WIRELESS AND MOBILE DATA NETWORKS

WIRELESS AND MOBILE DATA NETWORKS

AFTAB AHMAD

WILEY-INTERSCIENCE
A JOHN WILEY & SONS, INC., PUBLICATION

Published by John Wiley & Sons, Inc., Hoboken, New Jersey.
Published simultaneously in Canada.

Library of Congress Cataloging-in-Publication Data:

Ahmad, Aftab, 1961–
Wireless and mobile data networks / Aftab Ahmad.
p. cm.
Includes bibliographical references.
ISBN-13 978-0-471-67075-9 (cloth)
ISBN-10 0-471-67075-8 (cloth)
1. Wireless communication systems. 2. Mobile communication systems. 3. Computer networks. I. Title.
TK5103.2.A43 2005
621.382—dc22

2004025911

10 9 8 7 6 5 4 3 2

To Mahmooda

CONTENTS

PREFACE

As broadband access reaches more and more homes and businesses, paradigm changes are occurring in all aspects of data communications. Security in Wireless LANs is becoming an ever more important issue, cellular networks are geared toward a service-oriented design, broadband access does not necessarily imply 'fixed' networks and, above all, network architectures with a range of data rates for personal operating space have been specified. Various factors, both international and national, have impacted the interoperability endeavors and we see an unprecedented collaboration among operators, vendors and standardization agencies. A large number of wireless data technologies provide solutions for users of wireless data. Arguably, the 'secret ingredient' in all the new and traditional technologies seems to be the Internet Protocol (IP). Without IP, a networking technology, wireless or not, seems to be destined to . . . including IP. However, each of the various network architectures has its own place in the market. Each wireless network relieves its users from some restrictions, such as having a plethora of wires and, many times, provides the freedom to move while connected.

The kind of freedom that wireless networking has promised is not only irreversible, but is also subject to growth, in strides, that is. The depth of knowledge in wireless networking has gone to a point where we talk about changing and choosing modulation schemes from burst to burst, of mobility in excess of 200 kmph, and of license-exempt bandwidth topping 1.5 GHz. Putting it all together in one book is practically impossible without sacrificing one thing or the other. However, it is possible to have a book with a theme, for example, to give enough breadth that the knowledge gained covers sufficient types of networks, and enough depth that the knowledge obtained is not superfluous.

This book tries to meet this general goal of providing a breadth of the technologies in wireless data networks while requiring a respectable background in communications network architecture and some background in fundamental algebra. The emphasis is on data networks.

When we talk about 'data networks', we usually imply packet-switched communications, of which voice could be a very important application. Following this logic, we describe only the 'data' part of networks, where both voice and data parts exist. Also, 'multimedia' in packet-switched networks includes voice communications as well. Therefore, voice, such as in voice over IP (VoIP) is automatically a part of discussing data networks. However, while voice is QoS-intensive, it does not shine as a killer application for high-speed networks, including wireless networks. A killer application would ideally be the one that requires network capability to the fullest and would be in demand to the fullest as well. New architectures for cellular networks seem to have decided to deploy sufficient infrastructure and leave the question of killer application to future, thus providing the scope for third-party service development environment. Nevertheless, the question of a killer application does not really exist in all types of network architectures, specifically, the ones used for access or the ones designed for specific applications. The example of the former are the WLANs and broadband wireless access networks, and the examples for the latter are sensor networks designed specifically for a certain application. We have included a wide range of network architectures, along with chapters to enhance their understanding.

The first three chapters have the goal of enhancing the understanding of later chapters. First chapter gives a bird's eye view of various wireless and mobile network types. It ends with a discussion on the frequency spectra allocated for these networks. Chapter 2, in continuation, discusses the protocol architectures of various network types. Even though we classify networks as personal, local, metropolitan and wide area networks, their real classification is in terms of protocol planes. Chapter 3 discusses various components of wireless LANs. A wireless LAN is much more complex than the wired counterpart and utilizes many concepts that are relatively more advanced. Instead of explaining these concepts as a digression, we have included them in a separate chapter. Following Chapter 3, there are two chapters on WLANs: Chapter 4, on descriptions of the physical layer (PHY) standards, and Chapter 5, an account of the medium access control (MAC) layer standards. The material presented in these chapters is organized in a convenient sequence. Also, the chapter on components of a WLAN (Chapter 3) is kept in view while organizing Chapters 4 and 5. In a way, WLANs are for low-level mobility (link-level). The next step in mobility would be the wide area mobility for wireless data terminals. The next three chapters and Chapter 10 cover this topic.

In Chapter 6 we discuss the two main Internet protocols that bear the responsibility of wide area mobility provision, the mobile IP and the session initiation protocol (SIP). Mobile IP provides what is called macromobility and SIP provides signaling mechanisms for macromobility on a higher protocol

CHAPTER 1

WIRELESS DATA—INTRODUCTION

In studying the principles of data communications, the wireless spectrum is generally treated as part of communications media only. This may give the impression that the remaining components of a wireless data network were the same as those of a fixed, wired network. The reality, however, is quite different, thanks to a number of factors with varying degree of roles in wireless and fixed, wired networks. There are network components that exist in only one network type and not the other. There are also network components existing in both types, but playing a less significant role in one or the other. There are many sub-systems, such as antenna radiation and mobility management that do not surface in the fixed, wired networks. Wall connectors are not usually part of transmission systems in wireless networks. There are systems that do make an essential part of both network types, but with much less significance in one than the other. Examples of such systems are power consumption systems, data security, and privacy, by containing signal, signal-detection techniques and error-control techniques. Lastly, there are certainly many components that play equally important roles in both types of networks, such as switching and routing techniques, flow and congestion control mechanisms and call-control procedures. Thus a study of wireless data networks has its own scope, different from networking systems in general.

Wireless, however, does not imply mobility. There are wireless networks in which both ends of communications are fixed, such as in wireless local loops. In satellite communication systems, even though the satellite is always mobile, the mobility profile of the satellite is designed so as to provide a constant signal

Wireless and Mobile Data Networks, by Aftab Ahmad

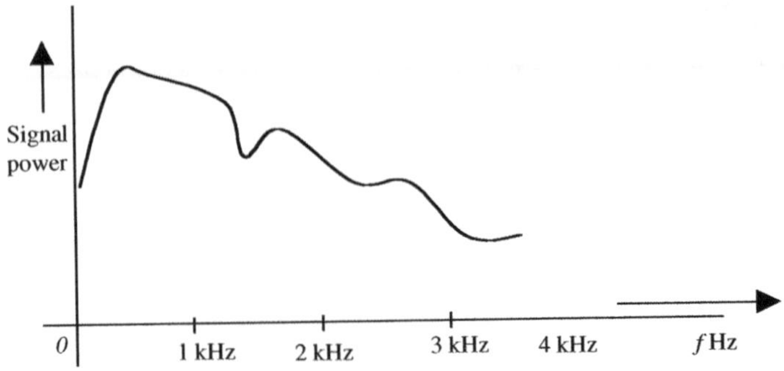

Figure 1-1. Power spectrum of a speech-like signal.

level to the connected terminals, thus emulating a fixed end. Wireless networks with mobility, however, provide the biggest challenge to the network designer.

We will devote this chapter to various types of wireless data networks that network engineers have to design and deal with. We will start the discussion with wireless voice communication, as the bulk of data in cellular networks is still voice. Also, most of the telecommunications developments have been in telephony.

1.1. WIRELESS VOICE

Before beginning a discussion on wireless data networks, a few words about the voice signal might be advisable. Even though the wireless data systems were the precursor of all electronic communications systems, most of the progress in telecommunications is a result of voice networks. In fact, most of the developments in cellular systems to date owe their existence to the voice signal[1]. Wireless voice poses somewhat relaxed requirements to system designers, which make it easier to make engineering decisions. Here are some examples of the characteristics of wireless voice.

1.1.1. Fixed Minimum Bandwidth

The voice signal has most of its energy within 300 Hz to 3400 Hz, giving a bandwidth of 3.1 kHz, such as shown in Figure 1-1. For typical digital transmissions, a nominal value of 4 kHz is assumed. Consequently, all channels with a bandwidth of 4 kHz or higher could ideally provide the same quality of transmitted voice if all other factors are kept constant. Digital speech is transmitted in one of the several standard coding forms, such as ITU G.711, G.721, G.722, G.723, G.728 and G.729. These standards are based on different mechanisms

[1] The same is true for wired networks, where PSTN has spearheaded the progress.

of speech digitization and compression, and produce a digital bit stream of either fixed (G.711, 721,) or variable but a known average rate (G.728, 729). PSTN uses G.711, which is based on 8-bit per sample PCM transceiver using one of the two quantization techniques (A-Law in Europe and μ-Law in North America and Japan), both resulting in a 64 kbps encoded voice bit stream. PCM is a waveform coding technique that deals directly with the speech signal for digitization and transmission purposes.

Other standards use model-based coding, which extracts certain parameters from the speech signals and transmits these parameters instead of the speech signal. These later systems, called Vocoders, result in bit streams anywhere from 16 kbps to less than 4 kbps. However, due to the inflexible nature of the PSTN, the 64 kbps standard is the one most used for voice transmission. For bandwidth-constrained systems, such as wireless networks, lower bit rate coding techniques have been considered as better alternatives. For example, the European GSM systems typically employ regular pulse excited hybrid voice coding (RPE), resulting in 13 kbps bit stream and the U.S. Department of Defense (DoD) uses 4.8 kbps code excited linear predictive (CELP) technique in federal standard FS 1016. In either case, once a network is designed to support a certain type of voice coding, the required minimum bandwidth is fixed. Such is not the case in data communications. Numerical, textual, or graphical data could be transmitted using any bandwidth without impairing its quality, as long as error-control mechanisms are employed to remove errors or retransmit lost packets and packets with errors. The channel bandwidth can only limit the speed of data transmission.

1.1.2. Vague Definition of Service Quality

A second characteristic of voice signal is a lack of a strict scientific definition of the quality of transmitted speech. The quality of voice transmitted is perception-driven and can't be adequately measured. Even though the International Telecommunications Union (ITU) standards based on scientific definition of quality perception allow for an automated measurement of voice quality, the most commonly used metric is still the mean opinion score (MOS), a subjective quality-determining mechanism in which listeners allocate a number between 1 and 5, where 5 is for excellent quality. The procedure for MOS is defined in ITU recommendation ITU-TP.800. A standard introduced for automated quality assessment was introduced in the early part of 2001. Called Perceptual Evaluation of Speech Quality (PESQ), it takes into account factors such as packet loss, delay and jitter. PESQ is defined in the ITU standard ITU-T862. Though its usability for Internet is agreeable, its validation, too, is done by comparing it to MOS.

In circuit-switched wireline networks, a fixed voice coding mechanism is employed, usually based on waveform coding. However, in connectionless Internet, neither fixed coding scheme must be employed, nor do the network characteristics remain constant. In wireless networks, the wireless channel is highly unstable, enhancing the vagueness of quality. In fact, despite strides in speech coding mechanisms, there is a discernable degradation in the quality of transmitted speech in cellular networks as compared with PSTN voice quality.

1.1.3. Delay Requirements

A third and perhaps the strictest characteristic of conversational speech is the stringent requirement on maximum delay. Due to its highly interactive nature, the conversational speech signal is required not to have more than a fraction of a second delay (250 msec maximum recommended by ITU). The variation in delay is expected to be even smaller by at least an order of magnitude. These requirements make the flexibility of packet switching somewhat less than ideal for voice communications. Therefore, the voice networks have traditionally been circuit switched. This applies to cellular wireless networks as well. Consequently, a voice network consists of a simple circuit-switched data part and a rather complex signaling system to monitor, supervise, and audit calls and resources.

In fact, the complexity and intelligence of the modern PSTN is due to its signaling systems. The contemporary cellular networks make use of the same signaling systems by adding a mobile part to it for mobility management and interaction with fixed PSTN. Future cellular networks, (termed as beyond 3G or 4G) are expected to circumvent signaling systems altogether and use connectionless packet switching for voice and other applications. This also leads us into a debatable definition of data networks. It is the our view that by data networks we imply packet-switched networks, such as IP networks. This is perhaps because such networks are ideally suited to bursty data applications, such as file and e-mail transfers, which can use store-and-forward mechanisms.

With the increase in demand for packet-switched data, the wireless data networks have evolved into many types, such as:

- Wireless LANs that provide wireless access just like the broadcast type fixed LANs provide access to fixed wide area networks. These wireless local area networks, relative latecomers as compared with their wired counterparts, are taking over the scene rather quickly. Their integration with the wide area cellular networks has become possible due to packet-switched third-generation (3G) systems.
- Wide area cellular systems, predominantly designed for voice, have incorporated packet switching all over the world from 3G and above. In fact, the precursor to 3G systems (sometimes dubbed as 2.5G) started packet data transmission before 3G technologies.
- Fixed wireless systems are becoming popular for broadband Internet access for ease of installation.
- Personal area networks (PANs) are the latest addition for short-range, serial-line-like wireless connections with limited mobility.

- Satellite-based data systems, though nothing new, are an essential part of the wireless and mobile networks.

We will look at the characteristics of some of these networks in this chapter. More detail will follow throughout the rest of the book.

1.2. WIRELESS LOCAL AREA NETWORKS (WLANs)

Protocols for wireless local area networks (WLANs) typically consist of specifications for the OSI-RM equivalent of physical and the data link control layers. The physical layer specifications deal with utilizing the indoor wireless channel for transmission and reception of wireless signal. These specifications have two types of limitations; the ones set by frequency regulation agencies, and the others set by the protocol specification agencies. Usually, the bandwidth and radiation amounts are regulated by the spectrum regulating agencies and the bandwidth utilization mechanisms (modulation, data rates) and power radiation mechanisms (direct, indirect, line-of-sight) are set by protocol agencies, according to the guidelines provided by the spectrum regulating agencies. The medium access control (MAC) specifications are set altogether by the protocol specification agencies. These specifications deal with issues such as channel access, synchronization of frames, power control, resource management for multimedia, and so forth.

The most popular WLAN standards, recommended by IEEE (we call these the IEEE 802.11 suite), use infrared and the unlicensed spectra. These spectra are allocated in many countries for research and developments in industry (I), science (S) and medicine (M)—therefore, called the ISM band. The IEEE standard PHY provides several mechanisms for the use of ISM band (and unlicensed national information infrastructure U-NII band), designed to combat interference from other sources of the same bands. This is necessary because the use of such a system does not require license from the government, which could result in numerous sources of interference. The infrared band specifies only one type of radiation, that is, indirect radiation reflected from a course surface (called diffused infrared). The medium access control mechanism specifies a distributed coordination function (DCF) for channel access, distributed referring to the fact that it is to be implemented in all participating wireless stations. It defines several device types, for example, a mobile station (STA), which is a user terminal, and an access point (AP), which relays data between two stations or a station and a terminal on a fixed LAN. This gives rise to two configurations of WLANs, as shown in Figure 1-2, the infrastructure WLANs and ad hoc WLAN.

1.2.1. Ad hoc WLAN

In an ad hoc or independent WLAN, two stations communicate directly with each other without an access point. Mobile stations for such networks may

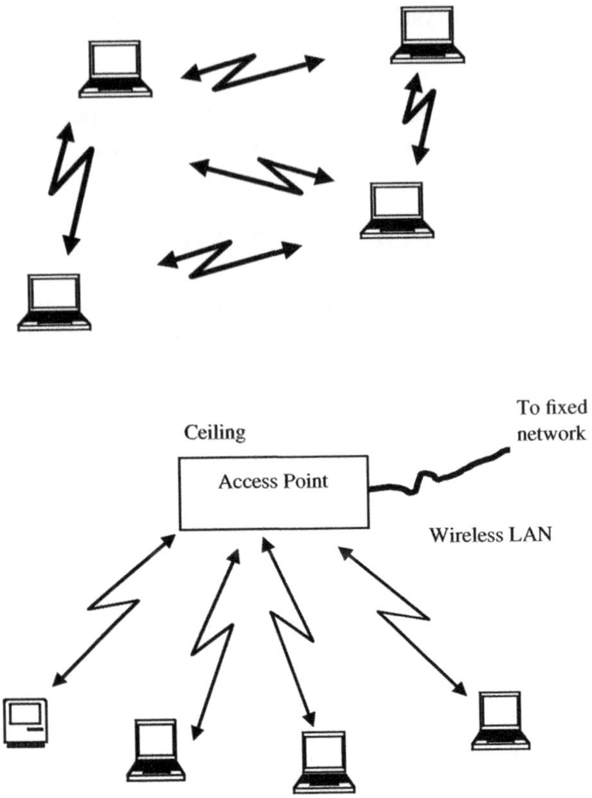

Figure 1-2. Ad hoc (*top*) and infrastructure WLAN with four computers and one access point (*bottom*).

require the capability to forward a packet, thus acting as a repeater. With this relaying capability, two mobile stations could exchange data packets even if they are unable to receive signals directly from each other.

Wide area networking in ad hoc networks is possible if one or more stations are connected to a wide area network, such as an IP network. However, this connectivity is not guaranteed and there is no guaranteed communication mechanism outside the ad hoc network.

1.2.2. Infrastructure WLAN

In an infrastructure WLAN, two stations exchanging data can communicate only through an access point. Figure 1-2 shows an access point connected to the ceiling with a cable connection to the wired network. The access point performs several functions in addition to relaying packets between stations in wireless and wired networks; such as implementing a point coordinating function (PCF) to allow reservation based communications for delay-bound traffic.

The MAC sublayer of the IEEE WLAN provides access-related mechanisms. For this purpose, it employs a mechanism similar to Ethernet. The Ethernet MAC (IEEE 802.3) uses carrier sense multiple access with collision detection (CSMA/CD). However, collision detection can't be efficient in wireless media, due to the rapid attenuation of the signal with distance. Instead of collision detection, a mechanism for collision avoidance is specified. Collision avoidance is implemented by requiring certain minimum time between any two packets transmitted. This time is called the inter-frame spacing (IFS). Due to the collision avoidance mechanism, the IEEE 802.11 MAC procedure is called CSMA/CA, carrier sense multiple access with collision avoidance. The subject of WLANs is as important as the application of such networks and it will be extensively discussed throughout the text.

1.3. WIDE AREA CELLULAR NETWORKS

Voice communication has been and continues to be the main application of cellular systems. These systems use PSTN-friendly infrastructure that employs circuit switching and signaling systems. However, with the wide spread of the Internet use, packet-switched services were introduced in enhancements of digital cellular systems. These included time division multiple access (TDMA)-based systems, such as GPRS (general packet radio service) and enhancement of code division multiple access (CDMA)-based systems IS-95B. The wireless standards for the new millennium that were internationally coordinated under the name of international mobile telecommunications 2000 (IMT-2000) (known from their air interfaces, WCDMA in Europe and cdma2000 in North America) have IP capability with data rates much higher than GPRS and IS-95B. The data networks of the first digital cellular generation were a result of defining new user terminals types, network devices and signaling system above the existing voice network, as shown in Figure 1-3 for GPRS. The latest generation wide area cellular networks provide access mechanisms for circuit- and packet-switched communication. For a true packet-switched cellular network an access mechanism similar to the WLANs could provide a better transport vehicle for data applications. Work is in progress in that direction and some countries already have WLAN access using the wide area cellular backbone for auditing and admission control purposes.[2] Such networks are expected to make broadband wireless access as ubiquitous as the Ethernet for the Internet. The next releases of the cellular networks could be a starting point for

[2] The WLAN access was initially seen by some countries regulating agencies as competition to cellular networks. There was some resistance to allow public deployment of such networks. At the writing of this book, this resistance is largely gone, even though WLANs do present competition in hot-spots, where they might provide a faster data vehicle than 3G system with no money spent on spectrum as against the costly cellular spectra.

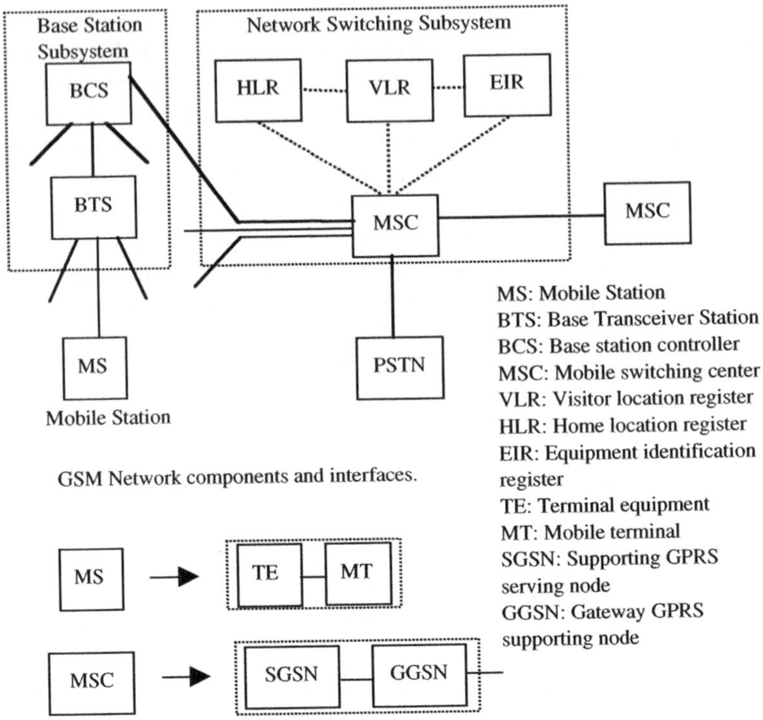

Figure 1-3. New components in the GPRS network.

this true merger of IP and cellular networking. At this time, broadband wireless is available in the form of fixed wireless networks only.

A universal installation of cellular systems based on 3G and above has been hampered by various technical, economic, and political factors. On the technical side, the world remains divided into groups based on evolution of their current systems. Two main camps are the European, supporting Wideband CDMA and North American, supporting cdma2000 evolution. The 3G partnership projects (3GPP for WCDMA and 3GPP2 for WCDMA) are destined to help actual implementation of 3G+ systems and take steps toward harmonization of the two camps.

1.4. FIXED WIRELESS NETWORKS

Started as a solution to carry subscriber's loop in hard-to-reach areas, the fixed wireless networks have emerged as a phenomenon unto themselves. This is owing to their ease of installation and the availability of broadband frequency spectra for this purpose. Toward the late nineties, it was obvious that short-

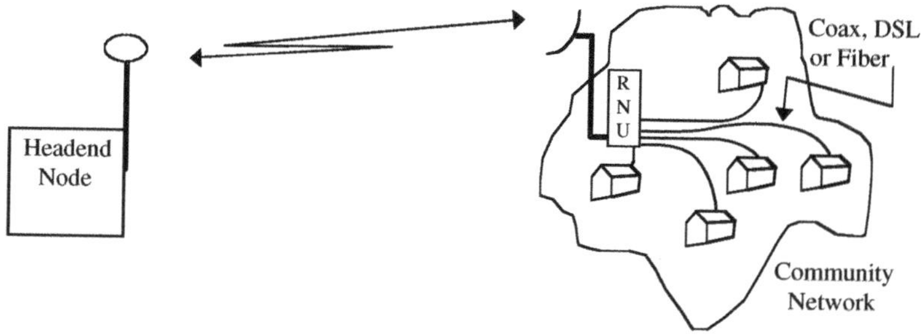

Figure 1-4. Fixed wireless network.

range, broadband, fixed wireless, line-of-sight networks would provide an excellent alternative to wired broadband Internet access. Many countries allocated a spectrum specifically for this purpose, in the millimeter wave range (around 28 GHz). Figure 1-4 shows an example of use of such networks.

Figure 1-4 shows a community network of digital subscriber's loop (DSL), coaxial, or optical fiber being fed by the Internet Service Provider (ISP) headend node from a radio network unit (RNU) via point-to-point fixed wireless connections. Systems such as local multipoint distribution systems (LMDS) could easily accommodate several hundred Mbps rates for similar radio connections. The IEEE standard 802.16 (Air Interface for Fixed Broadband Wireless Access Systems) is one of the latest arrivals in this arena. This technology is expected to revolutionize the way Internet service is provided. The standard, along with the enhancement IEEE 802.16a, is projected to lay foundations for across-the-board wireless data systems from desktop to Internet backbone. The IEEE 802.16 Working Group on Broadband Wireless Access also calls this standard as Wireless Metropolitan Area Network (Wireless MAN™). IEEE 802.16 is defined for line-of-site (last mile) spectrum of 10–66 GHz, while IEEE 802.16a extends its capability to non-line-of-site spectrum of 2–11 GHz. It provides service differentiation and bandwidth negotiation capabilities in order to be customized according to application needs and channel conditions. Fragmentation and packing at the MAC layer allows for efficient use of available channel resources. A separate Privacy Layer provides specifications for encryption. The MAC layer is designed keeping multicasting in mind. Due to the complex nature of the IEEE 802.16 architecture, and due to the fact that the IEEE standards need conformance testing, a compliance forum WiMAX (Worldwide interoperability for Microwave Access) has been formed by leading developers of equipment.

In addition to the fixed wireless access to packet-switched networks, a number of standards and products exist for wireless local loops for PSTN. These include a standards suite by ANSI called Personal Access Communications System (PACS) [12] , Qualcomm's QCTel, and base station to PSTN system by Lucent Technologies, called Wireless Subscriber System (WSS).

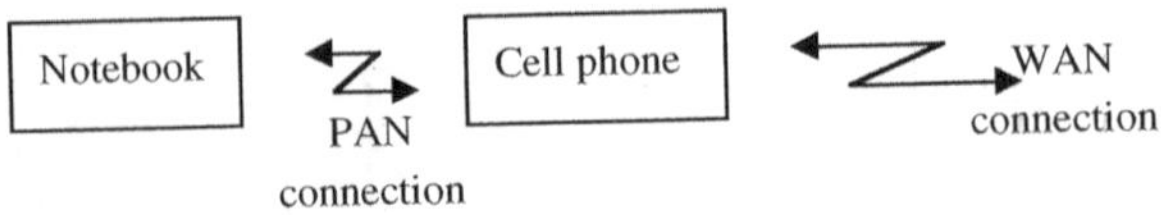

Figure 1-5. The personal area network connection.

1.5. PERSONAL AREA NETWORKS

Short-range wireless data networks, also called personal area networks (PAN)s have seen some recent standards activity on various fronts. These networks are more like fixed wireless networks than mobile networks, due to low mobility. However, they have an identity of their own due to several characteristics such as their use of the ISM bandwidth and limited span and mobility. These networks can be thought as a generalization of the television remote control mechanism to include cell phones, personal digital assistants (PDAs), and a host of other devices that use short-range cabled connections.

Figure 1-5 shows a use of such networks, where a notebook user is connected to a cellular network through a pocket-held cell phone. The connection between the laptop and cell phone forms a personal area network. Standards in this range of wireless communications are already available by HomeRF Working Group, infrared data association (IRDA), Bluetooth, and lately by IEEE, in the form of a new suite of standards IEEE802.15.

The IEEE 802.15 group has come up with a series of standards recommendations ranging from adopting Bluetooth v1.1 (802.15.1) to very low complexity, low speed and low battery solutions for sensors (802.15.4). The Bluetooth Special Interest Group (SIG) spearheaded the initiative of standards in the license-free spectrum for PANs. The standard recognizes the differing needs of various applications and accommodates these by defining profiles. Each profile is like a different protocol stack with some common layers, making it possible to have all or several profiles easily implemented. Four general profiles are, (1) generic access profile (GAP), (2) serial port profile, (3) service discovery application profile, and (4) generic object exchange profile [1]. Other profiles have been added in this and later versions to facilitate specific usage environments. Figure 1-6 shows an example profile of LAN access.

1.6. SATELLITE-BASED DATA NETWORKS

Many wide area mobile networks and some wireless local loops use satellite-based communications. Satellite systems have made their own niche due to the wide geographical area covered and the availability of a broad spectrum in the microwave range. The main disadvantages of satellite systems, that is, cost of increasing payload and significant delay, have been offset by the fact

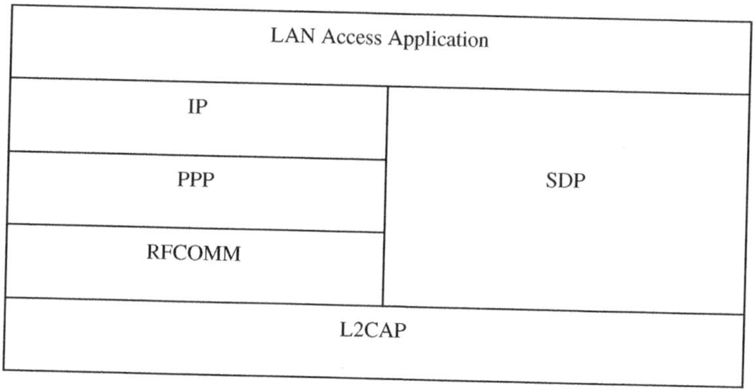

Figure 1-6. Bluetooth profile for LAN access.

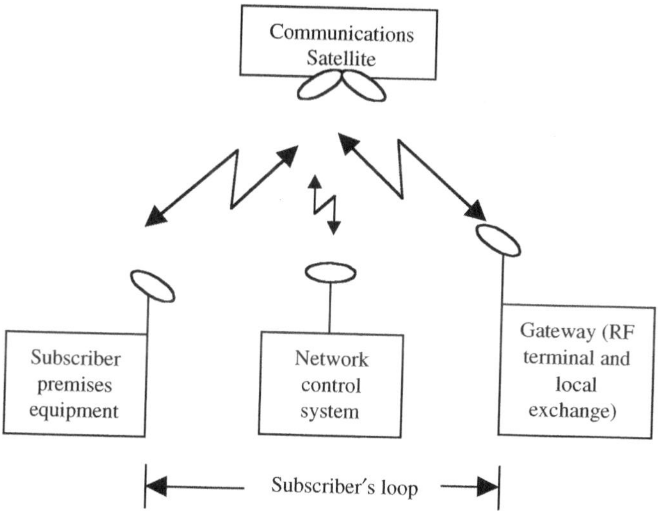

Figure 1-7. Satellite-based subscriber's loop system.

that plenty of solar energy is available in the orbits and that data services do not require critical delay bounds. Figure 1-7 shows a schematic of the use of satellite in a wireless local loop system. HNS Terminal Earth Station Quantum System uses a similar architecture.

One of the critical issues to be resolved in satellite system design is their location in the earth orbit. A satellite could be geosynchronous, low earth orbit (LEO), or a medium earth orbit (MEO). Some well-known systems, such as INMARSAT and EUTELSAT, consist of geosynchronous satellites. These satellites are located at an altitude of about 36,000 km and appear stationary with respect to the earth. LEO-based systems, set around 600–1600 km above the earth include Iridium and Globalstar. The MEO-based systems that are above the LEOs and below the geosynchronous satellites include Intermediate Circular Orbit (ICO) and a new INMARSAT proposal called INMARSAT-P. Usually the satellite systems based on LEOs and MEOs use a number of satellites (66 for Iridium, 48 for Globalstar). Ellipso is a satellite-based data system that employs elliptical orbits around the earth, sometimes called Highly Elliptical Orbit (HEO).

In spite of their wide range application to mobile and wireless data applications, the satellite systems are on a slow or receding course and some are even facing bankruptcy.

1.7. MOBILE IP

While mobility in wireless networks implies data transmission with the terminal in motion, it has a different connotation in mobile IP. Mobile IP is a protocol to be used with the existing Internet to allow a user terminal to be attached at a different point from its home network. This is essentially not a wireless network, but a portable fixed network protocol. Figure 1-8 shows various components of a mobile IP network.

The home network of the mobile terminal has a router that performs the function of home agent to keep the information of user's current location and forward, packets to that location. This packet forwarding is done with the help of a care-of address, required in addition to the permanent IP address of the mobile terminal.

When a mobile terminal enters a visited area, it requires the services of a foreign agent. The foreign agent provides registration and packet-forwarding services to the visiting terminals. Each mobile IP host uses one permanent IP

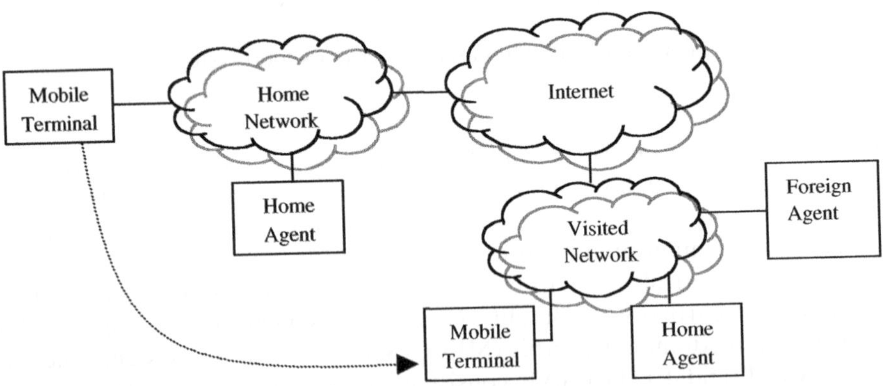

Figure 1-8. A mobile IP network consists of a home agent and a foreign agent, which perform the functions of mobility.

address (home address) and one temporary address (care-of address) if away from the home network. Thus, the IP packet exchange consists of the three mechanisms; namely, (1) discovering the care-of address, (2) registering the care-of address with the home agent, and (3) the home agent redirecting the received datagrams to the foreign network using care-of address. The mobile IP is not a wireless protocol, however, it could be employed for the IP infrastructure of the cellular networks. Several sources are available for a discussion on this important protocol, such as [8]. The Mobile IP Working Group of the Internet Engineering Task Force (IETF) was formed in 1996. The group has spearheaded the effort to come up with draft standards and recommendations for IPv4 and IPv6. The Cellular IP (cIP) project at Columbia University (http://www.ctr.columbia.edu/~andras/cellularip/) will use Mobile IP "to support migrations between Cellular IP Access Networks".

1.8. THE WIRELESS SPECTRUM

Wireless networks exist since before the concepts of frequency and bandwidth were understood. After over a hundred years of experience, significant strides have occurred in the understanding and management of wireless spectrum. Respective countries own their radio spectrum and most regulate it for up to 300 GHz. Since the actual usability of the spectrum is driven by its characteristics, generally similar frequency windows are allocated for same applications worldwide. International Telecommunications Union (ITU) has recommendations that could provide further ground for international cooperation. Unfortunately, a host of factors have resulted in some incompatibility of the allocation among various countries. In an era when wireless telecommunications is central to global trade, divisions exist in spectrum allocation. However, the coordination, too, is unprecedented and the world is not divided into as many blocks as the countries. ITU has divided the world into three regions, based on factors that influence wireless signal propagation. Region 1 covers Europe, Africa, parts of the Middle East and Northern Asia. Region 2 covers the Americas, Caribbean and Hawaii. Region 3 is everywhere else, that is, Asia, Southeast Asia, the Pacific Islands, Australia and New Zealand. Formation of international trade agreements, such as GATT (General Agreement of Trade and Tariffs), WTO (World Trade Organization) and OECD (Organization of Economically Developed Countries) have helped bring countries together in designing internationally interoperable goods and commodities, including wireless communications and the networking industry. Since our interest here is public and private data networks, we will have a look at some of the spectrum assignments in this area. For this purpose, we divide the spectrum according to the network scope, starting with the low power/personal area, through local area, to metropolitan area, to wide area to cellular networks.

The variability of the radio channel with frequency of signal is quite complex. The reason for this complexity is that the radio propagation characteristics are a function of many man-made and natural factors. Depending on the distance between the transmitter and receiver, the temperature, the surrounding material and mobility, we get a different profile for different conditions. The governments own and regulate the radio spectrum and, for reasons understandable, break it down according to some common characteristics of frequency bands. In general, the bandwidth of the wireless channel is extremely large (300 GHz regulated in most countries). However, due to breaks in performance at different frequency levels, only windows of continuity are available. Luckily, these windows can be suitable for different applications. This gives the government agencies the opportunity to allocate various windows for different applications. A company wanting to use a certain spectrum would normally require a license, which amounts to buying the spectrum. In order to provide a flexible platform for research and development, many governments have allocated parts of radio spectrum that don't require licensing.

1.8.1. Licensed and License-Free Bands

To allow new unhindered developments and research, parts of the spectrum are assigned to be used without a license. In the last 10–15 years, advancements in interference limiting signaling schemes have resulted in these unlicensed spectra being used for networking products. The use of the unlicensed band called ISM (Industrial, Scientific and Medical) has become so prolific that during late 1990s and early 2000s it created a false alarm of competing with cellular services (that are based on expensive licensed spectra). Interestingly, allocation of more spectra in both licensed and license-free arenas continues, while each is finding its niche. For the unlicensed band the regulating agencies usually regulate emissions.

1.8.2. Low-Power Wireless Data Systems

These systems include wireless personal area networks (PANs) and wireless sensors. Bluetooth™ (IEEE 802.15.1) and Ultra-Wide Band (UWB) are some of the technologies making use of spectrum for low power ranges. The bands used are UWB, ISM and Un-licensed National Information Infrastructure (U-NII).

1.8.3. Ultra-Wide Band (UWB)

Ultrawide band (UWB) is generally considered any band with a bandwidth of 1.5 GHz or higher [2]. The distinguishing characteristic of UWB is that emissions are at extremely low power (power density below −41 dBm in the United States). UWB signal is below the noise floor. It can penetrate through hard surfaces and provide very high data rates for very short distances. In the

United States, the bandwidth 3.1 GHz–10.6 GHz is regarded as UWB. It is sometimes broken into two main bands, the lower one 3.1–4.8 GHz and the upper one 6.1–10.6 GHz. The reason for leaving out the middle band it that it is allocated for Un-licensed National Information Infrastructure (U-NII).

Another distinguishing feature of UWB is the possibility of baseband transmission due to wide bandwidth. Instead of using sinusoidal signals, thin pulses are used in UWB transceivers. These thin pulses, with extremely low duty cycles, practically eliminate multipath. Multipath is the phenomenon used to describe multiple receptions of the wireless signal from reflections in the surroundings.

The UWB band has been in use since the 1980s for radar applications. However, its main projected uses are in Personal Area Network, IEEE 1349 and wireless universal serial bus (WUSB). The IEEE 802.15 Working Group is considering it for IEEE 802.15.3a PHY. In the United States, the commercial use of UWB has been allowed since February 2002. Other countries have not yet considered its use. This has led the industrial community into two groups, one advocating that the UWB standards be made, keeping in view that one day other countries may join in allocating the band. This group, with companies like Intel in it, favors a multi-band approach that could employ spread spectrum schemes, such as frequency hopping. The other group, with Motorola among its participants, originally advocated the use of the one big chunk of 3.1–10.6 GHz to be used with an alternative spread spectrum techniques, Direct Sequence Code Division Multiple Access (DS-CDMA).

1.8.4. The ISM Band

The ISM band is dedicated for the research and development of wireless devices for short-range applications in fields of industry, science and medicine. The actual location of the ISM band is not continuous and varies from country to country. However, the ISM band close to 2.4 GHz is allocated in most of the developed countries. Some variations of allocation are shown in Table 1.1.

The availability of the ISM band to anyone makes it an easy candidate for excessive interference. Spread spectrum communications techniques are used

TABLE 1.1. ISM Band Allocations

Geographical Area	ISM band around 434 MHz	ISM band around 900 MHz	ISM Band around 2.4 GHz
Europe	433.05–434.79	868–868.6	2.4–2.4835
USA		902–928	2.4–2.4835
Japan			2.471–2.497
France			2.4465–2.4835
Spain			2.445–2.475

TABLE 1.2. License-free Allocations Around 5–6 GHz

Geographical Area	ISM Band around 5 GHz	Remarks
Europe	5.725–5.825/5.825–5.875	5.15–5.35/5.47–5.725 reserved for HiPeRLAN
Japan	5.15–5.35, 4.9–5/5.03–5.091	
USA	5.15–5.35/5.725–5.825	
International Recommendation	5.15–5.35 Indoor 5.47–5.725 Outdoor	As per World Radio Conference (WRC) 2003.

in many devices for mitigating interference. In addition to the 2.4 GHz some unlicensed radio spectrums are allocated around 900 MHz.

1.8.5. U-NII Spectrum

About 300 MB of spectrum around 5–6 GHz has been allocated for high-speed digital wireless communications to provide broadband access to the information highway. This is also license free band. The United States and the European Union have slightly different allocations, because in the European Union, the 5.15–5.35/5.47–5.75 GHz was already allocated for High Performance LAN (HIPERLAN) [3]. Table 1.2 shows the U-NII allocations. IEEE 802.11a PHY is specified for this frequency band.

As shown in Table 1.2, Japan originally had 100 MB from 5.15 to 5.35 GHz. The remaining bands were added later on.

In the United States, FCC requires the companies only to follow radiation restrictions in this band. The European Union, however, requires the implementation of the HIPERLAN standard in part of the spectrum. WRC recommends the power limitation of 200 mW for indoor and 1 W for outdoor bands.

1.8.6. Cellular Systems' Spectrum

Table 1.3 shows different parts of spectra allocated to cellular systems. These range from generation I to III in the United States and II and III in the European Union. The 3G systems can, in general, use all the spectra allocated before them as long as they do not interfere with other systems. In fact, according to http://www.cdg.org/technology/3g/spectrum.asp, the 3G systems (cdma2000-based) are deployed in all bands, including 450 MHz, 700 MHz, 800 MHz, 900 MHz, 1700 MHz, 1800 MHz, 1900 MHz and 2100 MHz. Figures 1.9 to 1.12 taken from the website of NTIA, USA and show various worldwide plans for spectrum reserved for 3G systems.[3]

[3] http://www.ntia.doc.gov/ntiahome/threeg/3g_plan14.htm "Plan to select spectrum for third generation (3G) wireless systems in the USA," Oct. 2000.

TABLE 1.3. Cellular Bands in the United States

Band	Generation	Purpose
824–849 MHz	All	Uplink
869–894 MHz	All	Downlink
901–902 MHz	Narrowband PCS	Paging, Text msg. etc.
930–931 MHz	Narrowband PCS	Paging, Text msg. etc.
940–941 MHz	Narrowband PCS	Paging, Text msg. etc.
1850–1990 MHz	II and III	PCS, GSM (5 bands)
1710–1755 MHZ	III	mobile and fixed
2110–2150 MHz	III	mobile and fixed
2160–2165 MHz	III	mobile and fixed

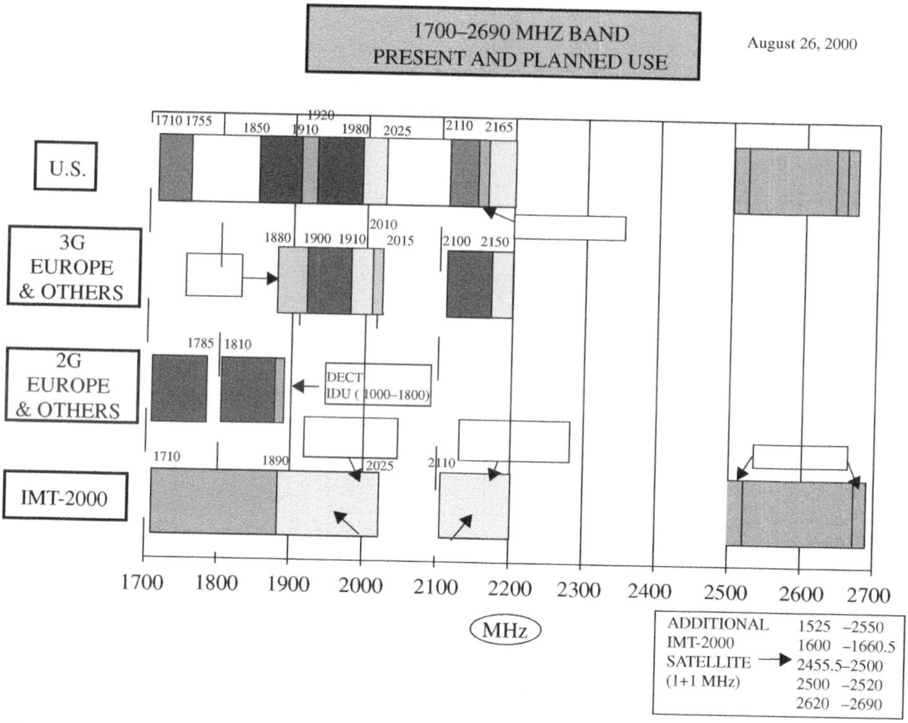

Figure 1-9. Comparison chart for the European Union and the United States for new allocations in 1700–2690 MHz range.

1.8.7. Fixed Wireless Systems

There are several types of fixed wireless systems (not mobile). Wireless local loops and broadband Internet access systems are of most interest to us. These systems, also referred to as last-mile technologies, are generally allocated spectra that have short ranges and are for line-of-site communications. Several

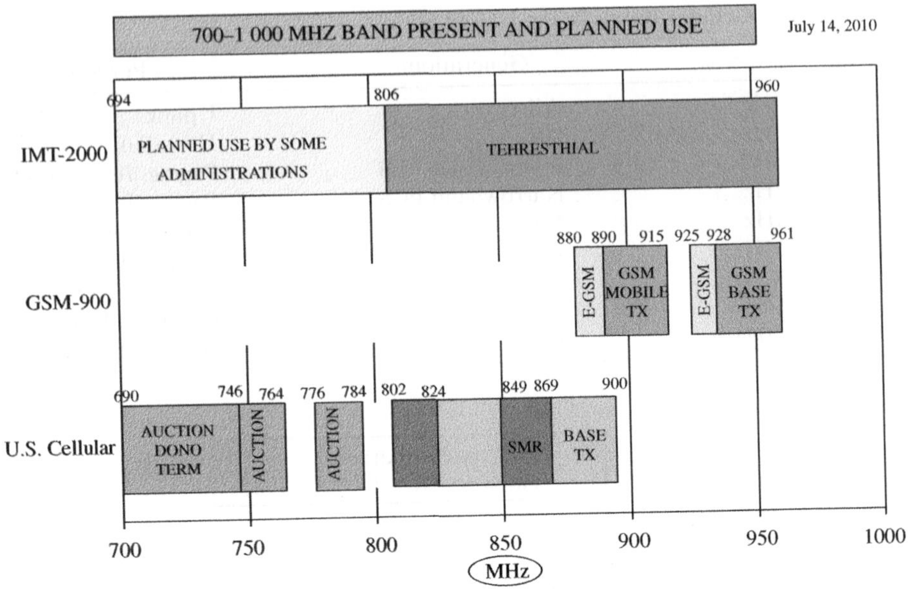

Figure 1-10. The lower bands (700–1000 MHz) allocation with proposed IMT2000 spectrum.

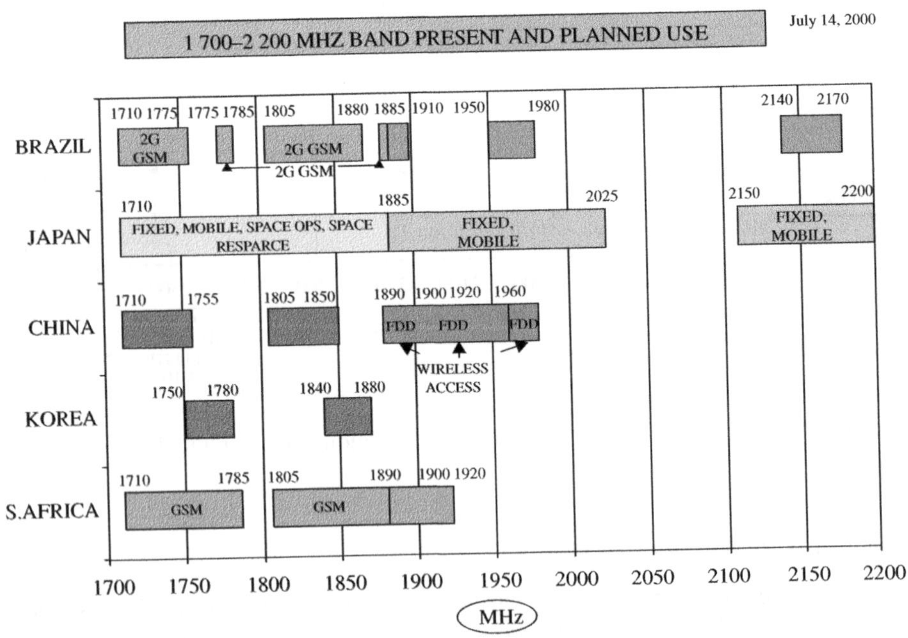

Figure 1-11. 1700–2200 MHz present and planned allocation by Brazil, Japan, China, Republic of Korea and South Africa.

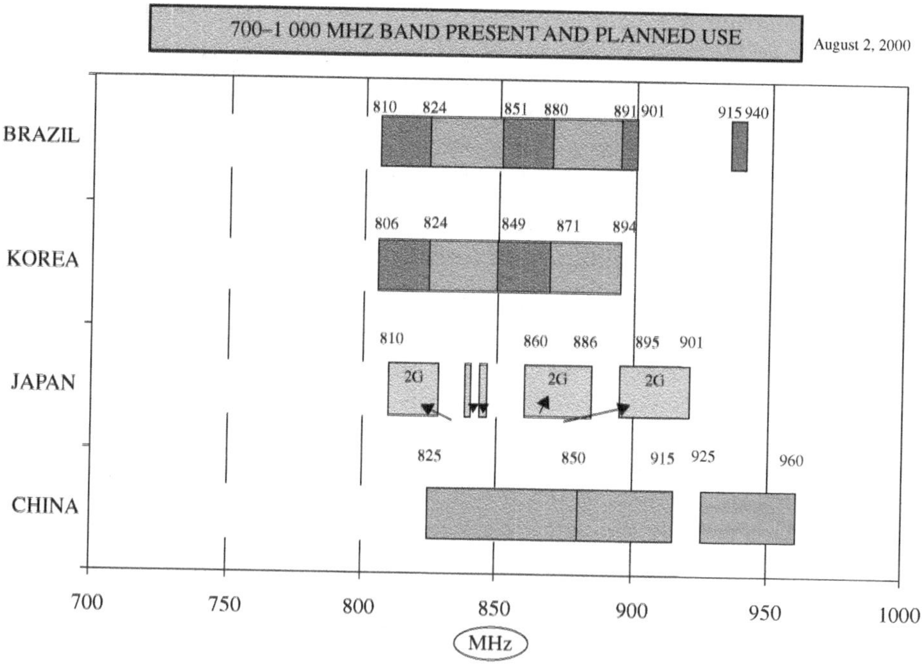

Figure 1-12. Lower spectrum (700–1000) allocation and plans by Brazil, Japan, China, Republic of Korea and South Africa.

existing PCS and 3G bands can also be used for this purpose (which will not have the line-of-site and last-mile restrictions). However, using PCS band for fixed wireless access (FWA) could be considered waste of its capability of supporting mobility.

The fixed systems have been used for analog communications, for example, a system called MMDS (microwave/multichannel multipoint distribution system) has been used to provide wireless 'cable' television in the New York area. New systems for digital Internet access are also under development in this frequency spectrum, generally below 10 GHz. Above 10 GHz, line-of-sight systems with 1–3 km range, for example, LMDS (local multipoint distribution system) can potentially provide over a Gbps Internet access. In the United States, LMDS uses spectrum around 28 GHz (total 1.3 GHz) while in Europe the allocated spectrum is around 40 GHz. Additionally, two companies, Winstar and Advanced Radio Telecommunications, were allowed to get license for some bandwidth around 38 GHz. Another 100 MHz have been purchased by Telegent in the United States around the 18 GHz band. FCC has also allowed digital transmission in the originally 'analog' MMDS band around 2.1 GHz and 2.5–2.7 GHz. A detailed account of spectrum allocation for broadband Internet access is given in [4] for various countries[4].

[4] The author, Roger Mark, is the Chair of the IEEE 802.16 Working Group on broadband wireless access.

Some spectrum has been allocated for narrowband data and voice over local loops. In [5] three categories of WLL are described as digital cellular, cordless, and proprietary. The digital category uses one of the cellular interface (GSM, TDMA or CDMA) with the second-generation and PCS bands and provide narrowband data access. Such local loops have been developed for both the digital cellular and PCS bands, albeit lower rates.

1.8.8. Wireless Metropolitan Area Networks (WMAN)

Various parts of the wireless spectrum between 11 GHz to 66 GHz have been allocated all over the world for broadband wireless Internet access. Unlicensed and licensed spectra below 11 GHz are also usable, making possible high speeds while connected. Several standards initiatives are in place to take advantage of these plans and allocations. Perhaps the most discussed among these are two IEEE initiatives, the fixed WMAN and Mobile WMAN (IEEE work Groups 802.16/16a and IEEE 802.20, respectively).

Ten GHz is roughly the cutoff for line-of-site communications. Therefore, non-line-of-site (mobile) broadband access systems are possible below 10 GHz. For example, IEEE 802.20 has targeted the spectra below 3.5 GHz for mobile broadband access, while IEEE 802.16a is designed keeping in mind the various bands in the 10–66 GHz range (e.g., 10.5, 25, 26, 31, 38 and 39 GHz). An amendment in IEEE 802.16a is considering lower spectra for non-line-of-site systems. An international interoperability forum WiMAX (Worldwide interoperability for Microwave Access) is working on developing interoperability principles for these standards and companies like Intel are among its participants.

1.8.9. Satellite Data Communications

In its frequency recommendations, the ITU identifies satellite communications bands along with the terrestrial bands. In spite of the overestimation in the success on lower orbit systems, satellites continue to play an important role in data communications. Communications satellites use spectrum in UHF, SHF, as well as EHF. Some of the operational bands are listed in Table 1.4. These are in addition to the proposed ITU recommendations for 3G technologies in 2 GHz and 2.2 GHz to be employed in Asia and Europe.

TABLE 1.4. Frequency Bands for Fixed Satellite Service Communications

Range	Name	Application example
1.5–1.6 GHz	L-band	Intelsat, INMARSAT, MAREC
6/4 GHz	C-band	Intelsat 806, SDRS 5
14/12 GHz	Ku-band	Telstar 12, PAS 6B, Echostar 3
30/20 GHz	Ka-band	Echostar 9

There are military bands designated as X-band (8/7 GHz), Q-band (44/20 GHz) and V-band (64/59 GHz). Part of Ka-band is also allocated for military communications. In addition to the above *Fixed Satellite Service* (FSS), there are many other types of satellite systems, for example, mobile-services satellite (MSS) and broadcast-service satellite (BSS). In the United States, spectrum in 12–18 GHz has been allocated to direct broadcast satellite (DBS), such as DirecTV and Echostar. DARS has been authorized to use spectrum in 2–3 GHz for BSS. Comsat is operating MSS in the 1–2 GHz range. Some spectrum below 1 GHz is also used by low earth orbit satellite (LEOS) systems, for example, Orbcomm. In general, the spectrum allocation for satellite communications has been much more a complex issue than terrestrial systems for a number of reasons. The possibility of sharing satellite spectrum with terrestrial ones is one big source of this complexity. There is a lot in the pipeline for satellites based on demands, and WRC/ IMT-2000 recommendations. The presentation [6] throws some light on the complexity of satellite spectrum along with mechanisms of sharing it with terrestrial systems.

REFERENCES

[1] Bluetooth, "Bluetooth Protocol Architecture Version 1.0", Riku Mettala, available from https://www.bluetooth.org/foundry/sitecontent/document/Protocol_Architecture

[2] Safecom, "*Emerging Wireless Technologies: Ultra Wideband Communications*", available from www.pswn.gov

[3] David Skellern, "The Future of Wireless LANs" Cisco Systems, CSIRO-Macquarie Technology Trends 2003 Seminar Series, February 2003, available from http://www.ics.mq.edu.au/business/events/2003/WLANsFuture.pdf

[4] Roger B. Mark, "Technical consensus in broadband wireless access technology", National Institute of Science and Technology.

[5] Michael Lee, "WLL in emerging markets: key development issues" Intelecon, available from http://www.inteleconresearch.com/pdf/WLLForum.pdf

[6] Edward M. Davison, "Spectrum Issues related to satellite communications, NTIA Office of Spectrum Management, available from http://www.its.bldrdoc.gov/meetings/art/art99/slides99/hat/dav_s.pdf

[7] Yi-Bing Lin, and Chlamtac, Imrich, *Wireless and Mobile Network Architecture*, John Wiley and Sons, New York, 2001.

[8] Charles E. Perkins, *Mobile IP: Design Principles and Practices*, Addison-Wesley *Wireless Communications Series*, Reading, MA, 1997.

[9] R. Bekkers, and J. Smits, *Mobile Telecommunications: Standards, Regulation and Applications*, Artech House Publishers, *Mobile Communications Series*, Boston/London, 1998.

[10] R. Ganesh, and K. Pahlavan, (Eds.), *Wireless Network Deployment*, Kluwer Academic Publishers, Boston, 2000.

[11] B. Bing, *High-speed Wireless ATM and LANs*, Artech House Publishers, Boston, London, 2000.

[12] Anthony R. Noerpel, Lin Yi-Bing, and Sherry Howard, "PACS: Personal Access Communications System—A Tutorial", *IEEE Personal Communications*, June 1996, pp. 32–43.

CHAPTER 2

REFERENCE ARCHITECTURES FOR WIRELESS DATA NETWORKS

The capabilities of a network can best be understood through its protocol architecture. A network protocol architecture typically groups protocol functions into layers, sublayers, levels, planes, and so on.[1] There are a number of ways to classify networks, for example, (i) backbone, interconnecting and access networks, that is, based on location with respect to the user; (ii) personal, local, metropolitan, and wide area networks, that is, based on general geographical span; (iii) campus, business, and office networks, that is, based on a circle of activity; (iv) cellular, mobile, hotspot, satellite networks, that is, based on technology, and so on, The list goes on. What is common among all these networks is that they all provide a reliable vehicle for information transportation from one point to another. Arguably, the best (by no means perfect) way of comparing two different networks is by comparing their protocol architectures. The open system interconnection reference model (OSI-RM) provides a generic mechanism for this purpose. The OSI-RM, albeit not comprehensive, defines the seven user data layers that describe a complete application to application communications function across a heterogeneous or homogeneous system of networks as laid down in the following paragraph.

First, the physical layer (PHY) provides mechanisms of physical transmission of signals and bits. These signals could be transmitted individually as well as in logical groups. Once the receiving PHY receives these signals, the data

[1] Synonymous with network architecture, protocol architecture, protocol stack architecture, or by using some specific prefix or suffix, e.g. WLAN architecture, and perhaps a lot more.

link control (DLC) layer above PHY treats groups of bits as a frame sent by the peer DLC of the sending station. Thus, through this understanding (*logical connection*) between peer DLCs, data can be exchanged between two applications or network (NET) layers above the DLCs across a single link. These data, delivered to the next receiving NET, may need to be forwarded to another NET before it can reach the destination application. The NETs use switching and routing mechanisms to push data all the way through the network, until they reach the destination *host* machine NET. At this time, the data are delivered to a transport protocol (TP) layer, which may make sure that all protocol data units (PDUs) or packets arrived in-sequence and without errors. A number of such packets could form a communication session to be managed by the peer *session* layer duo above peer TPs. The session layer, concerning itself only with groups of bits as session layer packets, may not be able to make sure that the codes, languages, syntax, etc. used by the communicating applications are well coordinated. This coordination is left to the *presentation* layer, which delivers a data format to the application layer in an understandable (and perhaps decrypted and decompressed) form. The application (APP) layer makes the data available to the user through the applications program. The role of each layer is complemented in the opposite direction.

The OSI communications paradigm provides a framework on which we can discuss and compare various networks. We will use this layered structure in this chapter to briefly look at the protocol functions provided by various wireless data networks.

2.1. BLUETOOTH™

Among wireless PANs, Bluetooth (also specified as IEEE 802.15.1) has received wide acceptance in industry. With the concept of profiles, it has emerged as a network that can take care of more than just personal communications equipment. The standard defines protocols corresponding to PHY and DLC of the OSI-RM. Applications can be designed directly above these layers, as shown in Figure 2-1.

Objectives of the Bluetooth have evolved over time. Initially it was expected to meet the following simple goals [1]:

- Low power;
- Low cost (US$5 per chip once fully developed);
- Low range (10*m*), extendable to 100*m*;
- ISM band.

In simple words, originally projected to replace serial cable, it has emerged as a lot more than short, point-to-point, connection. A Bluetooth device can work

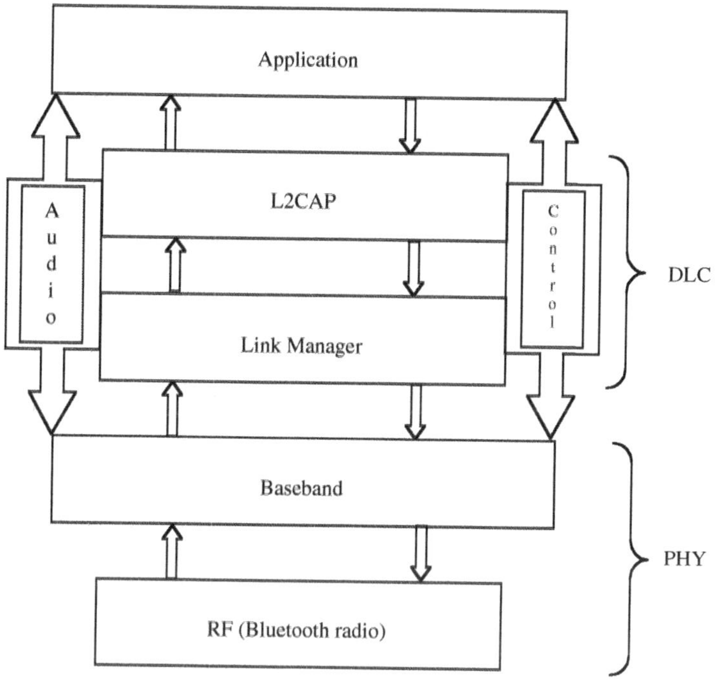

Figure 2-1. One way to look at the Bluetooth protocol stack.

with up to seven devices in a *piconet*, using a *master-slave* paradigm in which the master manages the connections. The master could be any one of the devices in the piconet with some restrictions. A slave device can belong to one or more piconets, while a master could also be a slave in another piconet, thus paving the way for routing mechanisms via a *scatternet*. A scatternet is a connected network of piconets. The Bluetooth specifications provide both connection capabilities, synchronous connection oriented (SCO), and asynchronous connectionless (ACL).

Through an agreement with the Bluetooth forum, IEEE project 802.15 adopted unchanged Bluetooth v1.1 as IEEE 802.15.1. Other enhancements for wireless PANs are defined in IEEE 802.15.2 (co-existence of IEEE 802.11 WLAN and IEEE 802.15.1 WPAN) and IEEE 802.15.3 (high-data-rate, low-power WPAN architecture) and IEEE 802.15.4 (for extremely low-power, low-data-rate applications). See the IEEE p802.15 website http://www.ieee802.org/15/ for detail. Following is the main function of each of the Bluetooth layer.

2.1.1. Bluetooth Radio

The radio layer performs PHY functions of proving an analog interface with the antenna and a digital interface with the link manager protocol (LMP) layer. It performs the RF functions relating to fast frequency hopping spread spectrum (F-FH-SS) modulation.

2.1.2. Baseband Layer

The baseband layer processes digital data for a number of functions pertaining to packetization, error control, security and channel filtering. It also performs a number of other functions related to frequency hopping spread spectrum technique.

2.1.3. Link Management Protocol (LMP)

LMP takes care of the physical mapping of the logical channels by resource management of the physical link. It assigns and terminates channels on the physical link in response to requests from the logical link control and adaptation layer (L2CAP) above.

2.1.4. Logical Link Control and Adaptation Protocol Layer (L2CAP)

This layer establishes, manages, and terminates logical connections for applications. It takes care of such details as QoS requirements of data, adaptation of data between application and baseband formats by providing segmentation and reassembly. L2CAP provides the capability of multiplexing by defining multiple channels, with each channel having a source/destination address pair, QoS tag and protocol. Bluetooth is a complete network architecture within its scope and specifies a mechanism for defining new application program interfaces (APIs). Two generic APIs provided by the network are telephony control protocol specification (TCS) and service discovery protocol (SDP) [2].

2.1.5. Bluetooth Profiles

The v1.1 defines 13 profiles [3] to imbed flexibility in the protocol architecture above the PHY and DLC layers. Later versions have standardized the incorporation of profiles so that a common framework is available for future profile development. Here's a brief definition of some important profiles.

2.1.5.1. Generic Access Profile (GAP). The GAP reflects the original goals and bare minimum protocol stack to be implemented in any Bluetooth-compliant device. Its purpose is to allow a Bluetooth device the ability to connect to another Bluetooth device without having any knowledge about the specific profiles available in that device. GAP is mandatory for Bluetooth compliance. Figure 2-2 shows a schematic for this profile.

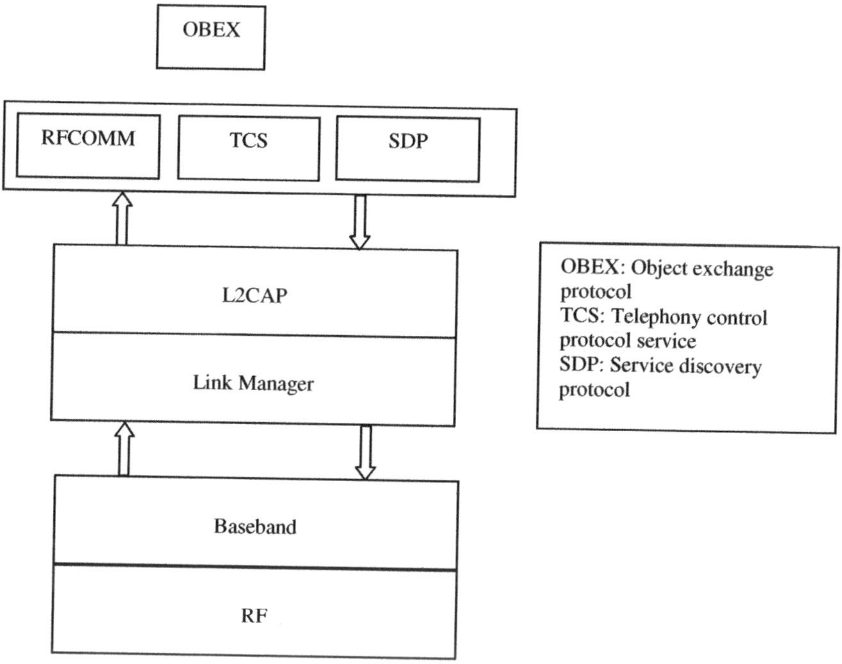

Figure 2-2. The generic access profile (GAP).

2.1.5.2. Service Discovery Application Profile (SDAP). This profile is designed for searching application services for Bluetooth devices. A required application, the *service discovery user application* uses SDP for general and specific known services search. Figure 2-3 shows a protocol stack for this profile [26]. Other profiles include the cordless telephony profile, intercom profile, serial port profile (emulating EIA-232 signaling with data rates of 128 kbps), headset profile, dial-up networking profile, fax profile, LAC access profile (using PPP above RFCOMM and SDP, a Bluetooth station that acts as a LAN station), generic object exchange profile (GOEP) using IrDA's OBEX and serial port profile, object push profile for small objects, file transfer profile, and synchronization profiles, used with GOEP. Figure 2-4 shows a grouping of various profiles [4].

2.2. IEEE 802.11

The IEEE 802.11 is an evolving standard and has a number of extensions at all levels. The basic network architecture as given in the original standard is depicted in Figure 2-5. The LLC is not part of the original IEEE 802.11 standard and is the same as Ethernet LLC, that is, IEEE 802.2.

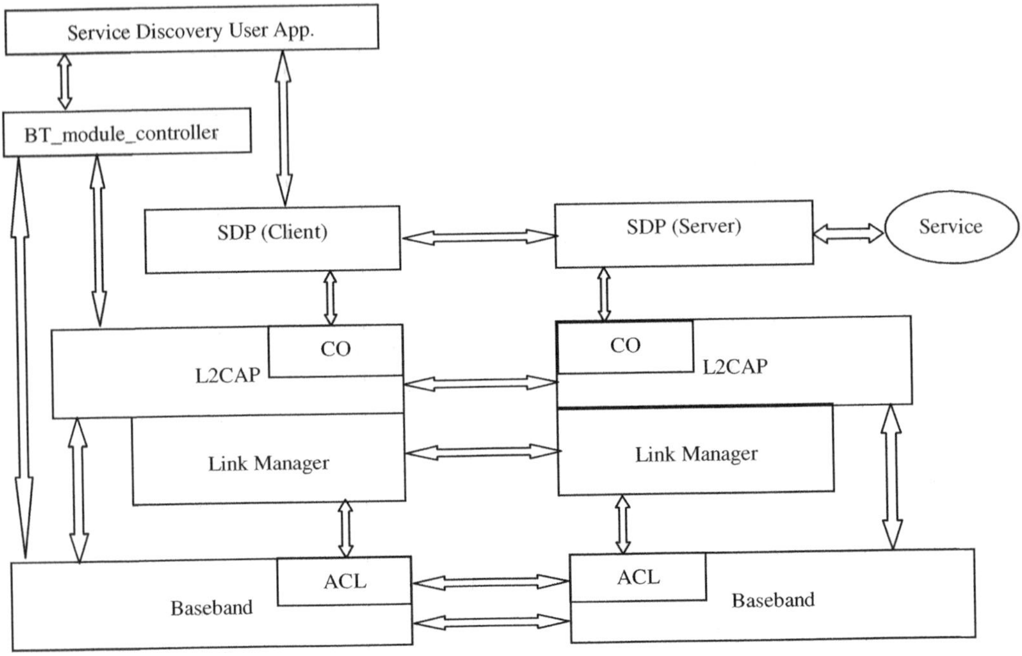

Figure 2-3. The SDAP profile.

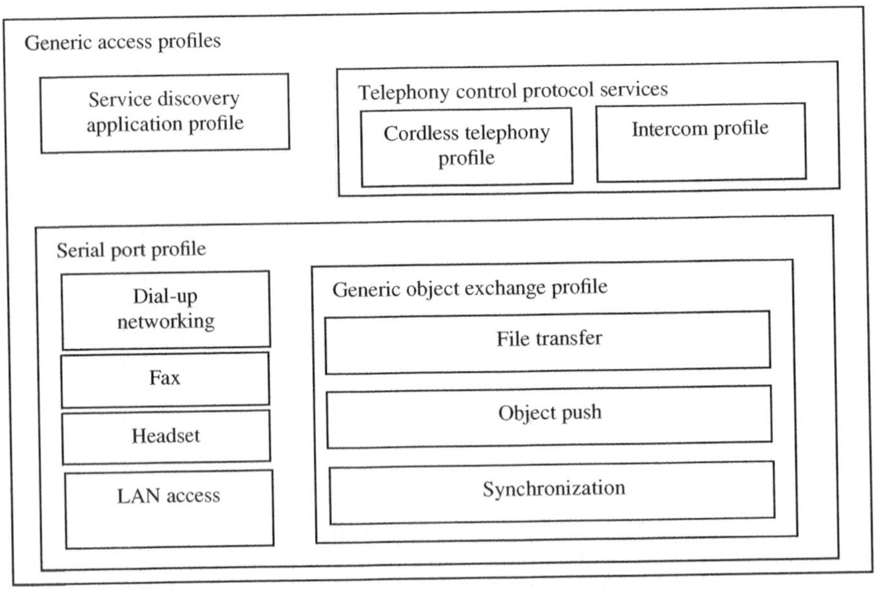

Figure 2-4. Relation among various Bluetooth profiles.

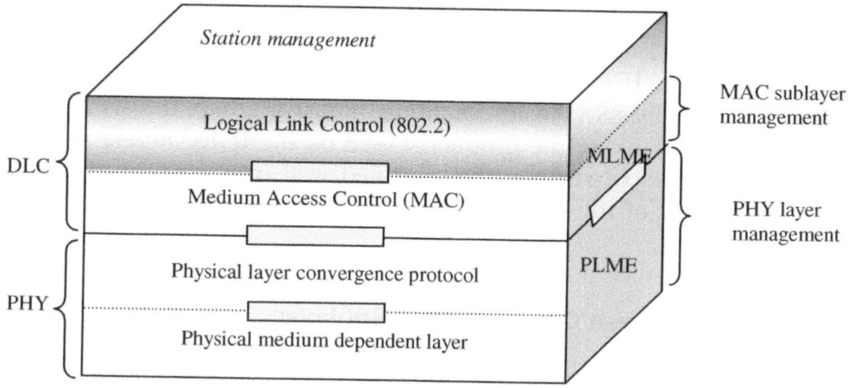

Figure 2-5. Protocol architecture for IEEE 802.11WLAN standard.

2.2.1. Physical Layer (PHY)

The PHY is divided into two sublayers, physical medium dependent (PMD) and physical layer convergence protocol (PLCP). PMD specifies functions, procedures and services for a variety of physical media (media here refers to spectrum *plus* signal conditioning) [5]. The original IEEE 802.11 defines three PHYs, two based on 2.4 GHz ISM band and one based on diffused infra-red radiation. All three provide a channel rate of up to 2 Mbps. Later extensions define three more PHYs: IEEE 802.11*b*, using 2.4 GHz spectrum for a maximum channel rate of 11 Mbps; IEEE 802.11*a*, using 5.7 GHz spectrum for a maximum rate of 54 Mbps; and IEEE 802.11*g*, using 2.4 GHz spectrum for a maximum channel rate of 54 Mbps.

2.2.1.1. Physical Medium Dependent (PMD) Sublayer. The PMD sublayer defines and specifies functions related to one particular PHY. Therefore, each of the six PHYs defined in this network architecture has an associated PMD. The PMD defines signal conditioning parameters, such as modulation and channel coding (if used) or pulse structure in case of baseband transmission. For PHY using spread spectrum technologies, PMD defines the exact spreading codes and spreading and despreading mechanisms. Transmission power levels are sensed by this layer if needed, for example, in clear channel assessment (CCA).

2.2.1.2. Physical Layer Convergence Protocol (PLCP). The PLCP takes care of the differences among various PHYs to provide a packet to the MAC layer so that the MAC packet format does not depend on a particular PMD sublayer. It might sound like an inter-operability function, but it is not. Inter-operability of two different PHY layers would require a function for converting signal format from one type of PHY to another. This is not the job of PLCP.

It simply liberates the MAC sublayer from having the knowledge of the PMD sublayer. PMDs are inter-operable if they use the same or compatible signaling schemes, that is, IEEE 802.11*g*, *b*, and IEEE 802.11, using direct sequence spread spectrum (DSSS). An international industry group, the *Wi-Fi* Forum, has evolved for certifying the products based on the IEEE 802.11 standards. It has been very successful and has certified well over 1000 products since March 2000. A list of the Wi-Fi certified products is available on its website, www.wi-fi.org

2.2.2. Medium Access Control (MAC) Sublayer

The MAC sublayer is defined as part of the IEEE 802.11 original standard. Later PHY versions use the same MAC. An enhancement, IEEE 80.11*e*, is in progress, for providing compatibility with QoS definitions of Internet and Ethernet (IEEE 802.1D). The original MAC was designed using the same channel access mechanism that Ethernet uses (CSMA). However, Ethernet employs a collision-detection mechanism, which cannot be implemented efficiently in a wireless medium. A slightly changed mechanism based on collision avoidance (CA) is specified. Collision is avoided by requiring a station to always wait for a time called interframe spacing (IFS) if the channel is idle. By allocating various values to IFSs, priority can be given to one station over the other.

The MAC standard requires CSMA/CA to be implemented in every station and access point. Thus CSMA/CA is a part of a distributed coordination function (DCF). Another coordination function, called point coordination function (PCF), can be implemented in an access point. An access point that uses PCF could implement channel access mechanism by reserving time for polling stations for data. The original intention for including PCF was to give reservation and priority mechanism for voice and other delay-sensitive traffic. When PCF is implemented and used, it defines a superframe of MAC protocols consisting of a repetitive cycle of collision free period (CFP) and contention period (CP). However, due to a lack of implementation of PCF and owing to many studies, the IEEE 802.11*e* group was created to come up with new specs for multimedia capable MAC. This task group is considering modified DCF, called enhanced DCF (EDCH), to provide service differentiation capability along with a hybrid coordination function (HCF) to replace PCF. More on the IEEE 802.11*e* is available in Chapter 5 on MAC for WLANs. Along with mechanisms for channel and multiple access, the MAC specifications provide mechanisms for privacy and security, power conservation, and mobility management.

2.2.2.1. Contention Windows. Even though the CSMA/CA is employed for contention resolution, the possibility of collision still remains. One or more stations sensing the channel simultaneously could find the channel busy. If they wait for the channel to become idle before they can initiate their IFS timers,

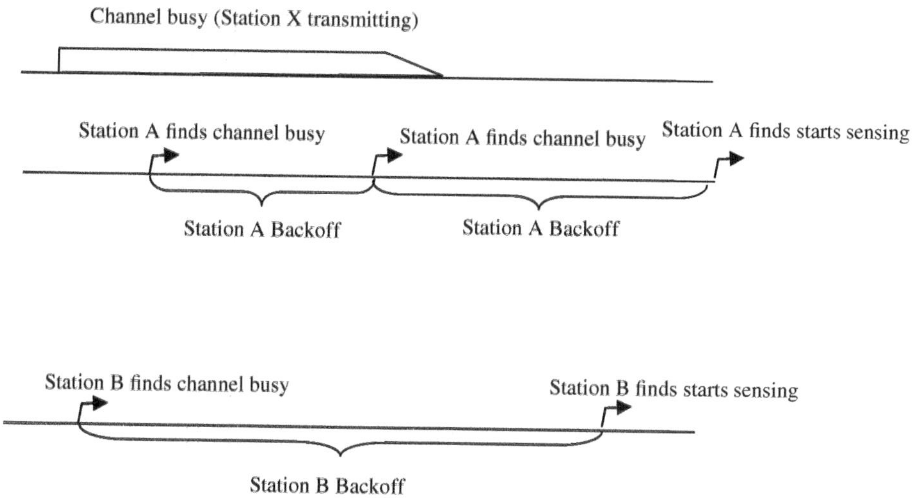

Figure 2-6. An illustration of the general concept of backoff. Stations A and B find the channel busy the first time they sense. Both are backoffed for a random time. Station A, after finishing the backoff period, finds the channel busy and is backoffed yet another time. Due to a different amount of total backoff, the two stations start idle channel sensing at different times.

chances are that some of the timers will expire simultaneously, allowing for simultaneous packet transmissions, causing a packet collision. To avoid such situations, the standard defines two concepts; random backoff and contention window. The random backoff time is also called *binary exponential backoff* and it is a random time between 0 and some maximum called *contention window* (CW). If a station has to backoff more than once, then every next time, the value of the contention window is doubled from the previous. This can happen between two limits, CW_{min} and CW_{max}. The selection of a random value of the backoff time randomizes the initialization of timers, hence reducing the possibility of collisions. Figure 2-6 shows this concept.

2.2.3. Layer and Station Management Planes

The layer management planes contain the layer management entities (LMEs) and management information base (MIB) for the respective layers. The MIBs have a list of layer parameters and value of different attributes. For example, the IFS is a function of PHY type and the MIB stores values of different IFSs for each type of PHY in PHY MIB and the values of different types of IFSs for a given PHY in the MAC MIB. Contention windows are also given values for their maximum and minimum attributes. The station management allows layer management entities to share data.

2.3. HIPERLAN/2

In this section, we will discuss the protocol reference architecture of HIPERLAN/2. HIPERLAN/2 is one of the four standards specified by ETSI to provide a concatenation of interoperable technologies from home or hotspot through cellular data networks (specifically, 3G) to an ATM backbone. The scope of HIPERLAN/2 is to provide an infrastructure or ad hoc wireless networks with low mobility (generally <1.5 *m/s*) and small radius (generally <50 *m*). Completed in 1996 by the ETSI committee RES 10, the standard is 'allocated' the license-free band in the range 5.15–5.25 GHz, with an extension to 5.35 GHz in some countries. This is in consonance with allocations in Japan and the United States for similar WLAN architectures.

One of the main strengths of its predecessor HPERLAN/1 is the support of isochronous traffic with low latency (32 kbps audio with 10 *ns* and 2 Mbps video with 100 *ns*). However, HIPERLAN/1 seems to be losing the steam in favor of IEEE 802.11. The HIPERLAN/2 is the second of the four standards completing the broadband radio access network (BRAN). The other three are: the original HIPERLAN/1, HIPERLAN/3 (Remote ATM access), and HIPERLAN/4 (ATM interconnect). As shown in Figure 2-7, and also explained in several papers [6], the standards go beyond the typical local area network definition and defines three layers. The following is a functional description of each layer and some of their components.

2.3.1. Physical Layer

This layer provides the standard OSI-RM PHY functions including RF interface to the medium modulation (OFDM) for combating multipath and several types of modulation for a variety of channel rates [7]. The PHY uses TDMA/TDD for bandwidth management [8]. Forward error correction helps the network meet more than one channel scenario. The PHY overview of the standard a given by ETSI is summarized at http://portal.etsi.org/bran/kta/Hiperlan/hiperlan2tech.asp

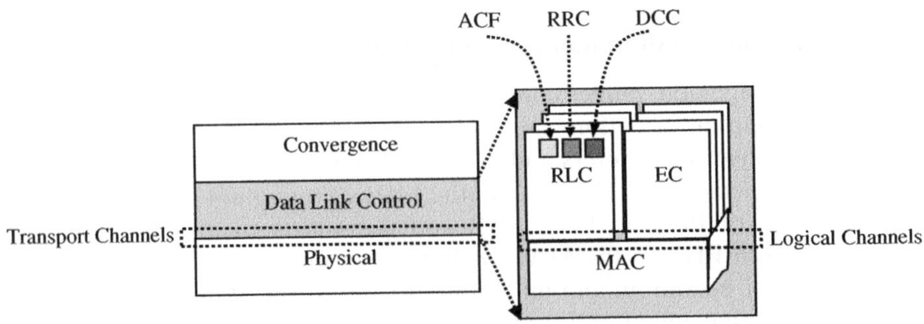

Figure 2-7. HIPERLAN protocol architecture.

In addition to the typical MAC functions, this sublayer also provides some functions unique to HIPERLAN/2. One example of such functions is link adaptation.

2.3.1.1. Link Adaptation. HIPERLAN PHY specifies mechanisms that can be used to design link adaptation algorithms. The standard allows for an MT and AP convey information about the channel clarity in an uplink or a downlink direction. The measurements can be made by MT and AP, but the decision to change the channel is in control of the AP.

2.3.2. Data Link Control Layer

HIPERLAN/2 has a rather complex DLC layer that is divided into many functions, as shown in Figure 2-7. Here's a definition of each sublayer and their functions.

2.3.2.1. MAC. MAC sublayer provides the main functions of channel access and multiple access. Channel access is provided through contention in channels dedicated for this purpose. Multiple access is provided through slotted frame, with a user data frame carrying 48 bytes of user data along with 6 bytes of overhead [9].

2.3.2.2. Radio Link Control (RLC). The RLC performs several tasks, such as association control function (ACF), radio resource control (RRC) function, handover support, power saving, and DLC connection control (DCC) [10]. ACF provides authentication and association capabilities. The standard requires the implementation of DES and it could optionally be used by ACF [11]. RRC is responsible for the efficient use of the available bandwidth. It provides measurement capability for signal-to-noise ratio (SNR) to be used for handover, association, link adaptation, and dynamic frequency selection (DFS).

2.3.2.3. Dynamic Frequency Selection (DFS). DFS is a carrier/channel selection mechanism employed by access point (AP) whereby an AP scans through all the available channels when it needs to allocate one to an MT. It allocates the clearest available channel, thus minimizing the overall interference among the active calls. Due to DFS, an MT does not have to look for a clear channel as in some other wireless systems. A group of APs can share channels and use DFS to provide the best possible channels to the users in the group.

Three types of measurements can be performed: (1) own channel quality, (2) measurement of interference (by shutting down own and MTs transmissions for a brief time), and (3) measurements of other channels to assist in handover. Each MT must have an RLC instance identified by a MAC ID. DCC is responsible for connection setup and release. A call reference ID is used in the DLC connection setup request message to create a relation with the higher level protocols.

2.3.2.4. Error Control (EC). EC specifies the use of selective reject ARQ. Additional forward error correction (FEC) can be added. Due to the special nature of selective reject ARQ, three types of messages are defined: (1) for accumulative acknowledgement, (2) for selective request for repeat, and (3) for accumulative discard (serial numbers of the message below which all messages have been discarded, which could happen due to some timer expiry or after certain maximum number of transmissions). Each EC instance is identified as part of a DLC connection ID.

2.3.3. Convergence Layer (CL)

The basic function of the CL is to provide HIPERLAN/2 DLC and PHY to other existing networks, such as ATM, IP, IEEE 1394 and Ethernet (IEEE 802.3). It does so by providing DLC services to higher-layer data and converting higher-layer packets into DLC data format for transmission within HIPERLAN/2 (segmentation and re-assembly). A CL is needed for every higher-level network, as shown in Figure 2-8.

Convergence layers for Ethernet, IP, ATM, 3G UMTS, PPP and IEEE 1394 (*firewire*) [12] have been specified. As seen in Figure 2-8, the CL is of two types, packet-based and cell-based. The packet-based CL provides convergence to and from PPP (for IP), IEEE 1394, and so on. The cell-based CL is for ATM connectivity. The packet-based CL is divided into two parts/sublayers. One of these sublayers provides functions common to all packet-based CLs and is called packet-based convergence layer *common part*. The other sublayer provides functions specific to one particular higher-level protocol, and consequently is called *service-specific convergence sublayer* (SSCS). A dif-

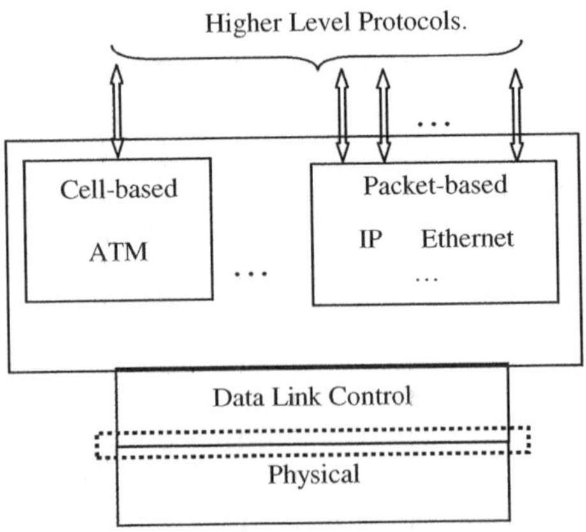

Figure 2-8. HIPERLAN convergence layer.

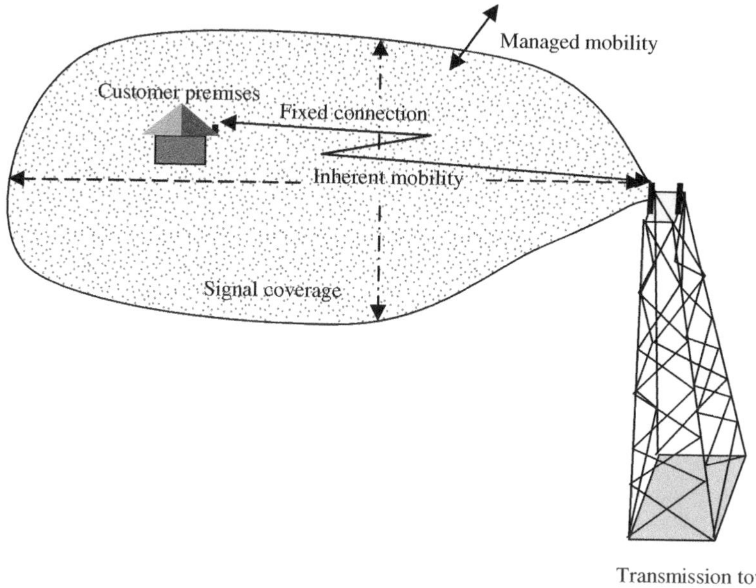

Figure 2-9. Fixed wireless networks may allow some mobility due to signal coverage.

ferent SSCS is needed for each higher-layer protocol to provide the relevant functions. For example, the SSCS for IEEE 1394 provides bandwidth reservation mapping, address mapping, channel mapping and self-id emulation from IEEE 1394 to HIPERLAN/2, in addition to IEEE 1394 clock distribution and time stamp processing. Similarly, the Ethernet SSCS adapts the Ethernet packets and QoS to HIPERLAN/2. Best effort is mandatory, while IEEE 802.1P prioritization is an option [13].

2.4. BROADBAND WIRELESS ACCESS NETWORKS

In this category, we have chosen the IEEE 802.16 and IEEE 802.20 recommendations.[2] The fixed wireless access implies lack of mobility, even though that is not the case in the broadband wireless access. In any wireless network, some mobility can be handled simply because the signal is available in a three-dimensional geographical area (shown as *inherent mobility* in Figure 2-9). A wireless terminal connected to another wireless terminal or a basestation tower can move around as long as both stay within the same antenna beam. The question of mobility becomes critical when a station moves out of the current antenna range from which it was receiving signal or to which it was sending signal. Figure 2-9 shows these two mobility types, the former as *inher-*

[2] In IEEE terminology, these are metropolitan area networks (MANs).

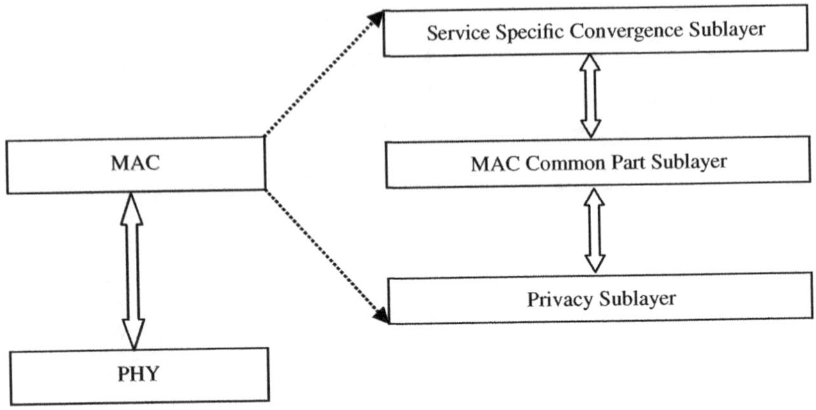

Figure 2-10. Reference architecture for IEEE 802.16*a*.

ent mobility and the latter as *managed mobility*. When we talk about fixed wireless networks, we usually imply the absence of managed mobility. The reason we call this type of mobility as managed mobility is because a network infrastructure other than signal coverage is required to handle this type of mobility, with functions such as registration, call routing, and forwarding and inter-system communications.

The IEEE 802.16 and IEEE 802.16s suite of standards provide point-to-multipoint and mesh architectures in the frequency range of 10–66 GHz and 2–11 GHz, respectively. With a data rate in excess of 120 Mbps, a large number of users can be simultaneously accommodated using TDMA in a large coverage area (about 50 *km* for IEEE 802.16a). A working group IEEE 802.16*e* is considering adding mobility to the standard for providing IEEE 802.16 services to vehicles within the basestation coverage area. An international forum (WiMAX) is aggressively addressing interoperability issues. IEEE document [14] provides some parameters for co-existence simulation for the cellular and mesh architectures in the 2–11 GHz range. The following discussion is based on [15] and Figure 2-10 showing the user plane of protocol architecture.

2.4.1. The User Plane

The data plane consists of the equivalent of PHY and DLC and defines three sublayers with the medium access control (MAC).

2.4.2. MAC Layer

The IEEE 802.16 MAC layer provides the dual capability of allowing *multiple network types* to use this layer to *share* the systems bandwidth in a *secure* fashion. The three italicized terms define three sublayers of MAC.

2.4.2.1. Convergence Sublayer (CS). *Called service specific convergence sublayer* in Figure 2-10, this sublayer is responsible for freeing the MAC common part from the details of the higher layer protocols using the IEEE 802.16 MAC services. ATM as well as packet convergence layers are included in this specification to provide a broad spectrum of higher-level connectivity.

2.4.2.2. MAC Common Part Sublayer (CPS). The CPS provides the core MAC services in the form of a connection-oriented point to multipoint transmission with channel access and bandwidth sharing. It employs time division duplexing (TDD) and frequency division duplexing (FDD) so that an FDD channel could be used by many stations in a sector.[3] The multiple access mechanism employed is TDMA and bandwidth is allocated in three ways, contention, polling and unsolicited reservation.

The MAC is designed for a multiple-sector network layout with all stations in a sector scanning all signals to check if they are the intended recipients of a segment of data.

2.4.2.3. Privacy Sublayer. As expected, this sublayer provides mechanisms for authentication, key-sharing and encryption.

2.4.3. PHY

The MAC sublayer is defined to work at the top of any PHY in the 11–66 GHz range. The standard requires two modulations (QPSK and 16-QAM) to be mandatory and 64-QAM as an option. Power density is limited to a maximum of 14 dBW/MHz for the base station and 30 dBW/MHz for the subscriber station. Since many different PHYs are possible in this range of spectrum, the actual PHY will have two parts: the PMD, for medium specific functions, and PLCP, for convergence to a common MAC.

2.4.4. IEEE 802.16*a*

This specification [16] is an amendment to the original standard. It proposes changes at both layers (PHY and MAC). It allows for the PHYs below 11 GHz (2–11 GHz), thus relaxing the condition of line-of-sight. However, due to losses at these frequencies, power management mechanisms are enhanced. The MAC enhancements deal with the need of ARQ to enhance the link reliability. A mesh topology is allowed for multipoint-to-multipoint transmission. The effect of mesh topology is that a distributed scheduling mechanism is required because the nodes do not communicate solely with the base stations. A subscriber station may take part in packet forwarding to act as a part of a neighborhood.

[3] Duplexing is multiplexing in opposite directions, e.g., uplink and downlink.

New PHYs for licensed and license-exempt bands at 2–11 GHz have been specified with the following characteristics: single carrier, OFDM and OFDMA for licensed bands, and one based on license-exempt band [17].

2.4.5. Mobile Broadband Wireless Access (MBWA) Network

The broadband fixed wireless access specified in IEEE 802.16 allows inherent mobility only. Another task group within 802.16, that is, IEEE 802.16*e*, is working on recommendations for adding managed mobility with up to vehicular speeds. This group, formed in December 2002, is expected to have the standard ready by the end of 2004. During this time, a new initiative, supported by IEEE by forming the group IEEE 802.20, has started looking for a new network architecture for mobile broadband access networks with very high mobility (up to 250 *km*). A second difference between IEEE 802.16*e* and IEEE 802.20 (other than magnitude of mobility) is that the latter is going to be specified for less than 3.5 GHz spectrum.

The reference architecture of the IEEE 802.20 may look similar to the IEEE 802.16 network and other IEEE 802 standards [18]. The main differences will be with PHY and mobility management.

2.5. CELLULAR DATA NETWORKS

Data transmission over cellular systems has seen major breakthroughs in recent years.[4] Over thirty years of mobile cellular networks advancements have culminated in the form of third-generation (3G) systems, with umbrella name of IMT-2000 under ITU banner and various other names in some other parts of the world. Even though the IMT-2000 systems were not all evolution, they are regarded so due to intermediate IP-based architectures between the 3G and 2G digital systems. It is understood by the cellular networking community that the most important part of the earlier generations core networks (signaling system) will be entirely removed in the near-future releases of the 3G systems. We will concern ourselves with the protocol architecture of only the latest or proposed, with a note on how they evolved. However, before we do that, we will discuss two points about cellular systems. The first point deals with the main difference between EU and North American specifications of cellular networks, and the second deals with the question of why we could not get as high data rates from the cellular network using voice-grade modems as we do from the PSTN, with the same signaling core.

2.5.1. North American and European Cellular Networks

Cellular networks consist of a core part and an air interface. Most of the technologies that we compare are really related to the air interface. This is no

[4] "What are the killer applications?", is still under debate!

coincidence, as the mobility and wireless nature of the medium impacts the air interface most. Roughly, the air interface as equivalent to the three bottom layers of the OSI-RM [19]. The core network is hidden from the user and is as secure and reliable as the PSTN, because traditionally, it has been PSTN.

In the United States, the specification of core network has been standardized as interim system 41 (IS-41, ANSI-41) and we have seen various releases of IS-41. Release B made major strides toward digital systems, while Release C is the current one and used in 3G systems. Throughout much of the North America, IS-41 is the open standard for the core network but there is no 'officially' standardized open air interface. Consequently, a number of air interfaces have been deployed, for example, analog based on FDMA (1G), digital based of TDMA (variety of standards, e.g., IS-54, IS-136, even GSM) and digital based on CDMA (IS-95). The European Union, however, partially learning from the American experience of IS-54, and partially due to international roaming as a major goal, decided to standardize both the core networks (GSM MAP) as well as the air interface. Together, the EU system has been called GSM up until 2G. Of course, there are other major differences, for instance, spectrum allocation, voice coding, and so forth. However, these are major philosophical differences in technology.

2.5.2. Voice-Grade Modems

We have seen some major breakthroughs in speeds of voice-grade modems used over PSTN, yet we do not have the same performance available for voice-grade cellular networks. There are two reasons for this, which can be considered critical. First is the voice coding schemes used in the two types of networks. PSTN internally uses 64 kbps PCM, which is a waveform coding and results in not only a clear vice reception (4.3 MOS), but also in a received signal that is a very good replica of the transmitted signal. By contrast, the cellular systems, focused on bandwidth, use bandwidth-saving coding, mechanisms with data compression, and low-bit rate modeling techniques. In some such coding mechanisms (CELP), the original signal waveform is not transmitted; instead one of the specified codes from a code book is transmitted. The resulting received signal is not as good a reproduction of the transmitted signal as PCM. A second major reason is that the wireless channel introduces multipath, fading, and interference, all non-existent in PSTN. This is in addition to Doppler effect due to mobile station and surrounding mobility. Equalization and additional error-control mechanisms consume part of the spectrum, resulting in much slower modems.

With the onset of 3G, we can now talk about data transmission of the order of several hundred kbps and even Mbps in fully mobile cellular networks. Therefore, we will focus only on 3G and beyond, sometimes referred to as 3G+.

2.5.3. Relative Look at Cellular Network Generations

Figure 2-11 [20] shows a relation of current 3G UMTS (also called WCDMA) air interface with past and future EU standards. Table 2.1 shows an evolution of data rates for various generations [21].

Two industry and standards organizations agreements, the third-generation partnership project (3GPP) and 3GPP2 to have interoperable implementations of the 3G cellular systems, have received much attention from standards associations, operators, and special interest groups. 3GPP is based mainly on

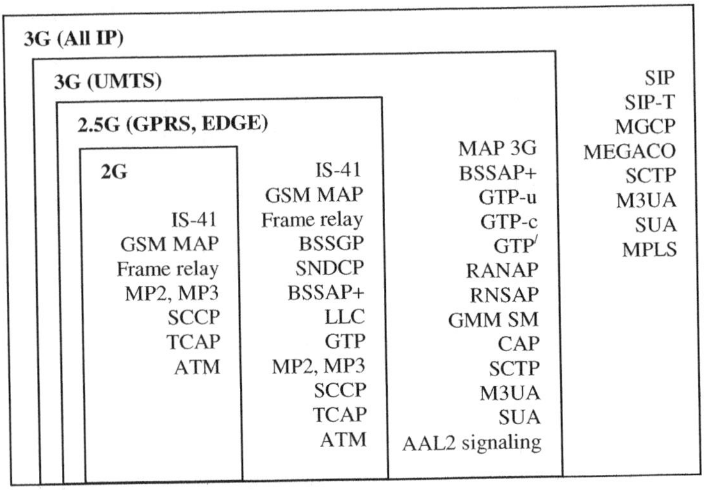

Figure 2-11. Evolution from second- to third-generation and beyond.

TABLE 2.1. Data Rates for Various Cellular Technologies

Technology	Data rate	Remarks
GSM	9.6 or 14.4 kbps	Circuit switched (C/S), 2G
GSM	28.8 to 56 kbps	High-speed C/S, 2G
GPRS (GSM)	Up to more than 150 kbps	IP and X.25 Packet Data
EDGE (GSM)	384 kbps	Packet data, 2.5G
IS-136	9.6 kbps	C/S, 2G
EDGE (IS-136)	384 kbps	Packet data, 2.5G
CDMA	9.6 or 14.4 kbps	C/S, 2G
IS-95B (CDMA)	64 kbps	Packet data, 2.5G
cdma2000 (1XRTT)	144 kbps	IP, 3G
WCDMA/WTDMA	2 Mbps	indoor, IP, 3G
cdma2000 (3XRTT)	144 kbps and 2 Mbps	IP, 2 Mbps indoor, 3G

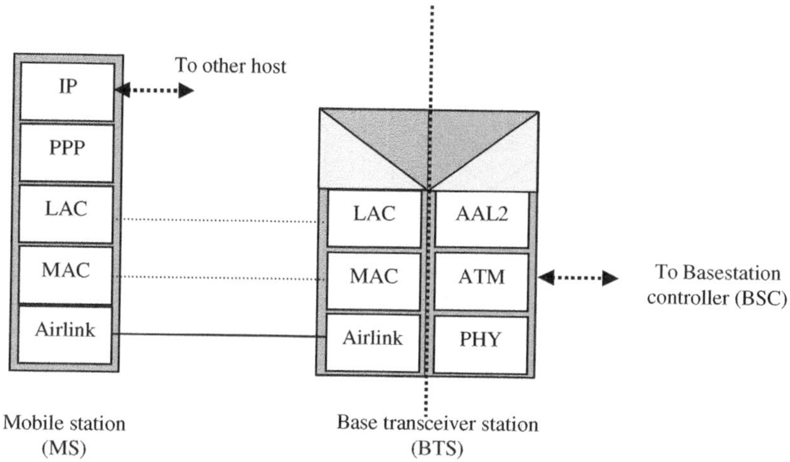

Figure 2-12. Air interface for 3GPP2 mobile station.

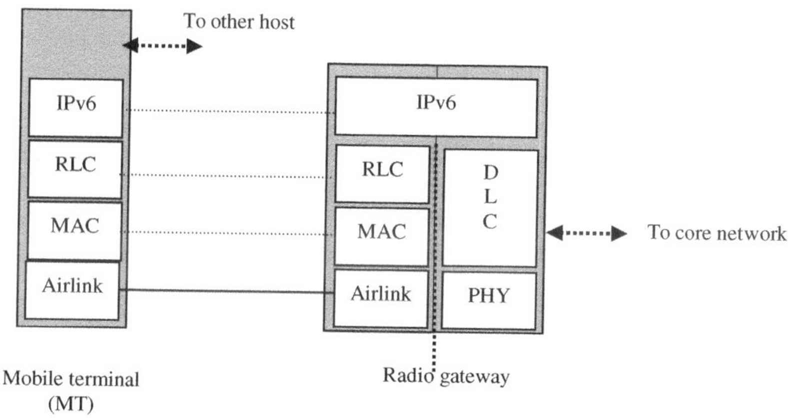

Figure 2-13. Air interface for 3GPP (WCDMA) based Moby Dick architecture mobile station.

WCDMA (the EU camp) and the 3GPP2 is based on the cdma2000 (North America) air interfaces. Figure 2-12 shows the air interface protocol stack for the 3GPP2 [22] all-IP option for 1xEV_DO.

Actual deployment (and even specification of an all-IP network is for the future, but there is general agreement that IP is the future and endeavors in this direction will continue [23]. Figure 2-13 shows the all-IP air interface for Moby Dick, a 3GPP architecture proposal [24]. Note that there is no final version of the all-IP architecture available at this time. The UMTS (GSM-based GERAN and the new W-CDMA) had four revisions already (R99, R3,

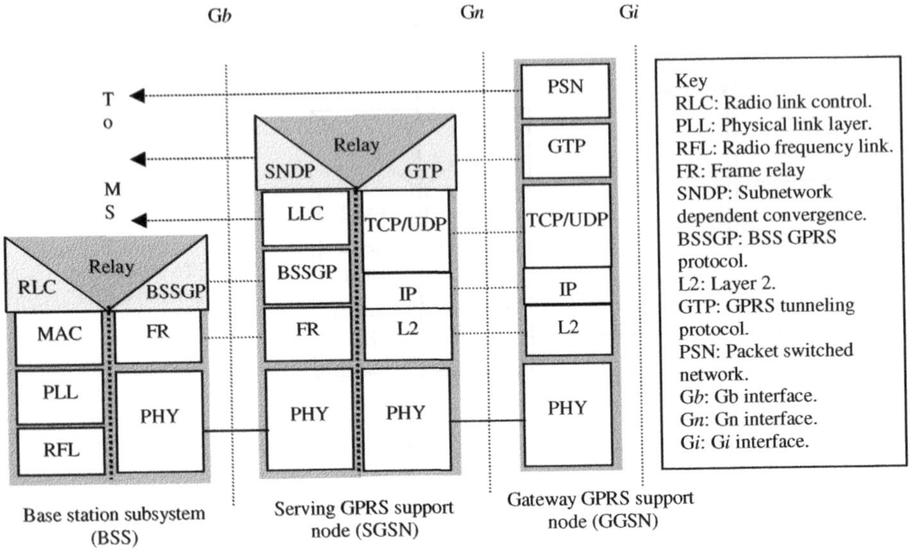

Figure 2-14. GPRS core network. *Gi* is interface between GGSN and the IP router.

R4 and R5) and several proposals for R6 are under review. It is expected to have R6 either entirely IP based or closest possible.

A Chinese proposal for TD-SCDMA (TDD-synchronous CDMA) is at a very elementary stage. Discussions on greater detail of 3G architecture can be seen in chapters 7 and 8.

2.5.4. Core Network

The core network for GSM and its advancements, originally based on an application layer (mobile application part – MAP), added to the PSTN signaling protocol the common channel signaling system number 7 (SS7). In North America, IS-41 is the core network, using the North American version of SS7 with mobile part added for cellular service. Due to its circuit-switched nature, the core network up until 2.5 generations (2.5G) has not been very efficient for data transmission.[6]

The 2.5G systems added packet-based service nodes to the 2G infrastructure to provide IP-like functionality with 56–150 kbps data rates. The general packet radio system (GPRS) is such a network. 3G core networks were initially an extension of the GPRS core network. Figure 2-14 shows its architecture for user data plane [25].

[6] Even though in the 1991 release of IS-41 (B) a document IS-41.5 for data communications was included.

New network databases (e.g., gateway location register – GLR) were defined in the 3G network, in addition to the existing databases in GSM and GPRS (not discussed in this chapter.

2.6. SUMMARY

In this chapter we gave a brief account of the protocol architectures of some wireless data networks. Following this chapter, we will discuss each type of the network in greater detail. The wireless data networks can now be deployed covering a very small area (WPAN) to a very large area (e.g., UMTS). Some latest developments, for example, the ultrawide band (UWB), have made it possible to have wide area wireless data network coverage from desktop to vehicular speeds without any cables. A desktop can be connected to its peripherals, via a wireless PAN, to other machines in the vicinity through wireless LANs, to the ISP through wireless broadband access, and can get mobility services through cellular data connection or, in the future, through a mobile wireless broadband access network.

REFERENCES

[1] Marcus Albrecht, Frank, Matthias, and Martini, Peter, "Bluetooth architecture and services overview: The Role of IP and Quality of Service Issue", *First Workshop on IP Quality of Service for Wireless and Mobile Networks (IQWiM'99)*, Aachen, Germany, April 1999.

[2] XILNK, "Bluetooth architecture: APIs, L2CAP, Link Management, Baseband and Radio", available from http://www.xilinx.com/esp/networks_telecom/bluetooth/pdf_files/bt_architecture.pps

[3] The Bluetooth SiG, *Bluetooth Specifications v1.1*, https://www.bluetooth.org/spec/

[4] Jennifer Bray, "What Are Bluetooth Profiles?", available from http://www.phptr.com/articles/article.asp?p=21336

[5] IEEE Working Group 802.11, "IEEE 802.11 architecture", *Document IEEE p802.11-96/49B*, March 1996. Available from http://grouper.ieee.org/groups/802/11/Tutorial/archit.pdf

[6] Goran Malmgren, Khun-Jush, Jamshid, Schramm, Peter, and Torsner, Johan, "HiperLan Type 2—An emerging world wide wireless LAN standard", *International Symposium on Services and Local Access*, 2002. Available from http://www.issls-council.org/proc00/papers/6_3.pdf

[7] Jamshid Khun, Schramm, Peter, Wachsmann, Udo, and Wenger, Fabien, "Structure and Performance of the HIPERLAN/2 Physical Layer". Available from http://easy.intranet.gr/paper_2.pdf

[8] W.C. Freitas Jr., de Almeida, A.L.F., Cavalcanti, L.F.P., and Lacerda Neto, R.L., "Physical layer performance of HIPERLAN/2 in measured indoor channels", XX

SIMP 'OSIO BRASILEIRO DE TELECOMUNICAC, ˜OES-SBT'03, 05-08 DE OUTUBRO DE 2003, RIO DE JANEIRO, RJ.

[9] Harol Teunison, "BRAN H/2 DLC update: Joint BRAN and 802.NW session". Available from http://nwest.nist.gov/mtg3/papers/teunissen.pdf

[10] "HIPERLAN/2 RLC", available at http://www.kbs.uni-hannover.de/~allert/hiperLAN/rlc0010.html

[11] Markus Radimirsch and Vollmer, Vasco, "HIPERLAN Type 2 standardization: An overview". Available from http://www.ant.uni-hannover.de/Forschung/Public/Kn/1999/RV1999.pdf

[12] Jamshid Khun Jush and Straub, Gilles, "ETSI project BRAN Hiperlan Type 2 for IEEE 1394 applications: system overview", available from http://grouper.ieee.org/groups/1394/1/Documents/br071r00.pdf

[13] Martin Johnsson, "HiperLAN/2 – The broadband radio transmission technology operating in the 5 GHz frequency band". Available from http://www.hiperlan2.com/presdocs/site/whitepaper.pdf

[14] IEEE P802.16, "System architecture for 2–11 GHz co-existence simulation", available from http://www.ieee802.org/16/docs/01/80216c-01_12.pdf

[15] IEEE P802.16, Part 16: Air interface for fixed broadband wireless access systems", available from http://standards.ieee.org/getieee802/download/802.16-2001.pdf under certain conditions.

[16] IEEE P802.16 WG, "Part 16: Air interface for fixed broadband wireless access systems—Amendment 2: Medium access control modifications and additional physical layer specifications for 2–11 GHz", *IEEE STD 802.16a™ –2003*. Available from http://standards.ieee.org/getieee802/download/802.16a-2003.pdf under certain conditions.

[17] Carl Eklund, Marks, Roger B, and Stanwood, Kenneth L., "IEEE standard 802.16: A technical overview of the WirelessMAN™ interface for broadband wireless access", *IEEE Communications Magazine*, June 2002, pp. 98–107.

[18] IEEE P802.20, "Requirements, topics and proposals as discussed at session #4 of 802.20", *IEEE 802.20-30/16r1*. Available from http://grouper.ieee.org/groups/802/20/WG_Docs/802.20-03-16r1.pdf

[19] Davis J. Goodman, *Wireless Personal Communications Systems*, Addison-Wesley Wireless Communications Series, 1997.

[20] Prasad Kallur, "3G Wireless: Evolution, new developments and challenges", *3GSM World Congress 2004, (3GSM'04)*, Cannes, France, Feb. 2004. Slides available from http://www.ccpu.com/pages/buzz/events/3GSM_04_CCPU.pdf

[21] Peter Rysavy, Rysavy Research, "The evolution of cellular data: on the road to 3G". available from http://www.rysavy.com/Articles/3G/3g.htm

[22] Mark Klerer, "Uniqueness and the MBWA PAR", *IEEE P802.20*, available from http://grouper.ieee.org/groups/802/20/SG_Docs/802m_ecsg-02-17.pdf

[23] N. Olaziregi, Micocci, S., Ravasio, G., Bachmann, J., and Berzosa, F., "Architecture of IP based network elements supporting reconfigurable terminals", *SCOUT Workshop*: IST Summit 2003 SRD Architecture. Available from http://www4.in.tum.de/~scout/

[24] Hans Eisiedler, Aguiar, Rui L., Jahnert, Jurgen, Jonas, Karl, Liebsch, Marco, Ralf Schmitz, Ralf, Pacyna, Piotr, Gozdecki, Janusz, Papir, Zdzislaw, Moreno, Jose

Ignacio, and Soto, Ignacio, "The Moby Dick project: A mobile heterogeneous all IP project", Advanced Technologies, Applications and Market Strategies for 3G (ATAMS 2001), Krakow, Poland, June 17–20, 2001, ISBN 83-88309-20-X, pp. 164–171. Available from http://www-ks.rus.uni-stuttgart.de/Publications/MobyDickATAMS_final.pdf

[25] Yi-Bing Lin and Chlamtac, Imrich, *Wireless and Mobile Network architectures*, John Wiley and Sons, 2001, Hoboker, NJ, Chapter 18.

[26] http://www.swedetrack.com/usblue4.htm

CHAPTER 3

COMPONENTS OF A WIRELESS LAN

The meanings of a network are different for different people. For a user, the application is the network. For communications engineer, the physical layer is the network. For network engineer, it is usually layers above the physical layer that constitute much of the networking function. When making standards to specify various components of a network, a standards agency has to keep all types of people in mind. Obviously, this applies to wireless LANs as well. These days, standardization of networking protocols is influenced more and more by user participation. Some standards agencies include users as members to get the essential feedback for the network applications. This has resulted in more comprehensive standards for modern networks as compared to the past. The standards for WLANs, for example, depart from their wired predecessors in many ways. As an example, prioritization of user traffic is an add-on for Ethernet, while it is built-in in many wireless LAN (WLAN) standards.

These and other reasons have rendered WLANs much more complex systems than their fixed counterparts. In this chapter, we will discuss generic components of a WLAN. Standards are, then, a projection to these generic components. Since the wired LANs, especially the Ethernet, are established already and many of their features are emulated in WLANs, a good reference point for discussion would be the wired LANs. We will begin this chapter with a brief discussion on the components of a wired LAN in the next section. This would be followed by sections outlining the differences and similarities between the wired and wireless LANs. Later sections will summarize the discussion into a list of components for a typical WLAN.

Wireless and Mobile Data Networks, by Aftab Ahmad

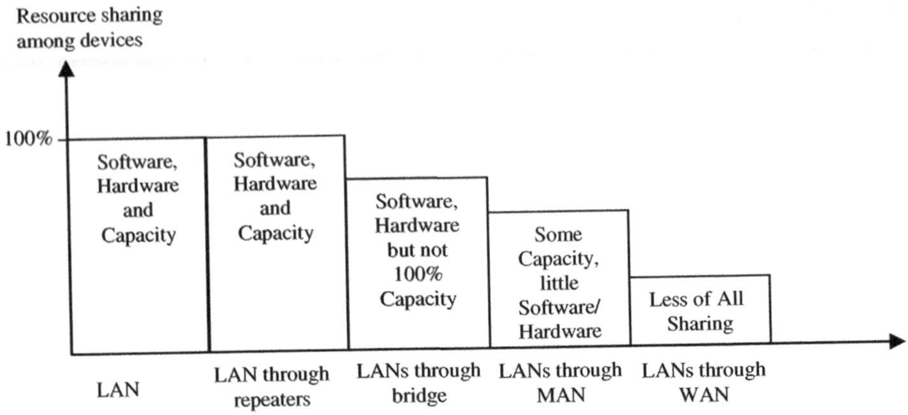

Figure 3-1. One way to discriminate LANs from other networks is by the amount of resource sharing among the network devices.

3.1. LOCAL AREA NETWORKS (LANs)

Local Area Networks [3,6] are networks interconnecting a few devices using software and hardware components and the network capacity. LANs can be extended in many ways. The simplest way to extend a LAN used to be by using repeaters, without changing its characteristics. There are usually a limited number of repeaters used for this purpose. Alternatively, they could be extended through a device called bridge, which could limit the capacity sharing capability between the bridged LANs. Thus the bridge would not allow a packet be transmitted across itself if it is not addressed to a station in the LAN across. These days, hierarchical LAN extension is the best way of expanding a local area network. In this mechanism, hubs are used as concentrators of networked stations as well as other hubs. Thus, a hub can be used to make one LAN connect to another hub with another LAN, and so on. LANs can be extended through metropolitan and wide area networks. This extension limits the sharing capability tremendously, by restraining much of the hardware sharing. The hierarchy of these extension mechanisms is shown in Figure 3-1.

As seen from Figure 3-1, the LANs are responsible for maximum sharing among the connected devices. When we translate this application of LANs into networking functions, we can get an idea that LANs require:

1. A way of interconnection, so that sharing and addressing is possible without a complex mechanism;
2. A medium specification, so that a certain minimum capacity is available for sharing;
3. Physical layer mechanisms, for signal representation and transmission of data;

4. Data link control (DLC) layer mechanisms, specific to the unlimited sharing mechanisms;
5. Either be interconnected via a WAN/MAN for allowing applications to exchange data, or have such capability; and
6. Have traffic differentiation capability if it is going to be used for multimedia.

These requirements translate into the following components.

3.1.1. LAN Interconnection (Topology)

The devices connected via a LAN could have many types of interconnection, such as a star, bus, ring or a tree. In a star LAN, all devices communicate through a central device, variously called a switch, hub, or a concentrator. This central device provides a path for signals from any device to any other device. If it is a switch, it performs the capacity management function as well. If it is an *active hub*, it performs the repeater function as well. If it is a *passive hub*, then it simply provides a physical interconnection without dealing with the signals. Figure 3-2a shows such as an interconnection.

A bus interconnection is rather simple. All LAN devices are connected to a single cable, and therefore exchange data through this cable. The two ends of the cable are terminated so that the signal does not reflect back to interfere with itself in the cable, as shown in Figure 3-2b. A ring interconnection is like bus, except that the two end of the bus are joined together to allow

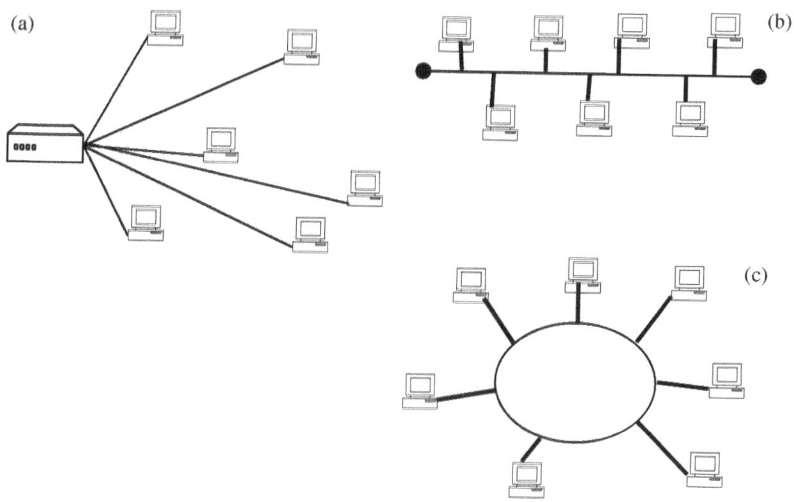

Figure 3-2. LAN interconnection. (a) Star connection; (b) Bus connection; (c) Ring connection.

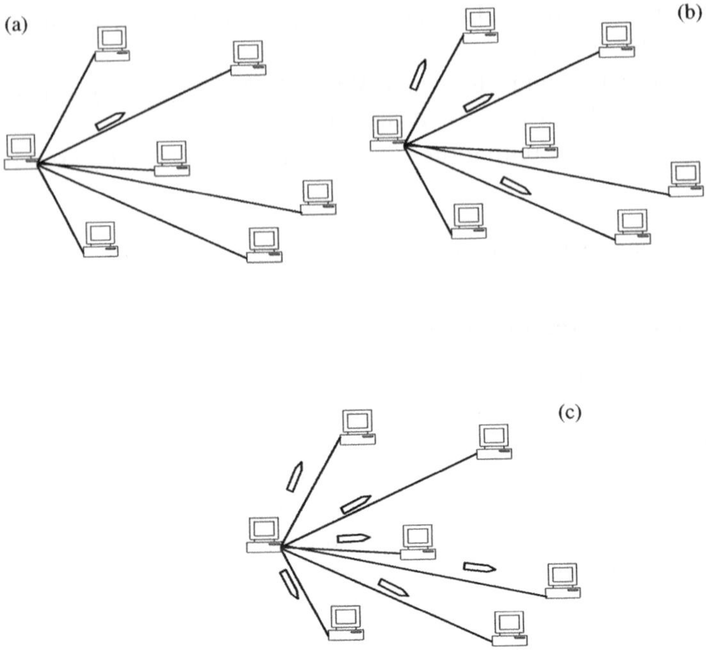

Figure 3-3. (a) Unicast addressing; (b) Multicast addressing; (c) Broadcast addressing.

circulation of packets, as shown in Figure 3-2c. Finally, a tree interconnection could be imagined as being a star of bus topologies, not shown in Figure 3-2.

3.1.2. Addressing Mechanisms

Each station needs an identification to recognize packets addressed to it. In fact, the actual number of addresses could be more than one. There are a minimum of two addresses required to identify each device on a LAN, a unicast address for packets addressed exclusively to the device and a broadcast address for packets addressed to everyone on the LAN. Additional multicast addresses may be needed if a terminal belongs to one or more multicast groups. Figure 3-3 shows the use of these addresses.

3.1.3. Medium Specification

LANs usually require high bandwidth media. Sophisticated signaling techniques are used in the transceiver design in order to get maximum data rate from the available bandwidth. Usually, a cable is part of a LAN standard so that a certain bit rate is guaranteed. Many times we recognize LANs from their cable types (twisted copper pair, coaxial, optical fiber) and sometimes from the bit rate (10 Mbps, 100 Mbps).

3.1.4. Physical Layer Mechanisms

The physical layer mechanisms of LANs are specified in terms of the PHY standards. Since there are many types of cables used, sometimes the PHY specification is broken down into two groups, one taking care of the medium-dependent characteristics and the other taking care of the transmission characteristics of the layer above the PHY. The former is called physical layer medium-dependent (PMD) sublayer and the later is called physical layer convergence procedures (PLCP) sublayer.

3.1.5. Data Link Control Layer

For a reliable logical connection between two stations on a LAN, a DLC layer specification is included in the standards. Part of the responsibilities of DLC layer is to allow fair and efficient bandwidth sharing among all stations because only one point-to-point connection is available at a time. In fact, this responsibility is the defining characteristic of LANs and is handled by the medium access control (MAC) sublayer. The result of a separate sublayer for medium access control is that LANs using MAC divide the DLC function into MAC and another sublayer for standard DLC functions. The other sublayer is called logical link control (LLC) sublayer in IEEE standards. This division of DLC layer into two sublayers allows the use of a common LLC above LANs with various MAC sublayers. The IEEE committee 802 [1] has defined MAC sublayers in recommendations IEEE 802.3 (Ethernet, CSMA/CD), IEEE 802.4 (Token Bus) and IEEE 802.5 (Token Ring) and many others. These MAC standards use a common LLC defined in IEEE 802.2.

3.1.6. Traffic Differentiation

As the demand for multimedia over LANs increases, several options have been brought about by new standards to be implemented in addition to the PHY and DLC layers of a LAN. These include call admission architectures, such as H.323, and user priorities definitions such as IEEE802.1P. Since these standards are not a part of any LAN, their compliance is not mandatory. We can safely say that the fixed LANs are not multimedia capable in the generic sense.

3.1.7. WAN/LAN Connection

Typically, the WAN/LAN connection is controlled by WANs. The Internet, for example, has a protocol defined for the routers to ask all LAN devices to register their LAN address with the router (router being a WAN device). This protocol is called the address resolution protocol (ARP) and could have its mirror image (called reverse ARP or RARP) to help LAN devices enquire their WAN address from a router. Enough intelligence does not exist in wired LANs to route packets among several LANs through a bridging function. A router is essential for wide area connectivity of several LANs.

3.2. WIRELESS LAN COMPONENTS

Wired and wireless LANs are similar in terms of their applications and network architecture. The main applications of both relate to hardware and software resource sharing, and high-speed access to some metropolitan or wide area network. The main architectural components consist of the capabilities of medium sharing, signal transmission and link management. However, there are subtle differences in the ways that these functions ought to be realized in the two LAN types. In this section, we discuss the components of a WLAN [2,4,5] with reference to their architecture.

3.2.1. Physical Layer Components

The physical layer is the most complex layer of any network architecture. It deals with the mechanical, electrical, functional, procedural, and signal-transmission characteristics of the physical interface between a terminal and network. These interface characteristics are specified in the form of many rules, functions, and components. Here is a list of some important components.

3.2.1.1. Station Types. Many wireless networks define terminals with specific characteristics. Among the factors that contribute in classifying a station are:

1. Whether it is a source/sink of data or a relay station or combination. The access point in IEEE 802.11 WLANs is an example of a station that is not a source or sink of data, but only a relay station.
2. What type of data are transmitted/received by a terminal, such as voice, video, store-and-forward (SAF) data. The SAF type of information is the one that could use variable delay to its advantage, such as in flow and error control procedures.
3. Whether it provides full functionality or partial.
4. Sometimes there are stations that have more information available about the resource/link condition than other stations. Such stations could be regarded as intelligent wireless terminals.
5. In wireless LANs, a terminal could temporarily disappear from some other terminals due to the wireless signal not being able to reach

everywhere in the covered area. Such a terminal is called as hidden terminal.

3.2.1.2. Channel Media. Channel media are an essential part of a LAN specification, as they define the bandwidth and geographical span of the LAN. The wireless channel media exist in the form of frequency windows allocated by regulating agencies. The WLANs are generally designed in the ISM bands. These bands are regulated for the amount of radiation per device. This makes such bands vulnerable to both involuntary and malicious interference. Even though radio frequency covers a wide range, the regulated spectrum is hardly above 300 GHz. Therefore, the use of an unregulated part of the spectrum, especially optical communication, is another possible medium. In summary, there could be four types of channel media, at least in theory:

1. The unlicensed ISM band.
2. Licensed bands, such as broadband access bands (LMDS, MMDS) and 18 GHz band, used in some commercial wireless products.
3. Unregulated bands, such as above 300 GHz.
4. Visible or invisible light, such as infrared.

Due to the popularity of the license-free bands, more than one band are allocated for ISM and more bandwidth could be expected in future. For example, ISM is allocated in the 900 MHz, 2.4 GHz and 5.7 GHz (USA) or 5.2 GHz (EU) ranges. Since anyone can generate radiation in these bands, special interference-combatting techniques are required for successful data transmission. There are many such techniques available in practice, each defining a different channel medium. Therefore, we can say that the channel media within the ISM band could be specified in terms of:

1. The ISM band, e.g., 900 MHz, 2.4 GHz or 5 GHz.
2. Interference rejection mechanisms, e.g., many types of spread spectrum communications. Figure 3-4 summarizes the above discussion.

3.2.1.3. Physical Link. Since the wireless channel is a shared medium, the physical link between two stations in a WLAN could be defined in more than one ways. Two such mechanisms were discussed in Chapter 1, in the form of ah hoc and infrastructure LANs. In ad hoc LANs the wireless link is established between the wireless terminals directly, whereas in infrastructure LANs, the link is established through a central point, the access point (AP).

3.2.1.4. Signal Conditioning. The signal conditioning is done to maximize the bandwidth utilization and make it possible for the receiver to easily recognize signals. In fixed LANs, baseband modulation schemes, such as NRZ-I, Manchester, and differential Manchester, can easily achieve the goals. How-

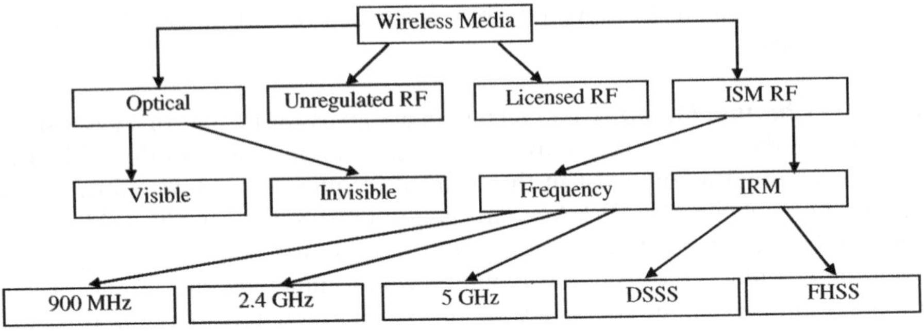

Figure 3-4. Summary of the choices for WLAN channel media.
RF = Radio Frequency Band.
IRM = Interference Rejection Mechanisms.
DSSS = Direct Sequence Spread Spectrum.
FHSS = Frequency Hop Spread Spectrum.

ever, wireless channels can't usually use baseband modulations because frequencies close to zero Hz cannot travel through the wireless medium. In fact, the interference-rejection mechanisms (IRM) mentioned in Figure 3-4 are part of signal conditioning. We will have a brief discussion on these methods.

3.2.1.5. Interference-Reduction Mechanisms

Method 1: Limiting Radiation Power. This method is trivial in the sense that it does not eliminate the problem of interference to a satisfactory level. It helps in one way, that is, the signal energy in microwave bands (1–40 GHz) deteriorates quickly. By fixing the maximum radiation power the regulating agencies make sure that a user will interfere only within a short area. In indoor communications, the path loss exponent is between 3 to 5. That means, for a distance of 100 times the reference distance, a 100 mW (20 dBm) signal will be reduced to $20 - 10n\log_{10}(100)$, which is -60 dBm or 10^{-6} mW for $n = 4$. This is the effect of distance alone. High microwave frequencies further reduce the power of the received signal.

Method 2: Using Spread Spectrum (SS) Techniques. Spread spectrum techniques make it possible to detect a signal transmitted with power level below the interference (or noise) power. It is made possible by spreading the signal bandwidth in a unique way, known only to the intended recipient. This serves more than just the interference-rejection properties; it also keeps outsiders away. In fact, these mechanisms were employed in military communications equipment to combat jamming interference. There are a number of ways to design SS techniques. Two of the more popular are direct sequence SS (DSSS) and frequency hopping SS (FHSS).

Direct Sequence Spread Spectrum (DSSS). In DSSS the spreading is done for each bit of data. A bit is sliced into many pieces along the time axis, each one called a chip. The number of chips per bit determines the additional bandwidth needed for spreading. This number is sometimes called as processing gain. The pattern of chips for a bit is fixed for all the bits of a user for a transmission session. This pattern is called pseudo-random noise (PN) code. It is known only to the intended user and appears as noise to others. Figure 3-5 shows the concept of DSSS.

Frequency Hopping Spread Spectrum (FHSS). In frequency hopping, spreading may be done by dividing the wide channel bandwidth (B_C) into many narrow channels, say, each with a bandwidth close to the signal bandwidth (B_S). The signal is transmitted using one of the narrowband channels for a short duration, after which this channel is replaced with another narrowband channel, and so on. Thus, by changing the narrowband channel in a pattern known only to the intended receiver, a frequency-hopping pattern is implemented. This hopping is called fast frequency hopping if the channel hopping speed is faster than the bit rate. In contrast, if more than one bit is transmitted per narrowband channel, then we call it as slow frequency hopping spread spectrum (SFH-SS).

The net effect of spreading is that the average signal power is divided all over the wide channel bandwidth, making the signal low level, as depicted in Figure 3-6.

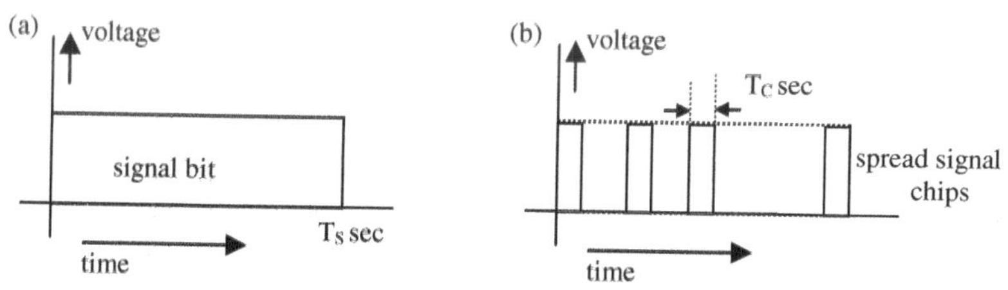

Figure 3-5. (a) Signal bit before spreading. T_S = bit duration; (b) Signal bit after spreading. T_C = chip duration.

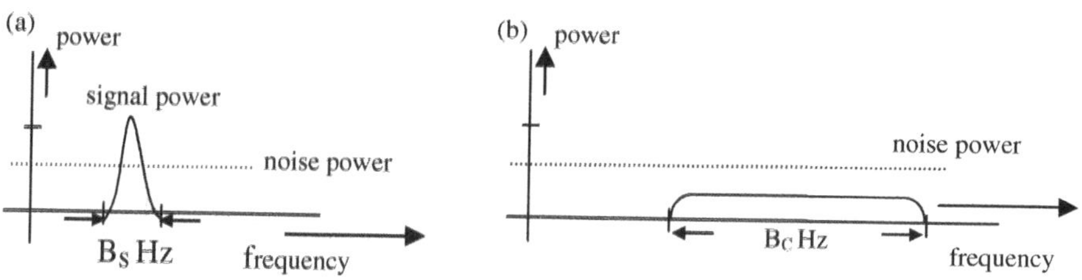

Figure 3-6. (a) Signal power spectrum before spreading; (b) Signal power spectrum after spreading.

3.2.1.6. *Modulation of Signals.* Since wireless bandwidth and media act to limit the WLAN data rates, efficient modulation techniques may be required for radio frequency ranges. Modulation of baseband pulses serves many purposes; namely,

1. Increasing the bandwidth efficiency by allowing multi-level modulation. If a modulation scheme allows to choose one of the M symbols to be transmitted in a certain duration, this equals to transmitting $k = \log_2(M)$ bits. In other words, the bit rate is k times the baud rate. M is chosen to be a power of 2, that is, $M = 2^k$, so that k is a whole number.
2. Baseband pulses attenuate more rapidly due to the presence of low-frequency components. The power transmitted in lower frequencies is wasted. By modulating signal to a carrier, power is shifted to carrier frequency and its surrounding spectrum.
3. Many modern modulation schemes allow imbedding error-control mechanisms in them. An example is the family of trellis coded modulation (TCM) schemes. This adds an error-control mechanism to modulation.
4. Modulation schemes are generally power efficient, which help prolong the battery life, a feature not that important in fixed LANs.
5. Another source of power waste in baseband transmission is the abrupt discontinuities in the pulses that induce power in very-high-frequency components. These components may be filtered out at the receiver, thus not being able to utilize energy they carry. Modulation schemes with continuous phase, such as minimum shift keying (MSK), leak minimum power to these frequency components.

3.2.1.7. *Data Transmission.* The physical layer data consists of frames, signals, and bits. In relatively new standards, only synchronous transmission is recommended on the physical layer. The physical layer frame could be defined separately for each media if several media could be used under a common physical layer. Since WLANs are relatively new technologies, it is expected that synchronous transmission of frames takes place between the two PHY protocols through the radio frequency (RF) interface as against the asynchronous transmission.

3.2.1.8. *Convergence Procedures.* Due to the possibility of having many types of channel media, a single WLAN architecture may require separate protocol mechanisms, one for each of channel media. Due to a variety of possible media, the convergence procedures are essential for any standard. Convergence functions are usually defined in the form of a convergence sublayer of the physical layer. It could have a frame format that takes care of the details of individual physical medium dependent sublayers.

3.2.1.9. *Rate Selection Capability.* Many physical layer protocols provide the capability of rate selection. This capability can be useful, as the wireless channel is generally unstable. Protocols can be made to fall back on lower rates during error bursts and then come back to the high rates when the channel is clear. If a station, such as a VoIP terminal, values quality more than the transmission rate, it can make the right choice.

3.2.1.10. Synchronization, Flow and Error-Control Capabilities. Even though PHY data exist at the signal level, the availability of a choice of channel media and the convergence requirements significantly add to the complexity of this layer. The resulting protocol data unit (PDU) consists of fixed or variable length frames. It is then possible to add the functions related to reliable frame exchange between two physical layers. In wireless media, this could be essential due to exaggerated channel errors as compared with wired media. In fact, placing the complexity at this layer relieves the logical link layer of added functionality for wireless networks. Consequently, a common LLC protocol can be employed between a station in WLAN and another station in fixed LAN.

3.2.1.11. Physical Layer Management. The provision of the above functions makes the physical layer a highly complex system. A large number of physical and logical elements, associated procedures, and their interaction may require many management functions to be included in a standard. These functions could be used to generate system state report, to set or get values of various parameters and even schedule changes in future. The physical layer management protocols facilitate management and control of physical layer parameters. It consists of a logical database defining management parameters and a protocol that could allow the use of this database. Figure 3-7 shows the physical layer management paradigm.

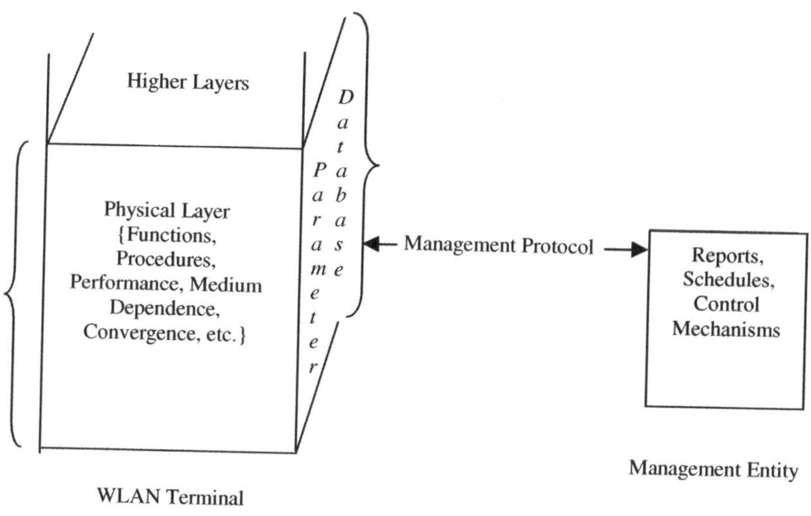

Figure 3-7. Physical layer management paradigms.

3.2.2. Medium Access Control (MAC) Layer Components

The MAC layer defines the 'personality' of a local area network since the main function of a LAN is to create a trust among the attached stations to share capacity and other resources. As mentioned earlier, the MAC layer is one of the two parts of what is called the data link control (DLC) layer in OSI terminology. Due to this reason, sometimes it is called a sublayer of the DLC layer. A MAC sublayer could consist of more than one sublayer. Any MAC layer has two primary functions: channel access and multiple access. Due to the wireless nature of the channel and due to the multimedia capability, other important functions may be added. Here's a brief list.

3.2.2.1. Network Configurations. Just as the physical layer defines station types, the MAC layer could specify various types of network configurations. These depend on the accessibility of the MAC layer protocol and the coordination among various stations. In wired networks, the network configuration is a function of topology, while in WLANs configuration is a logical concept to organize stations in identifiable groups. Though relatively new, this concept does exist in fixed LANs and WANs as virtual LANs/WANs, such as defined by IEEE 802.1Q standard for LANs and as virtual private networks (VPN) in IP networks. The IEEE 802.11 Working Group has defined several types of configurations that we will discuss in Chapter 5, on IEEE 802.11 MAC. These include basic service set (BSS) and extended service set (ESS).

3.2.2.2. Channel Access. Since LANs are generally shared media networks, a mechanism for channel access is required so that quick and fair access by all stations is possible with priority where needed. Most of the channel access mechanisms could be classified into one of the two types, contention based and controlled. In contention-based access, the terminals contend for access under a set of limitations. The mechanism with minimum amount of limitations is called Aloha, in which every station assumes that the channel is available whenever needed. Naturally, this could lead to lowering of channel utilization due to collision of data from simultaneous transmissions. A number of access mechanisms have been developed over the past three decades that relate with reducing the collisions and improving the throughput. Contention-based access (also called random access) not only lowers the channel utilization, but it also introduces variable delay that could result in jitter in real-time applications. Let's have a brief look at one of the random access mechanisms.

Carrier Sensing—Random Access. Ethernet uses carrier sense multiple access with collision detection (CSMA/CD), which gives fairly high throughputs as compared with Aloha and its variants. Carrier sensing ascertains that the channel is idle and collision detection helps notify the stations of an ongoing collision to avoid transmission. Though an extremely popular mechanism, it cannot be used efficiently for wireless LANs as collision detection

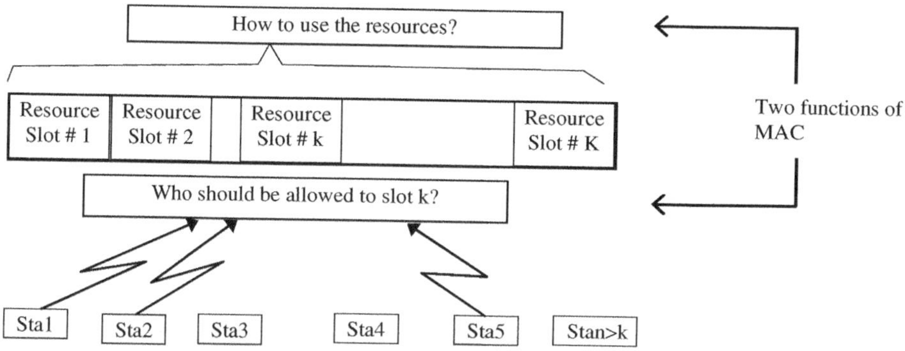

Figure 3-8. Two main functions of medium access control procedures.

requires energy levels higher than the peak voltage in the channel. Energy attenuates too quickly in the case of wireless media. However, CSMA and its other variants are possible through detecting the presence of the carrier signal.

Controlled Access. Many LANs use controlled access to ensure a minimum throughput or limited maximum access delay. This could be crucial in many cases, such as in manufacturing, where terminals could be robotic arms performing specific jobs in a given order at specific times. Such mechanisms are very useful when a large number of stations are active, as there are no collisions. However, their implementation and maintenance are complex.

Combined Random and Controlled Access. It's possible to combine the two access mechanisms into one. One way to do this is by randomly allocating waiting times whenever a station tries to access a channel. The random access results in success or failure of getting a waiting time. The amount of waiting time results in a controlled transmission. Such mechanisms could be a good candidate for delay-sensitive terminals.

3.2.2.3. Multiple Access. Multiple access mechanisms determine the extent of channel sharing. Once a station has accessed a channel, there is the question of how long or what part of channel resource should be allowed to the station. Figure 3-8 shows a relation between channel access and multiple access. There are a number of ways in which multiple access mechanisms are implemented. We will look at the most popular ones.

Packet Division Multiple Access (PDMA).[1] By far the most popular mechanism in LANs, it limits transmission to a single packet once a station has gained access to the channel. A maximum packet length is specified to limit

[1] This is not a standard term, and has been introduced here because it describes the concept clearly.

the amount of data transmittable in one transmission. If a station has a sequence of packets generated, all packets have to go through channel access procedure, thus possibly introducing jitter. This method is a natural partner of random access. Thus, in a LAN, data packet is used for both functions of MAC, namely, channel access and multiple access. The term packet division multiple access (PDMA) is not a standard one, and has been introduced here because it describes such mechanisms most appropriately.

Resource Reservation. An alternative to going through channel access procedures for every packet is to reserve channel resources for stations with successful access. Channel resources could be defined in many ways, such as a specified maximum data per access, a specified channel time, or a specified fraction of channel bandwidth. These result in various reservation mechanisms. They are frequency division multiple access (FDMA), time division multiple access (TDMA), and data division multiple access (DDMA). Let's have a brief look at each.

Frequency Division Multiple Access (FDMA). In this mechanism, the channel bandwidth is divided into smaller frequency bands, each one may be called as user (or control) channel. Once a station is successful in accessing the medium, one or more channels are allocated for the duration of transmission. It's been the mainstay of first-generation cellular systems. Figure 3-9 shows a schematic for FDMA.

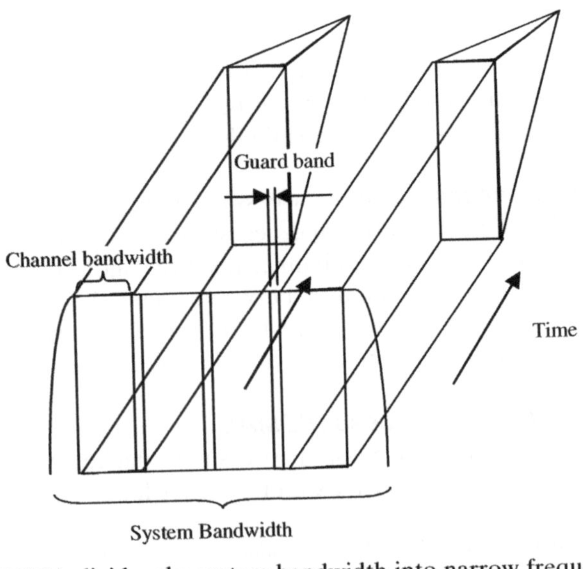

Figure 3-9. FDMA divides the system bandwidth into narrow frequency channels.

Time Division Multiple Access (TDMA). In this mechanism, the channel transmission rate is divided into smaller transmission slots (called time slots). Once a station is successful in accessing the medium, one or more slots are allocated for the duration of transmission. Second- and higher-generation cellular systems use this mechanism. In Figure 3-9, if we switch the Time and Bandwidth axes, it will be approximately the TDMA scheme.

Data Division Multiple Access (DDMA). This mechanism enforces a maximum amount of data that a station could transmit once it has succeeded in accessing the channel. A simple application of such a mechanism would be in centralized control where a station (called primary station in polling mechanisms) allows another station (secondary station) to send a specified chunk of data. The same amount of data from different stations does not take the same amount of transmission time, since each link to the station could be of different capacity. However, when a single link is used or all links have same capacity, it becomes similar to TDMA. If this mechanism is utilized in WLANs, then due to channel quality variation, it will remain different from TDMA.

In code division multiple access (CDMA), a mechanism similar to this, the stations are not restricted in the amount of data they can transmit, but in the amount of data rate at which they can transmit, which relates more to FDMA and TDMA. We will have a brief discussion on this in the next section.

Code Division Multiple Access (CDMA). In this novel DDMA mechanism, stations share the bandwidth by simultaneously transmitting their information at a bit rate much smaller than the total channel bit rate. Each user utilizes the whole channel bandwidth as against TDMA or FDMA. This is possible because user signal is spread using techniques such as direct sequence spread spectrum (DS-SS) technique. Each bit from each user is transmitted as a pattern of chips using unique spreading code. User signals could be detected through the knowledge of their PN-codes. The net result is the same effect as dividing channel capacity into smaller time slots as in TDMA or smaller frequency channels as in FDMA. However, one difference between CDMA and TDMA/FDMA is that the number of possible users is upper-limited in TDMA and FDMA, while it is not fixed in CDMA. By employing appropriate PN-codes and depending on channel conditions, the exact number of CDMA users for the same channel could be higher or lower than TDMA/FDMA. Different stations can be allowed to transmit at different rates under several conditions, such as:

1. By allocating more than one code to a single user, thus allowing multiples of a basic rate.
2. By accommodating more users for less data rates.
3. By reducing transmitted power of user signals, thus creating many subsets of a standard rate.

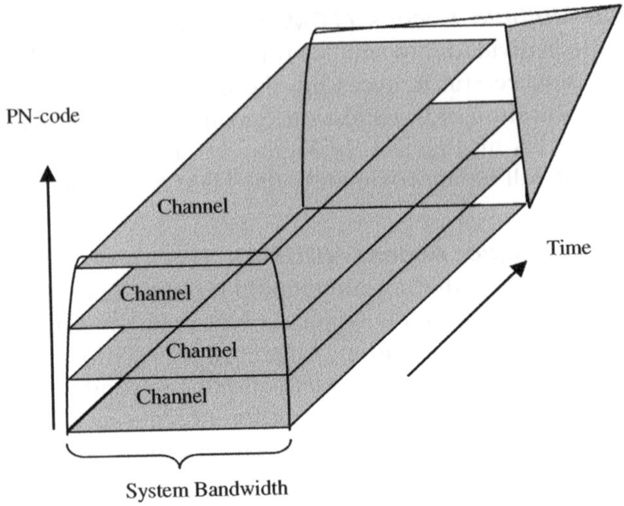

Figure 3-10. In CDMA, all users use all bandwidth for all the time. Unique PN-codes ar used to separate the users.

Figure 3-10 depicts the relation between time and frequency for a CDMA system. The PN-code axis in Figure 3-10 shows that each CDMA channel consists of the time/frequency plane that extends to the whole system bandwidth (sometimes called the CDMA channel) and to the call duration on the time axis. The figure depicts that user channels (planes) are parallel to one another. This is not in the literal sense. It is just to show that the PN-codes make these channels non-interfering to one another. In practice, all CDMA users interfere with all other users. PN-code properties make it possible to separate users as if they were non-interfering.

3.2.2.4. User and Data Privacy. Wireless medium is compromised by its very nature. Software and hardware sharing capabilities make a WLAN highly insecure. The problem worsens if an unlicensed band is used. Therefore, as a departure from fixed LANs, the WLANs may require multiple layers of user privacy. These could be roughly divided into three categories:

1. Network level, to permit resource access from other fixed or wireless LANs.
2. User and device level, to permit only legitimate users and devices.
3. Data level, to permit a user to hide data for security reasons. Using some public or private key encryption could hide data. In case a public key is used, a key distribution and confirmation mechanism may be required as well. The topic of security has seen many recent developments partly due to web services being used extensively. The wireless environment makes them only more crucial. A network standard could contain secu-

rity measures as part of its protocol stack, or could have a parallel stack for security protocols at all layers. What is beyond doubt is that security requirements are redefining the way we look at a network. A new plane has been added for security along with the user data and management planes.

3.2.2.5. Power-Management Mechanisms. Power-management mechanisms serve two purposes, to control interference, to enhance signal quality, and to increase the battery life, by transmitting at a minimum acceptable level. Both of these issues are central to a WLAN, especially if it uses the ISM band. The MAC layer could be used to provide this function once a physical link is defined between two stations.

Due to the unreliability of the wireless channel, the received signal strength fluctuates widely. Usually, there are three different situations in which power management mechanisms are needed. First, in order to help the wireless terminals to transmit at the minimum power levels, to save battery power. Second, to help the central point transmit at a minimum power level, to cause minimum interference. Third, to allow mobile terminals a chance to use minimum power consumption in an idle state. Therefore, a part of power-management mechanisms is to monitor the received signal strength. A second part of such a mechanism is to give feedback to the transmitter, so that it can increase or reduce the transmitted signal power. The frequency of feedback depends on channel type and conditions. Usually, channels with high mobility require higher frequency of feedback than the low-mobility channels.

A second mechanism of saving power is to allow wireless terminals to go into 'sleep mode' when they are inactive. The sleep mode could be implemented by many ways, such as:

1. Letting the wireless terminal use very lower power electronics to detect incoming calls. A fixed, controlling station generates warnings of incoming calls at a substantially high voltage level. After receiving this warning signal, the terminal could turn its receiving circuits on.
2. To allow the receiver to sniff through certain channels occasionally to check for incoming data. In this case, the terminal can be totally off for the rest of the time.

3.2.2.6. Fragmentation. Fragmentation results in a large packet broken down into smaller packets. It serves three purposes. First, it is possible that the packet is being delivered to another network or protocol that has a limit on the packet size that is smaller than the MAC frame. Second, fragments of a large packet could be transmitted as individual small packets, each requiring channel access procedures, thus giving chance to other stations to compete for channel access. Third, if a delay-bound or high-priority packet is generated during the transmission of fragments, it has some chance of getting channel access before the fragmented packet completes transmission.

3.2.2.7. Multimedia Service. Since the early 1990s, the demand for Internet-related multimedia services has been growing [7]. Several protocols and service architectures have been included in the Internet suite of protocols [8,9]. The main issue with multimedia is the variety of quality of services required for each type of information. In the broadest terms, we have the following types of services:

1. Delay sensitive services, which require a certain limit on end-to-end delay.
2. Loss sensitive services, which require a limit on packets lost during transmission.
3. Bandwidth sensitive services, which require a certain minimum guarantee of bandwidth end-to-end.

There could be applications requiring more than one service classes together. It turns out that the access network (LAN, WLAN, serial line) is more likely to become the bottleneck as the Internet backbone network shifts to ultra-high-speed wave division multiplex (WDM) links using multiple protocols label switching (MPLS). It then becomes an extra function of the medium access layer to combine access and multiple access mechanisms into services that could be used for multimedia applications.

3.2.2.8. Packet Forwarding. In ad hoc networks, a wireless terminal may be required to perform a relay function to forward a packet destined to a terminal in another location not reachable directly by the sender. This requires the routing capability, which could be adapted to the changing topology of wireless LANs. A mechanism may be required to allow wireless stations to create, update, and maintain a routing table of neighboring devices for packet forwarding. The wireless PANs are not projected to cover areas that would require routing in general. However, as their applications increase to include anything from cordless phone through wireless LAN to parking garage remote control, at some stage a routing function may become necessary or at least convenient. The 2.5G+ cellular networks have IP interfaces, while latest generations may be entirely based on IP infrastructure. Routing mechanisms that suite the wireless environment are being developed and tested in academia and industry all over the world.

3.2.2.9. Mobility Support. In addition to the routing capability, the medium access control must have functions to support mobile users when they cross the boundary between two wireless LANs. Due to haphazard radio coverage, the wireless domains are not clearly defined. Therefore, in a geographical area covered by more than one group of WLANs, a station may have to redefine its group while in motion. This new group could be redefined in one or both of the following ways: register with a new access point for infrastructure networking, or register with a new group of wireless terminals for ad hoc networking. Mobility requires a number of functions, including the following:

1. Channel sensing from more than one access points, for infrastructure WLANs;
2. Table creation and updating mechanism, for nearest neighbors in an ad hoc environment;
3. Packet or route forwarding mechanism in access points, for those terminals that have left their coverage area;
4. Registration mechanism, for terminals as belonging to a WLAN group and as visitors;
5. Resource re-allocation mechanism, for visitors who has reserved resources at the previous access point or ad hoc network during transition to the new WLAN territory; and
6. Direction determination mechanism, for the access point to warn the moving terminals before they cross the boundary. This warning could include information about the next possible access points and their resource-allocation principles.

3.2.2.10. MAC Layer Management. Just like the physical layer, this layer also needs a way of managing the connections, parameter values, and resources. Separate MAC management functions, procedures, and protocols may be required in a WLAN standard.

3.2.2.11. MAC Frames. The medium access layer may require a number of packets to perform the functions mentioned above. These frames could be the carriers of signaling, access, control, and user data. The protocol function can be realized in the form of fields in the frames and in the form of sequence of frames to be exchanged. Figure 3-11 shows an example of the use of two frames, one for data and the other for acknowledgement. As seen from this figure, there are various times involved in the exchange of data and ACK frames, such as propagation time (Prop.), processing time (Proc.), and protocol time (Prot.).

3.2.12. Teleconferencing Capability. Multicasting capability requires several functions in a protocol or network. These include the abilities to use a single packet for the whole multicast group and to adjust to the data rate of

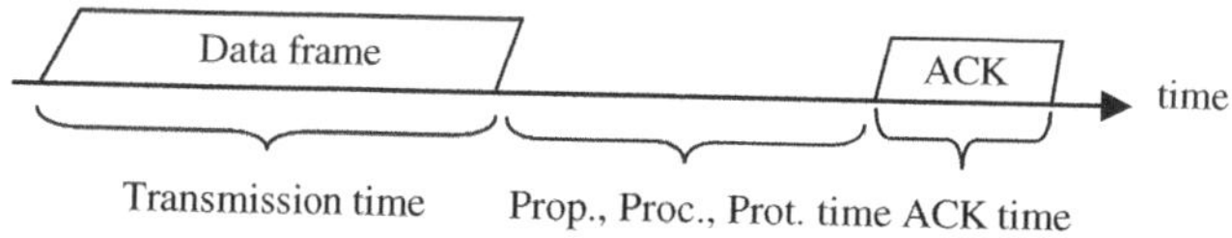

Figure 3-11. Exchange of data and ACK frames.

individual recipients. In addition to that, it requires signaling capability among various stations in multicast group as well as.

3.2.3. Logical Link Control (LLC) Layer

The LLC sublayer performs the same functions for wired and wireless LANs. There are benefits of using a single LLC on both types of LANs. These benefits relate to performance enhancements due to wireless station to fixed station LLC connection being possible without a need to translate the two. Also, the access point in infrastructure WLANs does not have to process LLC header either between two wireless terminals or between a wireless and a wired terminal. Since the main job of LLC is to allow reliable exchange of frames, it may be argued that WLANs require more robust LLC than fixed LANs. However, by adding reliability functions on PHY and MAC layers, a common LLC for wired and wireless LANs can result in more efficient interoperability.

REFERENCES

[1] IEEE, IEEE 802 LAN/MAN Standards Committee, available from http://grouper.ieee.org/groups/802/

[2] Matthew S. Gast, *802.11 Wireless Networks: The Definitive Guide*, O'Reilly Networking, O'Reilly and Associates Inc. Sebastopol, CA, 2002.

[3] James Martin, *Local Area Networks, 2nd Ed.*, Prentice-Hall, Upper Saddle River, NJ, 1994.

[4] Jim Geier, *Wireless LANs: Implementing Interoperable Networks*, Macmillan Technical Publishing, 1999.

[5] Benny Bing, *High-speed Wireless ATM and LANs*, Artech House, Norwood, MA, 2000.

[6] William Stallings, *Local and Metropolitan Area Networks, 5th Ed.*, Prentice-Hall, Upper Saddle River, NJ, 1997.

[7] Vijay Madisetti and Argyriou, Antonios, 'A Transport Layer Technology for Improving QoS of Networked Multimedia Applications', Internet Engineering Task Force, INTERNET-DRAFT, July 2002.

[8] R. Braden, Clark, D., and Shenker, S., 'Integrated Services in the Internet Architecture: An Overview', Network Working Group, Request for Comments 1633, Internet Engineering Task Force, June 1994.

[9] S. Blake, Black, D., Carlson, M., Davies, E., Wang Z., and Weiss, W., 'An Architecture for Differentiated Services', Network Working Group Request for Comments 2475 Internet Engineering Task Force, December 1998.

CHAPTER 4

WLANs: THE PHYSICAL LAYER

In this chapter, we will look at the physical layer components of some known wireless LAN architectures. The physical layer, by far, is the most complex layer in any communications network. The complexity and range of functions frequently requires this layer to be divided into sublayers; the physical medium-dependent (PMD) sublayer and the convergence sublayer. The PMD layer specifications relate to logical and physical transmission of higher layer protocol data units (PDUs). Logical functions relate to mapping bits into signals and back. Physical functions deal with signal coding and modulation to prepare it for transmission over the wireless medium. Convergence layer performs logical functions to prepare data from various PMD layers for higher layers. In addition to these components, the physical topology of local area networks is sometimes defined as part of physical layer. Topology has implications on both the physical and MAC layer functionalities. Due to particular characteristics of the wireless medium and an increasing demand for wireless multimedia services, various station types defined could also be associated to this layer. In the ensuing sections, we will look at these and other physical layer components of the IEEE 802.11 wireless LAN standards and ETSI HIPERLAN. We will try following the sequence of physical layer components from Chapter 3.

TABLE 4.1. Various Standards in the IEEE 802.11 Suite

Name	Purpose	Remarks
IEEE 802.11	Working Group. Also the 1, 2 Mbps WLAN standard.	3 PHYs, DSSS, FHSS and DIR and CSMA-CA based MAC. First two use 2.4 ISM band.
IEEE 802.11b	5.5, 11 Mbps extension	Uses 2.4 GHz band.
IEEE 802.11b-cor1	Corrigendum to the management information base (MIB).	MIB additions related to IEEE 802.11b.
IEEE 802.11a	6, 12, 24 Mbps extension	Uses 5.7 GHz U-NII spectrum. Optional 9, 18, 36, 54 Mbps.
IEEE 802.11g	22–54 Mbps	Uses 2.4 GHz band.
IEEE 802.11e	MAC enhancement	For Quality of Service
IEEE 802.11h	5.7 GHz band management	Spectrum Management for IEEE 802.11a
IEEE 802.11i	Addresses security concerns	Includes IEEE 802.1X etc.

4.1. IEEE 802.11 STANDARDS SUITE

The IEEE 802 (LMSC—LAN/MAN Standards Committee) has a series of standards at 2.4 and 5.7 GHz bands. These standards are prefixed as IEEE 802.11, the name of the Working Group for these standards. We address them collectively by calling them the IEEE 802.11 Standards Suite. Table 4.1 shows the definitions of the various recommendations/standards specified by various Task Groups.

Another standard expected to be used extensively with these recommendations is the IEEE 802.1X for port-based authentication. More details of this are delayed until Chapter 9. In this section, we look at the PHY components of the IEEE 802.11 suite.

4.1.1. Station Types

There are two ways of defining stations in a WLAN, from a general application point of view and from the standpoint of specific characteristics. Standard 802.11 takes both views into account to define the following three general types of station and some specific types of stations.

Access Point (AP). The AP is the central station in an infrastructure network. It performs network and communications control functions, such as authentication and association. It may optionally implement the point coordination function (PCF). If the PCF is implemented, the AP controls polling as well. An AP usually acts as a portal (see below) as well, providing interface with another LAN.

Portal. The standard defines portal as a device that provides bridging function between IEEE WLAN and another LAN.

Mobile Station (STA). The mobile station is the originator or destination recipient of the user data. In networks without AP, the STAs also need to have some of the packet relay functions. All devices, including mobile stations, implement the distributed coordination function (DCF). In order to use shared key exchanges, a STA must implement a wired equivalent privacy (WEP) or other algorithm.

Point Coordinator (PC). The point coordinator is a device that provides and controls PCF function. Typically, it would an AP.

CF-Pollable Station. A CF-Pollasle station is a station that has the capability to respond to the polling message of a PC during the contention free (CF) part of the MAC cycle. This could be a video conferencing terminal or a telephone.

4.1.2. Channel Media

There are two radio frequency bands and an infrared band employed in the standard. Table 4.1 shows a relation between the standards and media [2]. The bands could be used in various ways, discussed in the next section.

The 802.11b chipset (ACX1000) from Texas Instruments uses highly efficient modulation to provide up to double the bit rate. Similarly, the 'Turbo' chipset from Athero gives rates in excess of 72 Mbps for 802.11a.

The actual bit rates are lower than the projected maximum. In fact, the working group for the standards targeted a bit rate of about 3–8 Mbps for 802.11b [3] and above 20 Mbps for 802.11a [4].

4.1.3. Physical Links

The standard allows three types of physical links, as shown in Figure 4-1.

1. A point-to-point link between end stations. This is to be used for independent networks. At the routing level, this link could be either point-to-point or point-to-multipoint, depending on the type of routing. However, modulation techniques at the PHY allow for this link to be treated as a point-to-point link.
2. A point-to-point link between a station and an access point. The access point in this case acts as a relay station.
3. A diffused infrared link for infrared communications. In this case, all communications occur through reflection from a coarse ceiling.

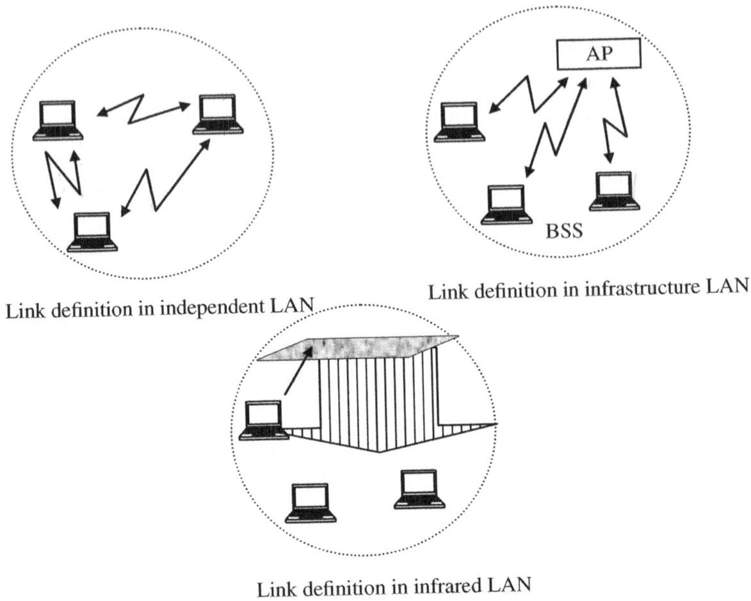

Figure 4-1. PHY link definitions in IEEE 802.11.

4.1.4. Signal Conditioning

The ISM band being licensed-exempt, WLANs are subject to unpredictable environment in terms of interference. IEEE 802.11 defines two standards for physical medium-dependent (PMD) sublayer to combat interference. These are both defined for the 2.4 GHz band. Based on spread spectrum technology, the frequency hop spread spectrum (FHSS) and direct sequence spread spectrum (DSSS) are described briefly in Table 4.2, based on a tutorial [5].

The characteristics of DSSS PHY for 2.4 GHz are given in Table 4.3 and 4.4. A more detailed tutorial [6] is available on the IEEE web site http://grouper.ieee.org/groups/802.11/tutorial/ds.pdf. See also [7] for a brief introduction on the 802.11b and 802.11a PHYs. Additionally, Table 4.5 describes characteristics of Infra-red PHY of 802.11, Table 4.6 lists some characteristics of IEEE 802.11a PHY and Table 4.7 shows the same for IEEE 802.11b.

4.1.5. IEEE 802.11g PHY

This standard is defined for the 2.4 GHz range. It provides upwards of 22 Mbps using OFDM. Its main attraction as compared to IEEE 802.11a (Table 4.6) is interoperability with IEEE 802.11b or earlier PHY standards. Since 802.11b is interoperable with slower versions (1 and 2 Mbps), the creation of 802.11g has made possible the interoperability of WLANs using stations from 1 Mbps to perhaps 54 Mbps data rates.

TABLE 4.2. Interference Rejection Characteristics of the FHSS PHY for IEEE 802.11

Parameter	Value	Remarks
Band	2.4–2.4835 GHz USA/EU 2.471–2.497 GHz Japan 2.445–2.475 GHz Spain 2.4465–2.4835 France	
Number of frequency channels	79 USA/EU 23 Japan 27 Spain 35 France	Minimum 75 recommended for USA, 20 for EU including Spain and France
Channel spacing	1 MHz	
Transmitted power	1 W maximum USA 100 mW EU, 10 mW/MHz Japan	
Minimum hop distance	6 channels	After channel x, next channel $x + \Delta$ satisfies $\|\Delta\| \geq 6$ Modulo 79.
Channels/hopping sequence	26	3 sets of a total 78 patterns.
Channel collisions	3 average, 5 maximum.	Over a hopping cycle.
Channel sequences for k^{th} hopping pattern	2.402+($b[i]$+k) Mod 79 US/most EU 2.473+[(I–1)×k] Mod 23 + 73 Japan 2.447+($b[i]$+k) Mod 27 + 47 Spain 2.448+($b[i]$+k) Mod 35 + 48 France	$b[i] \in \{0 \ldots 78\}$ and $i \in \{1 \ldots 79\}$ according to the hopping pattern tables for each territory.
Modulation Characteristics of the FHSS PHY for IEEE 802.11		
Passband Modulation	FSK	Gaussian shape (GFSK)
Baseband Modulation	NRZ	Data filtered by Guassian filter with bandwidth bit-time product BT = 0.5 or 500 MHz 3 dB bandwidth
Bit rate	1 Mbps	2 Mbps optional with 4-level GFSK
Scrambling polynomial	1001001	Scrambling
Receiver sensitivity	–80 dBm or lower for 1 Mbps –75 dBm or lower for 2 Mbps	For a frame error rate of ≤3% with a frame size of 400 octets.

TABLE 4.3. Interference-Rejection Characteristics of DSSS PHY Specifications IEEE 802.11

Parameter	Value	Remarks
PN-Code	11 chip Barker sequence $\pm\{+1,-1,+1,+1,-1,+1,+1,+1,-1,-1,-1\}$	Minimal sequence allowed in USA by FCC
Coding gain	10.4 dB	
Chip rate	11 Mcps	
Channel frequencies	$f(i) = 2412 + (i - 1)$ USA/EU $f = 2484$ for Japan	$i = 0,1, \ldots 10$ for USA $i = 2,3, \ldots, 10$ for EU

TABLE 4.4. Modulation Characteristics of the DSSS PHY

Parameter	Value
Modulation	DBPSK for 1 Mbps DQPSK for 2 Mbps
Bit rate	1 and 2 Mbps

TABLE 4.5. Characteristics of the Infrared PHY of the IEEE 802.11 WLAN

Parameter	Value	Comments
Wavelength	850–950 nm	
Radiation type	Diffused	LoS helpful as well
Range	10 m	20 m with more sensitive receivers
Transmitter	LED typical	
Receiver	PIN diode typical	
Modulation	Pulse position modulation (PPM)	4-PPM and 16-PPM
Receiver sensitivity (Minimum Irradiance)	2×10^{-5} mW/cm^2 for 1 Mbps 8×10^{-5} mW/cm^2 for 1 Mbps	For frame size of 512 octets and FER of 4×10^{-5}
Maximum radiated power	2 W ± 20% and 0.55 W ± 20% for two masks.	

4.2. INTERFERENCE REJECTION USING BARKER SEQUENCE, OFDM AND CCK

It is the use of orthogonal FDM (OFDM) and complementary code keying (CCK) as modulation schemes for interference rejection that has made it possible to have higher data rate PHYs for the IEEE 802.11 standards suite. It is, therefore, only fair to say a few words about these modulation schemes at this

TABLE 4.6. Characteristics of IEEE 802.11a PHY

Parameter	Value	Comments
Spectrum	5 GHz {5.15–5.25, 5.25–5.35, 5.725–5.825}	Unlicensed National Information Infrastructure (U-NII)
Interference Rejection Measure	Orthogonal Frequency Division Multiplexing (OFDM)	Using 52 subcarriers
Bit rate	6,9,12,18,24,36,48,54 Mbps	36 and above optional
Modulation	BPSK, QPSK, 16-QAM, 64-QAM	Each type used for two consecutive bit rates
Bandwidth	16.6 MHz	Occupied bandwidth
Central frequency for channel k	$5000 + 5k$ MHz	$k \in \{0 \ldots 200\}$
Error correction capability	½, 2/3 and ¾ code rates	Convolution codes

TABLE 4.7. Characteristics of IEEE 802.11b PHY (High Rate DSSS)

Parameter	Value	Comments
Spectrum	2.4 GHz	same as IEEE 802.11 PHY
Interference Rejection Measure	Complementary code keying (CCK) modulation Optional Binary convolutional coding	8 chip CCK, 11 Mcps chip rate
Bit rate	5.5 and 11 Mbps	
Modulation	DQPSK for 5.5 Mbps DQPSK+QPSK for 11 Mbps BPSK for optional PBCC	Packet binary convolutional coding (PBCC) could optionally replace CCK.
Interoperability	802.11 DSSS 802.11 FH with optional Channel Agility	Two types of preambles provided for interoperability with the 802.11 DSSS.
Receiver sensitivity	−76 dBm for 11 Mcps CCK	For a FER of 8×10^{-2} and frame size of 1024 octets.

point. We will be as simple as possible, due to the requirement of a background in communications theory for a full understanding of these schemes.

4.2.1. 11-Bit Barker Sequence

The 11-bit Barker sequence is a known sequence of binary numbers (+1,−1) that has some of the desired properties of the PN-codes. Two main desired properties are:

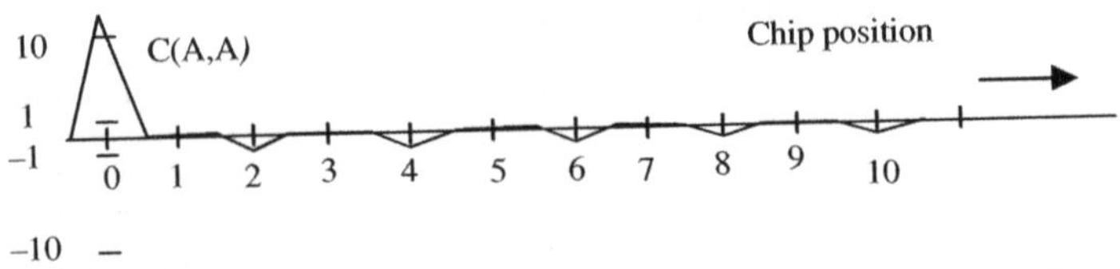

Figure 4-2. The autocorrelation peak in Barker code occurs only at the zero[th] chip position.

1. Orthogonality, demonstrated by high auto-correlation and insignificant cross-correlation of the sequences;
2. Fairly high number of sequence combination.

The 11-bit Barker sequence, specified in the 1 and 2 Mbps DSSS PHY, complies very well with the first requirement. However, due to the short size, it does not comply very well with the second requirement. This does not hurt the WLAN applications due to their limited range. The same sequences could be allocated in neighboring channels and rooms. Here's a description of how this sequence meets each criterion.

Correlation Properties. Correlation of two digital sequences $A = \{a_0, a_1, \ldots, a_{n-1}\}$ and $B = \{b_0, b_1, \ldots, b_{n-1}\}$ of length n is defined as another sequence $C(A,B) = \{c_0(a,b), c_1(a,b), c_{n-1}(a,b)\}$, whose j^{th} element $c_j(a,b)$ is obtained by

$$c_j(a,b) = \sum_{l=0}^{n-j-1} a_l b_{l+j}.$$

The above definition of correlation applies to auto-correlation as well as cross-correlation. If $B = A$, then the correlation is called autocorrelation of A; otherwise it is a cross-correlation of A and B.

For the 11-bit Barker code, the autocorrelation results in a peak at the zero[th] position (chip number zero) and no peaks off-zero positions as shown in Figure 4-2.[1] That simply means that Barker codes can be created simply by rotating a code one chip to the right or the left. The intended receiver has to know the amount of rotation and synchronize itself with it for correct reception of data bits.

$$\begin{aligned} Let\ A &= \{+1,-1,+1,+1,-1,+1,+1,+1,-1,-1,-1\}; \\ B &= \{-1,+1,-1,+1,+1,-1,+1,+1,+1,-1,-1\} \end{aligned}$$

Apparently, B has been obtained by anti-clockwise rotation of A by one bit.

[1] In other words, the inner product of a Barker's sequence with itself is much higher than with a version obtained by shifting bits.

Using the expression $c_j(a,b)=\sum_{l=0}^{n-j-1} a_l b_{l+j}$, the auto- and cross-correlations are given as:

$$C(A,A)=\{11,0,-1,0,-1,0,-1,0,-1,0,-1\} \textit{ is the autocorrelation of } \mathrm{A}.$$
$$C(A,B)=\{-1,10,-1,0,-1,0,-1,0,-1,0,-1\} \textit{ is the cross-correlation of } \mathrm{A} \textit{ and } \mathrm{B}.$$

From the calculations of C(A,A) and C(A,B) it is apparent that if a receiver is synchronized to the zeroth chip position it can detect signals only from the transmitter that uses the same code as the receiver. It also makes sure that sufficiently high power peaks occur at this position to discriminate between the correlation signal and interference signals.

4.2.2. Orthogonal Frequency Division Multiplexing (OFDM)

OFDM is the choice of modulation scheme for the IEEE 802.11a PHY. Here, we attempt at a general, easy-to-follow description of the scheme. There are two concepts that must be understood before we describe OFDM. These relate to (i) the actual frequency response of a channel versus its capacity, and (ii) inverse multiplexing.

We consider the channel bandwidth as the range of signal frequencies that could be transmitted through a channel without significant deterioration. Ideally, we would like to see a channel with brick-like frequency response, shown in Figure 4-3 as dotted.

In reality, a channel response is anything but flat with well-defined corners. Actual channels have a coarse and slowly varying frequency profile, shown by solid line in Figure 4-3. Bandwidth can be arbitrarily defined by taking a range of frequencies with attenuation less than (or amplitudes above) a certain fraction of the maximum amplitude. A commonly used bandwidth in electronic filters is the half-power (3 dB) bandwidth, which includes the frequency range with power at least half of the maximum value. If the channel response were flat, we would perhaps have a simple and predictable relation between channel frequency and bit rate for a given modulation scheme. But due to asymptotic channel behavior, the lower power components cannot carry data at a speed

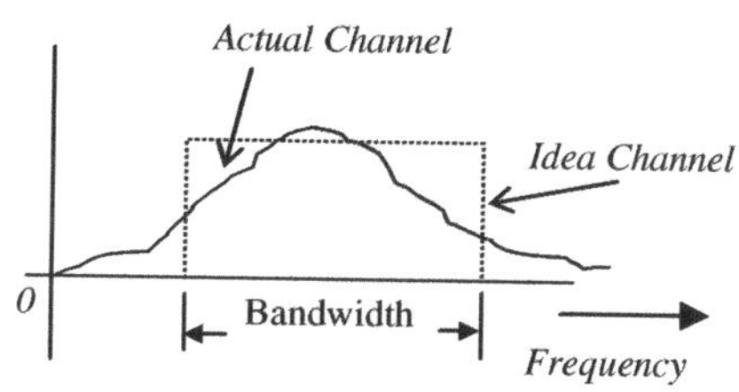

Figure 4-3. Difference in frequency response of idea actual channels.

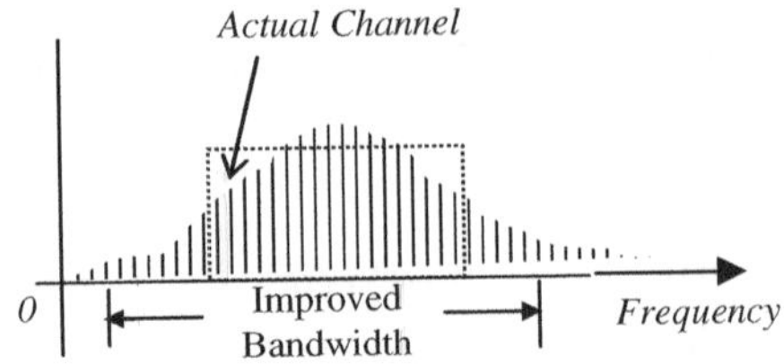

Figure 4-4. OFDM divides a channel into a large number of thin bands.

as high as the higher amplitude components. Due to this, some fraction of the bandwidth always remains unused, to provide sufficient power for the information-carrying signal and a cushioning between adjacent channels.

OFDM redefines channel response as consisting of a large number of thin frequency bands, as shown in Figure 4-4.

Each band acts as a different frequency channel and could carry data independent of other narrow bands. Thus, a narrow band (called OFDM carrier) could have a data carrying capacity according to its amplitude, higher data rates for higher amplitudes, and so on. In essence, an OFDM signal consists of a large number of carriers, each carrying a different amount of data (in general) from the same source. Thus, the data from a fast source could be divided among these channels, according to the channel amplitudes. This is opposite to multiplexing, which is why it may also be called *inverse multiplexing*. As seen from the figure, OFDM results in efficient channel utilization. However, it would be much more complex than using channel as a single frequency window. High-powered digital signal processing chips are available to implement such modulation schemes these days.

OFDM reminds one of Fourier series, in which a signal is represented by a series of components with frequencies orthogonal to one another. In fact, OFDM channels are orthogonal to one another due to their non-overlapping co-existence. Therefore, inverse Fourier analysis is used as OFDM transmitter and Fourier analysis as the receiver. The limitation of Fourier and Inverse Fourier analyses is that there would ideally an infinite number of carriers as input or output. This problem has long been resolved in the design of numerical techniques for Fourier analysis to be used in computers and DSP chips. One efficient mechanism, called Fast Fourier Technique (FFT), is employed in OFDM transceiver design, with Inverse FFT (IFFT) at the transmitter and FFT at the receiver. The total bit rate of the OFDM stream is the sum of bit rates in all narrowband channels. These narrowband channels are also called tones and OFDM is also called multi-tone modulation. Each tone in OFDM can use a different modulation inside it, for example, BPSK, QPSK, and so forth.

4.2.3. Complementary Code Keying (CCK)

CCK is, in effect, an MPSK with the symbol coding that has interference-rejection capabilities similar to Barker's codes. Using two layers of modula-

TABLE 4.8. Definition of $\{\phi_k\}$

Bit Combination (b_0 = LSB)	ϕ_k
$\{b_1,b_0\}$	ϕ_1
$\{b_3,b_2\}$	ϕ_2
$\{b_5,b_4\}$	ϕ_3
$\{b_7,b_6\}$	ϕ_4

TABLE 4.9. Phase Allocation for DQPSK

Bit Combination $\{b_{i+1},b_i\}$; $i = 0, \ldots, 6$	Value of ϕ_k in radians
00	0
01	π
10	$\pi/2$
11	$-\pi/2$

tions, for example, a CCK signal encodes 8 bits giving a baud rate of $\frac{1}{8}$ of the bit rate. Thus, for a bit rate of 11 Mbps, the baud rate is 11/8 = 1.375 Mbaud. An 8-bit per symbol encoding results in a bit rate of 5.5 Mbps. In the IEEE WLAN specification, the first layer of modulation results in determining values of four phase angles using Tables 4.8 and 4.9. The second layer of modulation results in determining the value of a multiphase symbol, which has eight components in case of 802.11b.

The 802.11b standard, that uses CCK, specifies the use of following multiphase symbol for encoding:

$$c = \left(e^{j(\phi_1+\phi_2+\phi_3+\phi_4)}, e^{j(\phi_1+\phi_3+\phi_4)}, e^{j(\phi_1+\phi_2+\phi_4)} - e^{j(\phi_1+\phi_4)}, e^{j(\phi_1+\phi_2+\phi_3)}, e^{j(\phi_1+\phi_3)}, -e^{j(\phi_1+\phi_2)}, e^{j\phi_1}\right)$$

The values of $\{\phi_k\}$'s are not unique and actual value of each depends on the bit pairs of the 8-bit symbol sequence and a DQPSK coding table as shown in Table 4.9.

In one of the actual implementations of CCK in a chip by Intersil, two bits corresponding to ϕ_1 are coded separately to shift (rotate) the signal phase created by b_7 through b_2.[2] Bits b_7 through b_2 are used to select one of the 64 phase values possible with 6-bit combinations according to the equation given by the 802.11b standard. This is possible because ϕ_1 is present in all eight components of c, and thus results in a phase shift for all components.

4.2.4. PHY Data Transmission

The modulated signal from one type of various PMD PHYs (IR, DSSS(802.11,*a*,*b*,*g*), FH) could generally be received only by the same type of

[2] http://www.intersil.com/data/an/an9/an9850/an9850.pdf

PMD PHY. This is due to different capabilities of each media type. These differences are to be hidden from the MAC layer. Therefore, data from each PMD PHY are framed by a convergence sublayer before they are exchanged between the MAC and PMD. The physical layer convergence procedures (PLCP) sublayer of PHY takes care of the capabilities of a PMD and define the PHY PLCP frames for each type of PHY media.

4.2.4.1. PLCP Frame Format for 802.11 Series. Figures 4-5 through 4-9 show the PLCP frame for each of the PHY discussed above:

4.2.4.2. Meanings of Frame Fields

Synchronization (Sync). This field helps the receiver to synchronize its operation at the frame boundaries. Different PHYs have different lengths of

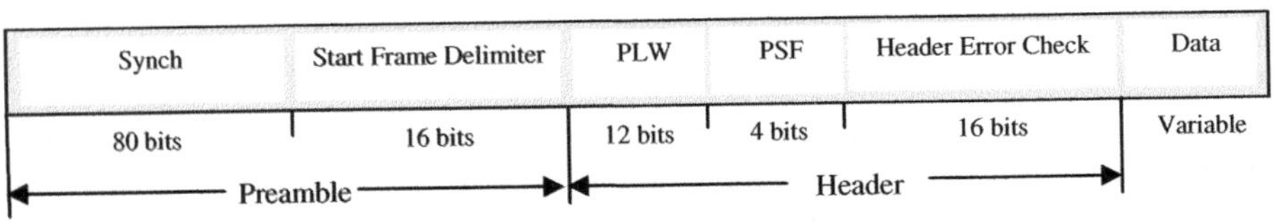

Figure 4-5. FH PCLP frame.

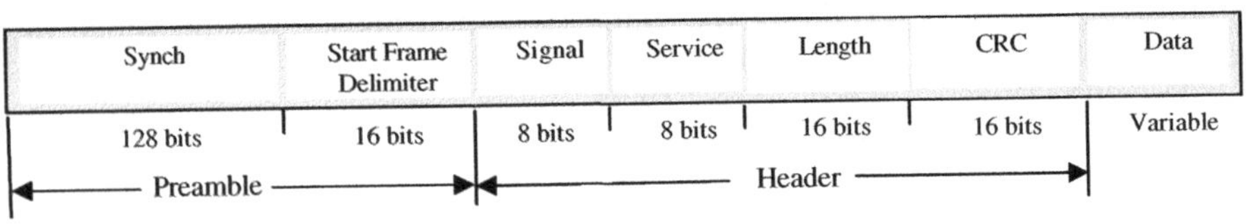

Figure 4-6. DSSS PCLP frame.

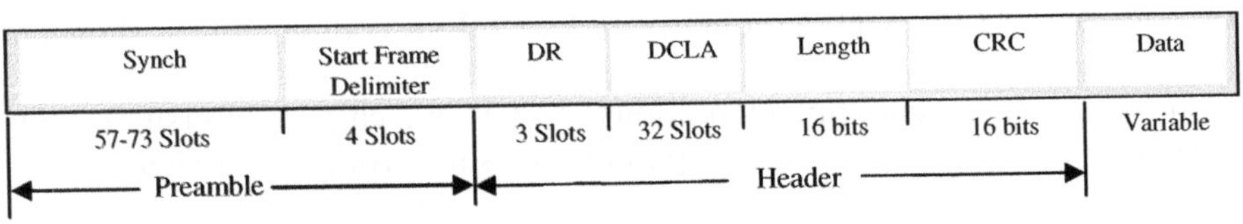

Figure 4-7. IR PCLP frame.

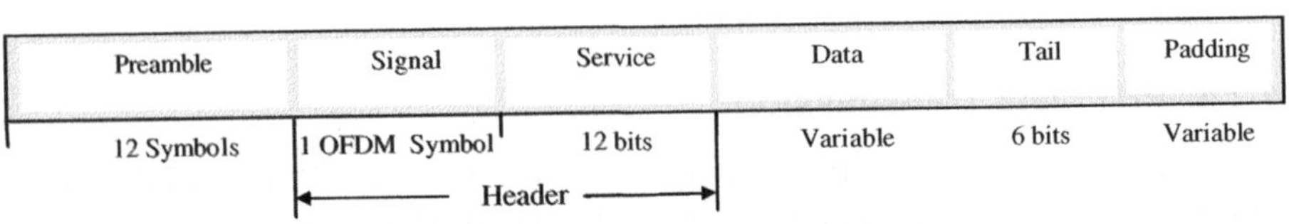

Figure 4-8. 802.11a PCLP frame.

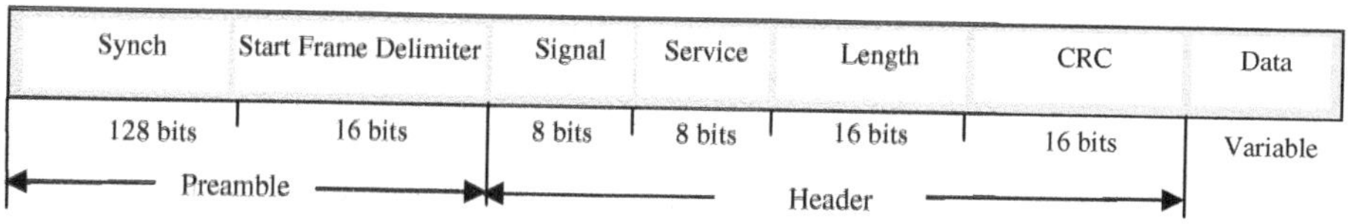

Figure 4-9. 802.11b PCLP frame.

the sync fields. This is in part due to different bit rates and in part due to different units of transmission times. For example, the infrared PLCP specifies the sync field in terms of slots, one slot equal to 250 ns. Another observation is that the sync fields for 802.11 and 802.11b have the same length, laying the foundation for compatibility.

Start Frame Delimiter (SFD). This field exists in all except the high-speed 802.11a PHY. This field indicates the start of the PMD parameters within PLCP. Following this field is the header that contains the parameters.

PLCP_PDU Length Word (PLW). This field is the number of data octets in the data field. The data field is the MAC-PDU.

PLCP Signaling Field (PSF). Specifies the data rate to be used in current frame transmission/reception.

Header Error Checksum and CRC. These fields are used for error detection using CRC.

The PHYs for the IEEE 80.11 suite have management functions defined in the layer management plane. Also, the MIB stores the values for various parameters for the PHY. We are not considering layer or station management issues and therefore will skip the discussion and go on to the PHY for the next WLAN.

4.3. HIPERLAN PHY

The HIPERLAN (High Performance Radio LAN) has been specified by European Telecommunications Standards Institute (ETSI) technical committee Radio Systems and Equipment 10 (RES10). Standardization of HIPERLAN was completed before IEEE 802.11, but continues due to its projected usage as an interface to next-generation cellular systems. HIPERLAN is one of a family of standards including another three. Following is a brief description of each:

HIPERLAN 1. Defined in the CEPT allocated band of 5.15–5.25 GHz for very short range (50 m), low mobility (1.4 m/s) for independent networks, as well as interface technology with high-speed data (Ethernet rate) and low-delay audio (32 kbps, 10 ns) and video (2 Mbps, 100 ns).

HIPERLAN 2. Uses the same spectrum as HIPERLAN 1, with some addition and is primarily used for interfacing with ATM backbone. Data rates in access of 50 Mbps are possible at PHY.

HIPERACCESS. HIPERACCESS is for outdoor, high-speed access (25 Mbps) to provide interface for wireless data to the cellular networks. It is a part of ETRI's third-generation partnership project (3GPP). HIPERACCESS is not projected to provide mobility services. It will work both in ATM and IP environments. It uses a number of licensed bands.

HIPERLINK. HIPERLINK is an interconnection technology between HIPERLAN/2 and HIPERACCESS, the former being in the license-free spectrum and the latter in the licensed spectrum. Together with HIPERACCESS and HIPERLAN/2, it completes the ETSI broadband radio access network (BRAN). Figure 4-10 shows the application of the BRAN 'family'.

Since HIPERACCESS and HIPERLINK constitute part of the access network, we will discuss them under broadband wireless access networks. Here we will continue with the discussion of HIPERLAN and HIPERLAN2 (together HIPERLAN/2) PHY.

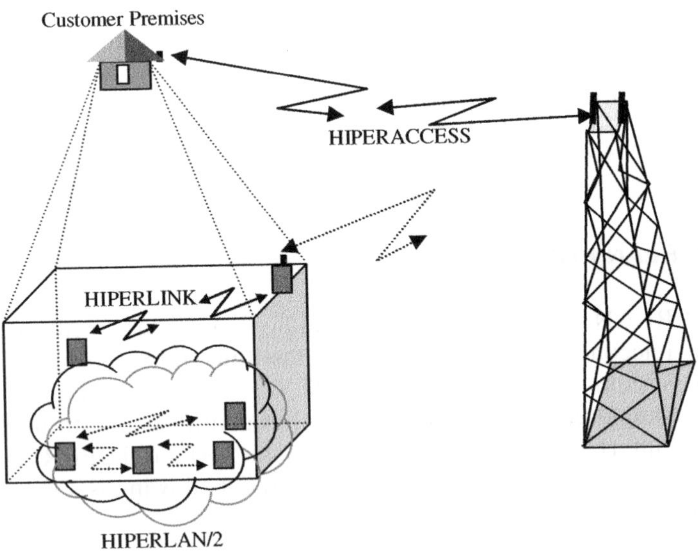

Figure 4-10. Various parts of the BRAN.

4.3.1. Station Types

Since HIPERLAN MAC is designed to provide multimedia services, voice, video, and data terminals all constitute HIPERLAN stations. The term mobile terminal (MT) is used to describe the HIPERLAN user terminals. The standard also uses access point (AP) with the same role as IEEE 802.11 AP. The HIPERLAN APs perform what is called *dynamic frequency selection* (DFS). The DFS algorithm results in automatic selection of the clearest channel. One difference between IEEE 802.11 and HIPERLAN is the use of a central controller (CC) in a group (cell) of HIPERLAN terminals [8]. The CC is an ad hoc assignment and any station could act as one. The main job of CC is to allocate time slots to various stations to emulate a connection-oriented protocol. This eliminates the hidden terminal problem in HIPERLAN at the cost of added complexity.

4.3.2. Channel Media

HIPERLAN 1 was designed for the unlicensed 5 GHz spectrum. The original allocation by ETSI was the 5.15–5.25 GHz band, even though a later extension to 5.30 GHz has been made in some countries. This band is largely for indoor usage. For HIPERLAN2, extra spectrum in the range of 5.470–5.875 GHz has been added for indoor plus outdoor usage. Part of the extra spectrum (5.725–7.875 GHz) is not license-free.

Both infrastructure and ad hoc configurations are possible with connection-oriented channel allocation. In case of ad hoc configuration, a group of MTs is called a cell. One of the MTs acts as a central controller to coordinate allocation of time slots.

4.3.3. Signal Conditioning

HIPERLAN uses OFDM due to its strength in mitigating multipath. IEEE 802.11a also employs OFDM (64 point FFT) in a manner compatible with HIPERLAN. In HIPERLAN, bandwidth is divided into 19 channels of 20 MHz each. Fifty-two sub-carriers per channel provide highly efficient utilization. Guard intervals of 800 ns are mandatory with an option of 400 ns. The symbol duration is 4 μs.

A large number of bits per OFDM symbols result in having slow symbols yet keeping a high channel bit rate. Error-control coding further enhances the number of bits per symbol allowing space for errors.

4.3.4. Modulation and Coding

There are four modulation schemes for seven channel rates (BPSK for 6 and 9 Mbps, QPSK for 12 and 18 Mbps, 16QAM for 27 and 36 Mbps, and 64QAM

TABLE 4.10. Modulation Parameters for HIPERLAN PHY

Channel Rate (Mbps)	6	9	12	18	27	36	54
Modulation	BPSK	BPSK	QPSK	QPSK	16QAM	16QAM	64QAM
Coding rate	½	¾	½	¾	9/12	¾	¾
Bits/symbol (data)	24	36	48	72	108	144	216
Bits/symbol (channel)	48	48	96	96	192	192	288

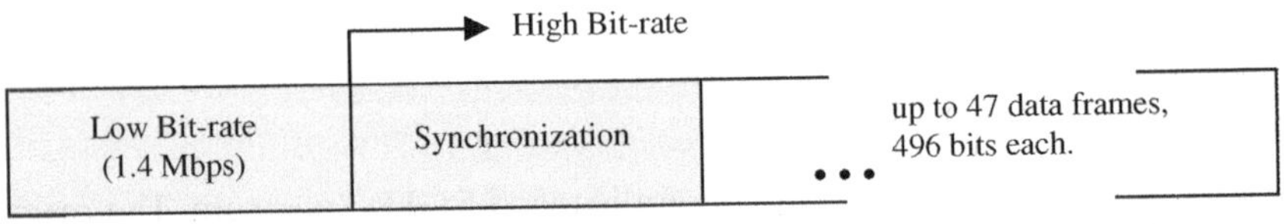

Figure 4-11. HIPERLAN PHY frame consists of low-bit-rate and a high-bit-rate parts.

for 54 Mbps option. The increase in channel rate is due to modulation and different error code rates. For BPSK and QPSK modulations, code rates of ½ and ¾ provide two different channel rates. For 16QAM, code rates of 9/16 and ¾ provide two different rates, while 64QAM employs a code rate of ¾.

Table 4.10 lists various parameters of modulation for HIPERLAN.

IEEE 802.11a specifies all the channel rates of HPERRLAN except 27 Mbps (instead of 27, IEEE 802.11a has 24 Mbps channel rate).

4.3.5. Data Transmission, Convergence and Rate Selectivity

Unlike IEEE 802.11 suite, the HIPERLAN has a single spectrum, therefore eliminating the need for PHY convergence. One of the strengths of HIPERLAN is the capability of link adaptation. A link adaptation algorithm automatically selects the appropriate transmission rate. This helps applications meet QoS requirements.

Figure 4-11 shows a HIPERLAN frame.

4.3.6. PHY Management

The MIB for PHY provides a number of parameters to configure the layer attributes. These parameters relate to, among others, Topology Information Base (TIB), Neighbor Information Base (NIB), Alias Information Base (AIB), and Route Information Base (RIB).

4.4. SUMMARY

This chapter looked at various PHY options for the two WLAN standards. Even though HIPERLAN 1 could be considered a competition of the IEEE 802.11, with the onset of HIPERLAN 2, which is very similar to IEEE 802.11a in terms of PHY, any PHY advantage is essentially eliminated. Due to the connection-oriented nature of HIPERLAN, it performs better for multimedia applications. The ATM interface of HIPERLAN 2 does not seem to be promising, as IP is looking more and more the only wide area technology. Also, with the introduction of 2.4 GHz IEEE 802.11g, even the high-speed IEEE 802.11a does not look so promising. All in all, IEEE 802.11, b and g seem to be destined to be universalized.

What is still a big strength of BRAN is the fact that it caters to end-to-end connectivity and integrates with next-generation cellular technology in the form of HIPERLAN 3 (HIPERLINK) and HIPERLAN 4 (HIPERACCESS). For all the right reasons, there is more and more cooperation between IEEE and ETSI, in order to make standards that are not far removed from or interoperable with each other.

REFERENCES

[1] Bob O'Hara and Petrick, Al, *The IEEE 802.11 Handbook: A Designer's Companion*, IEEE Press, 1999.

[2] Joseph Moran, 'Wireless Home Networking—Part II—Wi-Fi Standards', available from http://www.80211-planet.com/tutorials/article.php/1495031

[3] IEEE 802, 'Project Authorization Form, IEEE Standard Board, IEEE P802.11, Working Group for Wireless LANs Assigned Project Number 802.11a', available from http://grouper.ieee.org/groups/802/11/PARs/par80211bapp.html, 9 Dec. 1997.

[4] IEEE 802, 'Project Authorization Form, IEEE Standard Board, IEEE P802.11, Working Group for Wireless LANs Assigned Project Number 802.11a', available from http://grouper.ieee.org/groups/802/11/PARs/par80211aapp.html, 16 Sept. 1997.

[5] Naftali Chayat, 'Frequency Hopping Spread Spectrum PHY of the IEEE 802.11 Wireless LAN', Presentation to IEEE 802, March 1996, available from http://grouper.ieee.org/groups/802/11/Tutorial/FH.pdf

[6] Jan Boer, 'Direct Sequence Spread Spectrum Physical Layer Specification IEEE 802.11', Presentation to IEEE 802, March 1996. http://grouper.ieee.org/groups/802/11/Tutorial/ds.pdf

[7] John Hansen, '802.11b/a—A physical medium comparison', *RF Design*, Feb 1, 2002, available from http://rfdesign.com/ar/radio_ba_physical_medium/

[8] D. Hollos and Karl, H., 'A protocol extension to HiperLan/2 to support single relay networks', *Proceedings of First German workshop on Ad Hoc Networks*, Ulm, Germany, March 2002, pp. 91–808.

[9] Jamshid Khun-Jush, Schramm, Peter, Wachsman, Udo, and Wenger, Fabien, 'Structure and Performance of HIPERLAN/2 Physical Layer', IEEE VTC'99 (fall), Amsterdam, pp. 2667–2671.

[10] Angela Doufexi, Armour, Simon, Karlsson, Peter, Nix, Andrew, and Bull, David, 'A Comparison of HIPERLAN/2 and IEEE 802.11a', *IEEE Communications Magazine*, May 2002.

[11] ETSI, *Broadband radio access networks* (BRAN)s: HIPERLAN type 2 technical specifications: Physical (PHY)layer, August 1999. DTS/BRAN-0023003, V0.k.

[12] Ji Wang, Khokhar, Ashfaq, and Garg, Vijay, 'Video communications with QoS Guarantees over HIPERLAN/2', IEEE Fourth International Symposium on Multimedia Software Engineering (MSE'02), 2002.

[13] Ken 'ichi Ishii and Aghvami, A.H., 'An evaluation of HIPERLAN/2 scalability for mobile broadband systems', *European Wireless 2002*, Invited paper. available from http://www.ing.unipi.it/ew2002/proceedings/H2002.pdf

[14] Barry Forde, 'The Future of Wireless LANs' UCISA-NG Wireless LANs Conference, Lancaster University, Information Systems Services, Feb. 2001.

[15] Romain Rollet and Mangin, Christophe, 'IEEE 802.11a, 802.11e and HIPERLAN/2 goodput performance comparison in real radio conditions', *Proceedings of the IEEE Globecom*, San Francisco, Dec. 2003. available from http://www.mitsubishi-electric-itce.fr/English/scripts/Publications%20pdf/2003/rollet_GlobeCom03.pdf

CHAPTER 5

WLANs: MEDIUM ACCESS CONTROL

Sometimes, we differentiate among various LANs based on their MAC protocols. The two main functions to be provided by any MAC sublayer are the channel access and multiple access. In WLANs, the wireless nature of the channel adds its own conditions. Provisioning of multimedia requires even more functions. While developing the WLAN standard, many service-related facts were known to the developers, such as new developments in the Internet protocol suite, user demands, spectrum availability, new modulation and coding schemes and user attitudes toward wireless technology. All this added to the existing technical experience from wired network standards and resulted in wireless-friendly, multimedia-capable medium access control protocol. IEEE specified a MAC procedure for WLANs under the project number IEEE 802.11 and ETSI in Europe defined MAC sublayers for HIPERLAN 1 and 2. The channel access method of IEEE 802.11 is also called CSMA/CA, as a reminder that the main channel access procedure employed is carrier sense multiple access with collision avoidance (CSMA/CA). It is distributed among all participating stations, and thus called the *distributed coordination function* (DCF). Distributed implies that the procedure is to be implemented in all participating stations, while coordination function implies that it is a channel access and multiple access procedure based on cooperation from all stations. As we will see in this chapter, IEEE 802.11 MAC is not a single or a few protocols or functions; it is a set of functions and procedures designed to deliver some target performance for multimedia wireless communications

Wireless and Mobile Data Networks, by Aftab Ahmad

among a coordinated group of wireless terminals through or without an access point. The ETSI standard HIPERLAN 1 uses a MAC procedure called elimination yield non-preemptive multiple access (EY-NPMA) that integrates within itself the capability of assigning priorities to the contending stations. EY-NPMA is also distributed, but allows for frame relay capability. We will have a look at both of these protocols in this chapter.

We will follow the sequence of functions given in the MAC Components section in Chapter 3.

5.1. IEEE 802.11 MEDIUM ACCESS CONTROL

Specified as part of ANSI/IEEE Std 802.11, 1999 Edition, the medium access control layer of this radio LAN architecture is the closest to the Ethernet that the wireless medium could have, to deliver similar performance. It has several functions for security and multimedia coverage and is expected to continued to be enhanced for interacting with future cellular networks, as they are driven more and more toward data applications with the commonality of the Internet Protocol (IP). We will look at various aspects of this protocol in this section.

5.1.1. Network Configurations

The IEEE 802.11 standard views wireless networking mainly as a distributed function among a few stations in a Basic Service Area (BSA). The group in a BSA is called a Basic Service Set (BSS). Stations in a BSS are connected via a common central point (if at all). More than one BSS could be grouped through a Distribution System (DS) to form an extended service set (ESS). All stations in a BSS have common values for a set of data rates (Basic Rate Set) and use a common set of MAC PDUs called *Station Services* (SS). ESS is defined only for the infrastructure topology. For ad hoc or independent topology, the BSS is called Independent BSS or IBSS.

5.1.2. Channel Access in IEEE 802.11

IEEE 802.11 MAC uses a type of packet division multiple access mechanism (PDMA) as the main channel plus multiple access mechanism. It's called the *Distributed Coordination Function*. Carrier sense multiple access is used as part of distributed coordination function. Two mechanisms are provided for sensing the carrier, a physical mechanism clear channel assessment or CCA (defined at PHY, but it is relevant to MAC transmissions), and a virtual mechanism. Since collision detection is not efficient for wireless channels, a mechanism to avoid collisions is used. Due to the combination of CSMA and collision avoidance, the mechanism is called *carrier sense multiple access with collision avoidance* (or CSMA/CA). Collision avoidance is implemented through interframe spacing (IFS). Let's look at the channel-sensing and collision-avoidance mechanisms.

5.1.3. Channel Sensing

Channel sensing is carried out by either sensing the presence of a carrier signal in the wireless medium or checking the value of a parameter, called *Network Allocation Vector* (NAV). The MAC layer gets help from the physical layer electronics to implement the former. The PHY function to sense channel is called *clear channel assessment* (CCA). NAV results in what is termed *virtual sensing*, as it does not involve a physical signal-detection mechanism. In virtual sensing the NAV value is checked that was set by the station on detecting a short packet exchange (called *handshake*) between the intended transmitter and receiver. The transmitter neighbors set the NAV parameter value on detecting a Request to Send (RTS) type of MAC packet. RTS is sent by the transmitter to inform the intended receiver of the reservation request and to ask its permission. On receiving RTS, the receiver responds with Clear to Send (CTS), indicating its permission, as well as informing its neighbors of the transmission. On detecting CTS, the stations in the neighborhood of the receiver set their NAV value, as shown in the timing diagram in Figure 5-1.

This gives the transmitter and receiver pair the exclusive right to use the channel for the duration set by NAV. As seen in the figure, the receiver has to wait for a time called interframe spacing (IFS) before issuing a CTS. In fact, IFS is one of the main features of the IEEE 802.11 MAC to help avoid collisions. All transmissions must sense channel for the amount of IFS, even if the channel is idle. The actual amount of IFS depends on the type of data to be transmitted. However, no amount of IFS guarantees collision avoidance due to the fact that many stations could be sensing an idle channel simultaneously,

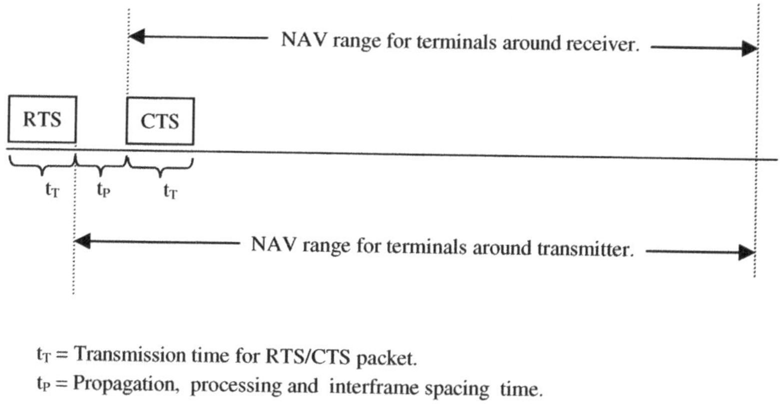

Figure 5-1. Handshaking to set up NAV reservation.

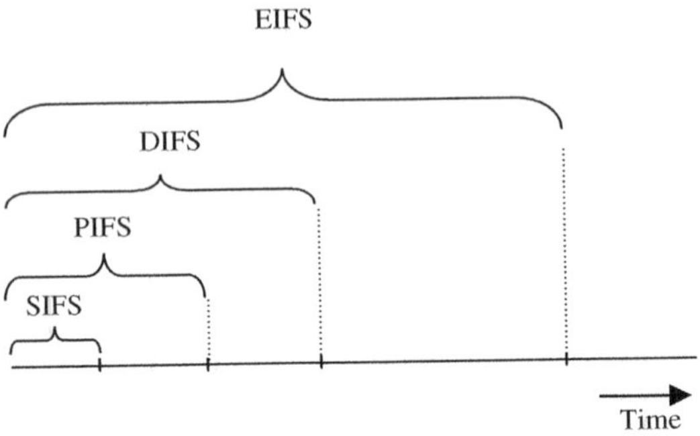

Figure 5-2. Relative difference among various IFS times.

and after the requisite IFS, more than one may transmit together. Therefore, collision-avoidance contains many elements that we discuss next.

5.1.4. Collision Avoidance

Since IFS may not provide sufficient guarantee for collision avoidance. The following elements are added to it for improvement.

5.1.4.1. Prioritizing IFS. The first step is to allocate different amounts of IFS times for different types of packets. The standard specifies four priority levels with respect to the IFS times. These are: (1) Short IFS (SIFS), (2) Distributed Coordination Function IFS (DIFS), (3) Point Coordination Function IFS (PIFS) and (4) extended IFS (EIFS).

The SIFS is for the highest priority (short) packets. CTS and acknowledgement (ACK) are examples of such packets. DIFS is the 'normal' IFS. To initiate a transmission of data or RTS packet, a station must find the channel idle for an amount given by DIFS. PIFS is for real-time applications, and has a value between SIFS and DIFS. The EIFS is to relieve the network from congestion and is used on occurrence of errors. Figure 5-2 shows a relation between various IFSs.

Actual values of the IFS are physical layer specific. The physical layers have a parameter called the slot time. The PIFS is one slot time bigger than the SIFS, while DIFS is one slot time bigger than the PIFS.

5.1.4.2. Random Backoff. In the case of a station finding a channel busy, it has to wait for some time before it can restart sensing. This time is called backoff time. The backoff time is selected randomly in IEEE 802.11. This ensures that two stations, which might have backed off together, should not start and finish to sense the channel together again. The time is generated from a random number generator every time a station has to backoff.

5.1.4.3. Discouraging Multiple Transmissions. If a station has multiple packets to be transmitted, the protocol spreads their transmission over time by requiring all packets, subsequent to the first one, to have a minimum of one random backoff, regardless of the availability of the channel. Thus packet number 2 will have to wait for a minimum of DIFS plus the random backoff time. Another station with a newly generated packet will have to wait only for the DIFS, giving it priority over the station with packet series. This is shown in box 7 in the state diagram for CSMA/CA procedure in Figure 5-3.

5.1.4.4. Binary Exponential Backoff. Despite the above factors, collisions may occur due to simultaneous sensing and transmission of the same priority packets. Occurrence of collision is determined from a lack of acknowledgement within a specified time. The state diagram in Figure 5-3 assumes the absence of collision. On determining a collision, a station generates a random number between 0 and $x \times 2^{n-1}$, where n is the number of collisions for the same packet transmission. Thus, the first collision generates a backoff time, between $\{0,x\}$, the second collision generates a number between $\{0,2x\}$, third between $\{0,4x\}$, and so on. Due to this increase in the upper limit, it is called *binary exponential backoff*.

5.1.4.5. Contention Window. The binary exponential backoff results in allowing a time window within which the station with collisions is allowed to contend for. This window is called *contention window*. Its value varies between a minimum C_{min} and a maximum C_{max}. It increases with the number of collisions until C_{max}. The value of C_{max} is an important parameter and determines the maximum access delay. This access delay is important for real-time applications. For example, voice packets should be delivered from application to application within a quarter of a second. By selecting a contention C_{max} larger than a few milliseconds, we risk allocating resources to a packet that would be useless even if it attains the channel. In the previous section, the value x is, in fact, C_{min}.

As can be seen from the above discussion, channel access is a complex procedure for IEEE 802.11 LANs. In reality, there is more to channel access than discussed. As an option, the standard allows an access mechanism for real-time or delay-bound packets as well. We will discuss this next under multiple access.

5.1.5. Multiple Access in IEEE 802.11

Packet division multiple access (PDMA) is the general multiple access (MA) mechanism adopted in IEEE 802.11. It is recommended as a Distributed Coordination Function (DCF). An optional part is included for centralized control, called Point Coordination Function (PCF). PCF is to be implemented at the top of DCF. Together, DCF and PCF define a multiple access cycle. The PCF is for delay-bound applications. During this time, the access point (AP) could

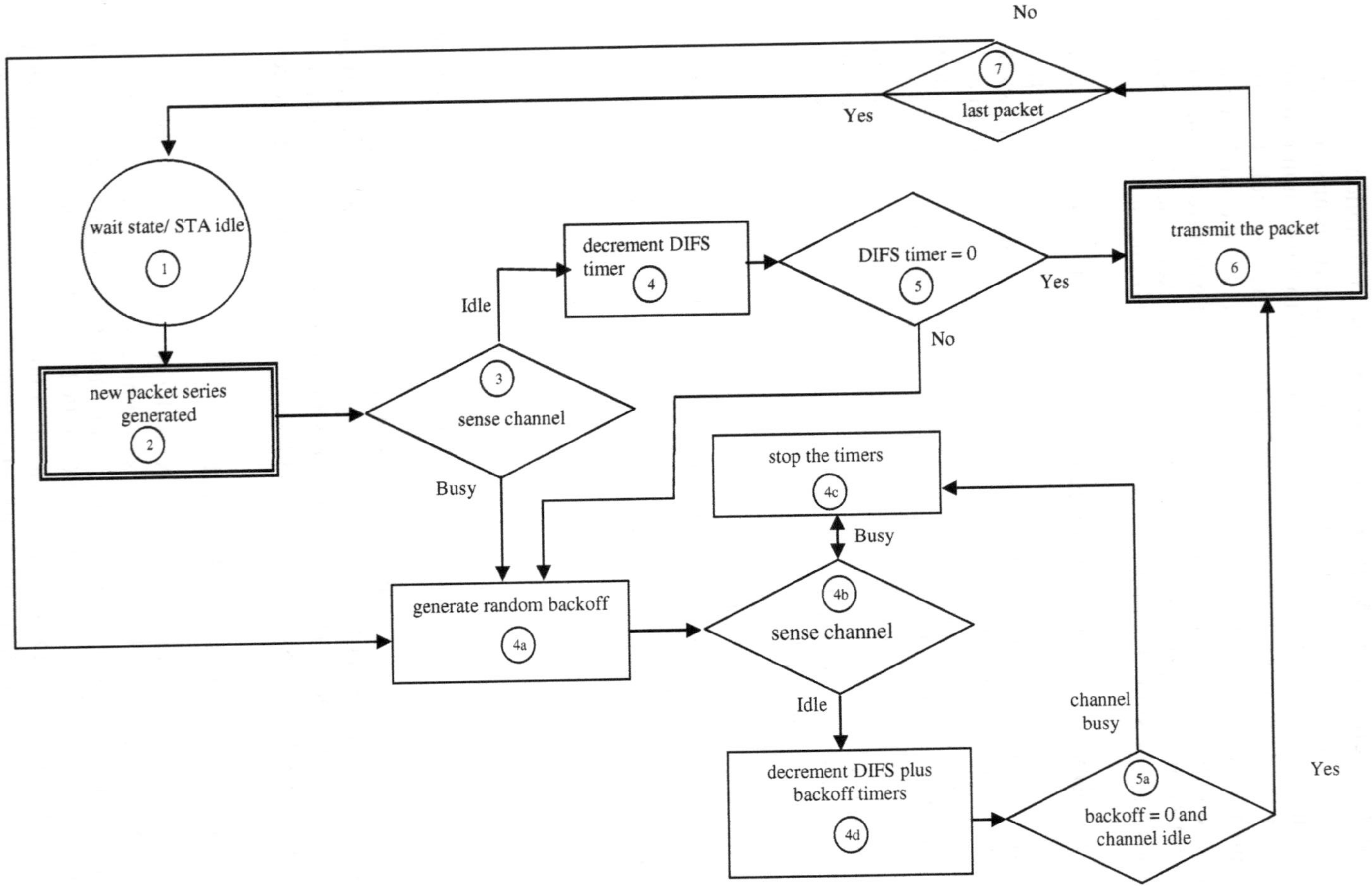

Figure 5-3. State diagram of the channel access and multiple access procedures in IEEE 802.11 MAC.

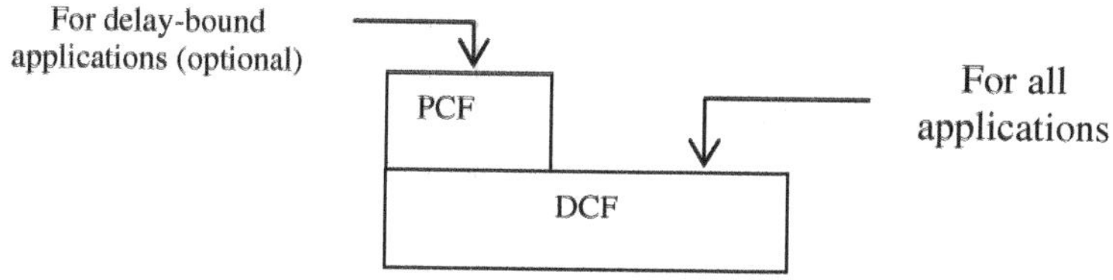

Figure 5-4. Delay-bound applications use DCF through PCF.

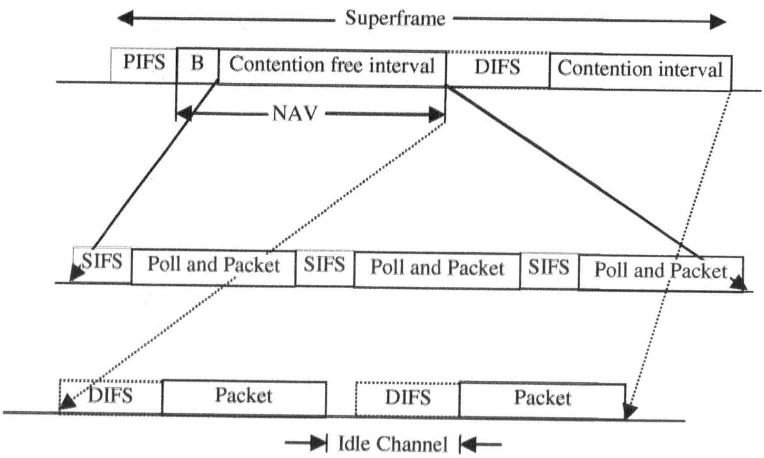

Figure 5-5. An example of PCF/DCF operation. On receiving the beacon (B) from access point, stations set their NAV variable to allow contention free communication.

solicit packets from stations using a polling mechanism. Figure 5-4 shows a protocol relation between DCF and PCF. Figure 5-5 shows the multiple access cycle.

5.1.6. DCF Transmission

The state diagram in Figure 5-3 shows a sequence of events for successful transmission of a series of packets from a single station. As seen from this figure, the multiple access mechanism tends to spread transmission of multiple packets from a single station by two mechanisms; first by requiring channel access procedure for each packet, and second, by adding a random backoff to the DIFS of all subsequent packets[1]. The situation changes if the series of packets are, in fact, fragments of a single MAC frame. In such a case, each subsequent fragment gets the highest priority, requiring only SIFS. Another excep-

[1] Fragmentation also helps in spreading bandwidth among stations.

tion is when a frame is not acknowledged. In this situation a binary exponential backoff is generated together with EIFS if transmission errors are detected. It may be noted that the standard specifies a single backoff procedure for many situations, such as busy channel and non-receipt of an ACK.

5.1.7. PCF Transmission

Within the PCF interval (called *contention-free period* or CFP), the point coordinator polls stations for contention-free data transmission. This is shown by enlarging the multiple access cycle in Figure 5-5. The term *superframe* describes a complete multiple access cycle. During the CFP, multiple access is enforced by allowing a maximum of a single PDU by each station participating in CF transmission. These frames are acknowledged within the CFP. However, if a frame is not acknowledged, the point coordinator does not backoff. The acknowledgements could be piggybacked in case of a full-duplex transmission.

5.1.8. User and Data Privacy

Due to the inherent disadvantage of the wireless media not being contained, the standard uses the concept of wired equivalent privacy (WEP) service. The idea is to protect the users from casual eavesdropping. There is a set of services provided for this purpose, including user authentication and data encryption. Network authentication is not provided. Thus, a user with legitimate user authentication parameters is eligible to belong to a BSS or IBSS, but the user cannot know if the access point is legitimate.

5.1.8.1. User Authentication. The standard provides two types of authentications, open authentication and shared key authentication. These authentication mechanisms consist of special MAC frames exchanged between a station and an access point for infrastructure network and between two stations for an ad hoc network.

Open Authentication. A station requesting open authentication uses the frame type called 'Authentication' that includes its station ID. The receiver of this request responds positively if it accepts open authentications or negatively if it does not accept open authentications, as shown in Figure 5-6.

Shared Key Authentication. The term 'key' is used to describe a known string of digits incorporated in encryption algorithms. It is used to encrypt and decrypt information, so that it is very unlikely that the information can be decrypted without the key. When a group of users uses the same key for encryption, it is called shared key. In shared-key algorithms, it is assumed that the key-allocation procedure was secure permanently. One of the mechanisms used to protect the shared key is by using 'write-only' circuitry for it. In this

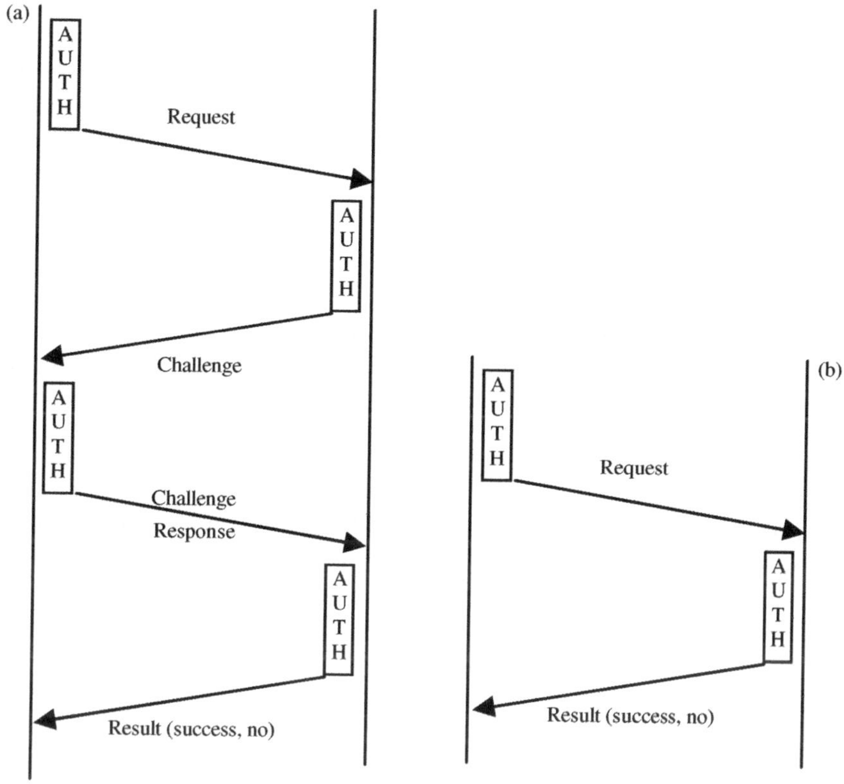

Figure 5-6. (a) shared-key authentication; (b): open authentication.

way, once a key has been entered in a terminal, it will be used only as part of the imbedded algorithm, without anyone being able to 'read' it.

Shared-Key Authentication. Figure 5-6 also shows a sequence of frame exchange for shared-key authentication. The first two frames are exchanged with the WEP OFF. The first frame is a request for authentication frame including the station ID of the sender. The recipient sends the second frame to test whether the sender of frame 1 has shared key or not. It does so by including some randomly generated data as 'challenge'. On receiving frame 2, the sender uses the WEP algorithm to encrypt this data. Then, it sends this data in frame 3. The authenticating station should be able to decrypt the data if the two stations are using a shared key. Thus, it decrypts the data and, if the decrypted data resembles the one sent in frame 2, the authentication is a success, otherwise it is a no-success. This result is transmitted back to the sender in frame 4.

5.1.8.2. Data Encryption. The same WEP service with shared key is used for data encryption as well. RC4 PRNG (Pseudo Random Number Generator) by RSA Data Security is recommended for WEP service. The algorithm takes as input the MAC PDU and generates encrypted version (ciphertext). It transmits ciphertext along with two more fields, the initialization vector (IV) and integrity check value (ICV). The IV adds up to the secret key and provides the seed for the decryption algorithm. The ICV acts as a parity field and helps in determining the integrity and accuracy of the received data. CRC-32 is used for ICV.

5.1.9. Power Management

The IEEE 802.11 specifies a mechanism for allowing stations to go to a power save mode when idle. This results in significant saving in battery life[2]. This mechanism could be implemented for infrastructure as well as independent BSS. Here is a list of components of the power management.

For BSS/ESS:

1. Buffers at the AP to store packets addressed to stations in sleep mode.
2. Mechanism of transmitting a periodic short message by AP, informing the stations of incoming packets stored at the AP. This message is called beacon. The beacon message contains traffic indication map (TIM). In TIMs, the stations learn if they have a packet stored for them by looking at these TIMs.
3. A synchronization mechanism, for the terminals to receive beacons.
4. A polling mechanism, to allow a station to request the stored packet from the AP.
5. A discard mechanism (timer) of packets stored in the AP buffers on non-receipt of a response to beacon.

For IBSS:

1. A mechanism for alternating the transmission of beacon among stations.
2. Synchronization mechanisms for receiving beacon and broadcasting intention to send. The intention to send is indicated in an announcement TIM (ATIM) window. ATIM follows the beacon.

For multicasting:

1. A special delivery TIM (DTIM) is defined to notify of the incoming multicast packets. Multicast packets are stored at the AP and are transmitted immediately following the beacon after DTIM message.

[2] Check http://www.novocomp.de/prod/wirl/WLAN/Bilder/Download/TB-022.pdf for an empirical study on the battery savings due to the IEEE 802.11 power-management mechanism for various devices.

5.1.10. Fragmentation

Fragmentation and reassembly is part of the IEEE 802.11 MAC recommendation. If a MAC PDU is fragmented, all fragments must be of equal size. The last fragment may be of a size smaller than the rest. If WEP is used, then initialization vector (IV) and integrity check value (ICV) would be added to each fragment. Since packet division multiple access (PDMA) is used, a fragmented packet is considered to be a single packet for multiple access purpose. Thus, stations have to wait for idle channel, albeit only for SIFS, before fragment transmissions. For reassembly, mechanisms of resequencing and signaling the end of fragments are included in the standard.

5.1.11. Multimedia Support

Multimedia support using PCF is optional in IEEE 802.11. PCF provides a contention-free period in which SIFS is used by the station to transmit packets. PCF guarantees shorter channel access times, but does not provide an absolute guarantee for the end-to-end delay. It has been shown that for higher loads, PCF is not very reliable, even in terms of channel access time. Therefore, research is open for the provision of multimedia with several proposals, such as efficient polling mechanisms [1] and distributed DiffServ MAC extension based on eliminating burst mechanism [2].

The IEEE 802.11 Task Group e is working on recommendations IEEE 802.11e as an enhancement to IEEE 802.11 MAC. The purpose of these enhancements it to provide an infrastructure for multimedia traffic over the IEEE 802.11 network. Since MAC is common to all PHYs (three IEEE 802.11, IEEE 802.11b, IEEE 802.11g and IEEE 802.11a), so is IEEE 802.11e. The incorporation of IEEE 802.11e will bring the MAC on par with enhancements in Ethernet (IEEE 802.1P, IEEE802.1Q and IEEE 802.1D). Also, the Differentiated Services (DiffServ) of Internet protocol suite use similar QoS nomenclature.

5.2. IEEE 802.11e FACTOR

The introduction of IEEE 802.11e in a BSS makes it enhanced BSS (QBSS), makes a station an enhanced station, and changes a point coordinator (PC) into a hybrid coordinator (HC). The MAC Protocol Data Unit (MPDU) can employ forward error correction (FEC) using Reed-Solomon (RS) codes in data (224, 208) and header (48,32). In other words, new fields for parity bits are added, as shown in Figure 5-7 [3]. Thus, the changes are done at station, AP, frame and protocol. Let's have a brief account of some of these changes on equipment and protocol.

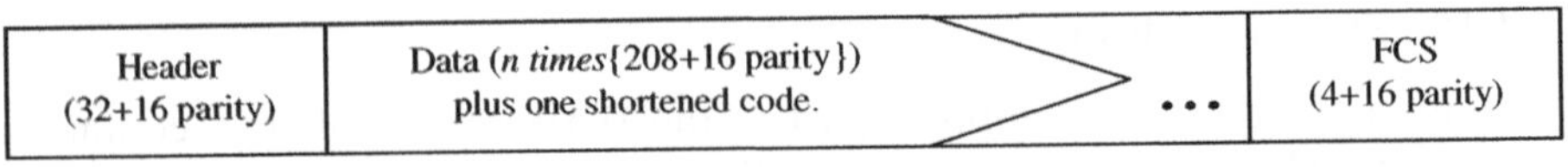

Figure 5-7. MAC frame with FEC. Fields have units of octets.

5.2.1. Enhanced Station

An enhanced station could have up to eight differentiable traffic classes (TCs), each corresponding to one of the priority levels. These traffic classes are managed through different queues. In a situation when a station has more than one non-empty queues, the traffic in each queue contends for the station independent of other queue. The priorities are allocated by the station and resolved by itself in case of a collision[3]. The station sends the information of each queue state to the access point, along with the QoS requirements of each. However, for maximum flexibility, the access point does not decide which queue will be picked up for transmission.

5.2.2. Hybrid Coordinator

The 'enhanced access point', or the hybrid coordinator (HC) implements several functions, new and modified. The new function hybrid coordination function (HFC) is enhancement of point coordination function (PCF) of IEEE 802.11 super frame. During this period, the HC allocates transmission opportunity (TXOP) to a station just like polling.

5.2.3. Enhanced DCF (EDCF)

In the enhanced version of DCF, new interframe spacing (IFS), called *arbitration ISF* (AIFS), is introduced. The AIFS is for internal arbitration of a station among various priority classes. Its value is allocated by the station and is at least equal to DIFS. The higher the priority of a TC, the lower will be the value of AIFS. If TC0 is the highest and TC7 the lowest priority traffic class and $AIFS_k$ is the AIFS for the TC_k, then the general principle is that $AIFS_7 > AIFS_6 > AIFS_5 > AIFS_4 > AIFS_3 > AIFS_2 > AIFS_1 > AIFS_0 \geq DIFS$. [4]. Each priority level is associated with one of the four access categories (AC). Each of these ACs contend for channel just like a DCF using a contention window generated from the time slot and a pseudo-random number between 0 and CW such that ($CW_{min} \leq CW \leq CW_{max}$) [5]. The contention windows limits (CW_{min} and CW_{max}) are a function of AC. In other words, the generic DCF for AC_k is replaced by $EDCF(AC_k)$ [6] and the (DCF, DIFS, CW_{min}, CW_{max}) is replaced by {$EDCF(AC_k)$, $AIFS_k$, $CW_{min}(AC_k)$, $CW_{max}(AC_k)$ }.

[3] It is easily imaginable that in case of a collision an enhanced station will pick up the highest priority queue for transmission and randomize the sensing of the rest of the queues.

5.2.4. Hybrid Coordination Function (HCF)

The HCF in IEEE 802.11e replaces the point coordination function (PCF) of the original standard. The optional PCF has not been implemented as widely as DCF due to its inability to differentiate among traffic types. With the new priority levels defined, HCF can now help stations make polling decisions based on these priority levels. In IEEE 802.11, PCF defines a super frame, which is basically a cycle consisting of pure contention-based part (called contention period or CP) and a polling part built using PCF (called contention-free period or CFP). In case of HCF, the cycle is still defined as consisting of EDCF and HCF with the difference that HCF works throughout the cycle. The HC does that by using the concept of transmission opportunity (TXOP).

5.2.4.1. TXOP. The TXOP is a permission granted to a station by HCF either during CP or CFP. During the contention period, TXOP is earned by a station either when its AISF plus backoff timers count down to zero (while the channel is sensed idle) or when a station receives a poll (QoS CF-Poll [7]). The difference of use of TXOP between CP and CFP is that during CFP, contention-based TXOP is disabled. In effect, HCF provides QoS throughout the network operation but contention based delivery only during the contention period. The QoS CF-Poll is given priority even during the CP by requiring HC less IFS for polling. HC waits for the channel to be idle only for PIFS as compared with contending stations having $AIFS_k$ ($k = 0, 1, \ldots, 7$) plus backoff time. Recall that AIFSs are all greater than or equal to DIFS, which is greater than PIFS. Figure 5-8 shows a successful TXOP given to a station during CP.

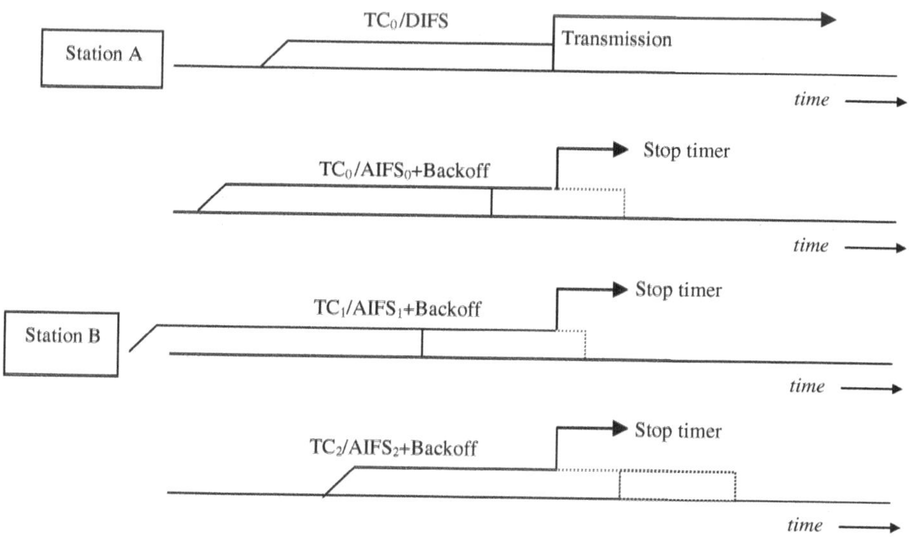

Figure 5-8. Two stations, A and B are contending for TXOP during CP. Station B has three queues. Station A gets TXOP using DCF.

5.3. ROUTING AND MOBILITY SUPPORT

A routing or relay mechanism is not a part of IEEE 802.11 MAC. The topic of routing is, however, central to large independent networks and will be covered as a separate chapter (Chapter 10). The standard introduces the concept of distribution function to be implemented in case of multiple infrastructure BSSs. The distribution function is not specified, and can be any of the several interconnecting devices (bridges, routers) [8]. Additionally, IEEE has published a document on recommended practices for access points' interoperability. These include management-level procedures for successful handoff [9]. This draft describes an inter-access point protocol (IAPP) that could be invoked for handoff across distribution systems. The *Association* function provides a mechanism for a station to be associated with an access point. A *Reassociation* MAC function allows for changing the AP of association. Additionally, a *probe* function allows a moving station to request channel scanning for handoff initiation.

With respect to mobility, three profiles exist in the IEEE 802.11 paraphernalia. These are discussed in the next section.

5.3.1. No Transition

This implies that a mobile station remains within the BSS in which it was associated. This does not require any mobility management functions, as stations can communicate with the BSS. It might occur that a mobile station is physically closer to another access point than the one to which it is associated. However, this does not warrant a handoff, as long as reception on the current association is acceptable.

5.3.2. BSS Transition

If a moving station loses the required signal strength to or from its associated AP, and has a clearer and acceptable signal to and from another AP, it may use the probe and reassociation functions to be transitioned to the new AP. A new BSS ID will be given to the station corresponding to the new AP. The ESS will remain the same, unless it is an ESS transition.

5.3.3. ESS Transition

If a moving station is closer to another AP, which is part of another ESS, it may require the distribution system (DS) to perform an ESS transition. The standard does not provide support for ESS transition. A proprietary mechanism may be employed, in which case the BSS and ESS IDs on the frames originating from and destined to the station will be changed. Alternatively, a deassociation followed an association with the new AP could form a simple solution for ESS transition. Apparently, this will cause transition delay.

5.4. MAC LAYER MANAGEMENT

A management information base (MIB) is part of the standard that stores parameters for the layer operation. These parameters correspond to all functions that the MAC layer provides, such as slot time, IFSs, power-management parameters. The MIB serves the dual purpose of designing the protocol as well as designing tools for managing the protocol functions.

Just like data types in a programming language, the parameter values could be one of the many types. For example, the attribute WEPOn is of Boolean type and dot11WEPKeyMappingLength is of integer type.

5.5. MAC FRAMES

The MAC layer receives higher layer PDU as service data units (SDUs) and makes an MPDU by attaching a header/trailer to them. Additionally, the layer may generate and process some frames by itself, for example, for bandwidth reservation and acknowledgement purposes. A number of control packets are defined for providing the functions numerated in paragraphs above. Between turning ON and fully operating, an IEEE 802.11 station goes through three stages with respect to Authentication and Association (all but Unauthenticated, Associated state). Each of these states allows for a given set of MAC frames to be used. The standard defines three classes, as shown in Table 5.1.

TABLE 5.1. Classification of MAC Frames

Class	Frames Allowed
Class 1	Data: Data frames in both directions. Control: RTS, CTS, ACK, CF-End, CF-End+ACK Management: Probe (Request/Response), Beacon, Authentication, Deauthentication, ATIM
Class 2	All Class 1 frames Plus Control: Association (Request/Response), Reassociation, Disassociation
Class 3	All Class 1 and 2 frames Plus Data: Data subtypes Control: PS-Poll Management: Deauthentication

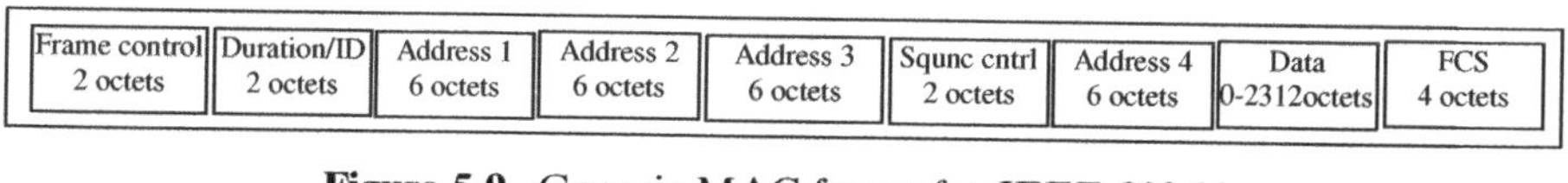

Figure 5-9. Generic MAC frame for IEEE 802.11.

In addition to the BSSID, there are included addresses of the source (SA), destination (DA), immediate transmitter (TA), and receiver (RA) in the MAC frame. The actual use of addresses depends on the combination of two bits (To DS/ From DS) in the frame control field. The frame control field defines frame types. Data (called *frame body* in the standard) contains data for the type of frame. FCS uses 33-bit generator polynomial $x^{32} + x^{26} + x^{23} + x^{22} + x^{16} + x^{12} + x^{11} + x^{10} + x^{8} + x^{7} + x^{5} + x^{4} + x^{2} + x + 1$. Sequence control provides a mechanism to identify each PDU and fragment with a sequence number (4 bit fragment number and 12 bit sequence number).

5.6. MULTICASTING CAPABILITY

Since MAC layer for multicasting only requires a multicasting address, the receiver address (RA) can be employed for this purpose. A multicast PDU may be treated differently from a unicast one, for example, fragmentation is not applied to multicast data. The multicast (and broadcast) data is transmitted with less reliability than the unicast data. This is due to the reason that ACKs are not practical in this case. Also, reservation (RTS/CTS) is not used for multicasting when To DS = 0.

Addressing capability is one of the many needed for multimedia teleconferencing type services. Other capabilities are provided by enhancements, like IEEE 802.11e, and queue management algorithms that might be needed for certain applications.

5.7. HIPERLAN MAC

The HIPERLAN 1 was designed to emulate Ethernet like capability over the wireless medium. It departs from the IEEE 802.11 MAC significantly due to the capability of prioritizing channel access and inherent multihop routing [10] [11].

5.7.1. Network Configuration

HIPERLAN 1 defines a fully distributed MAC with every station requiring to have the capabilities of packet forwarding and receiving in infrastructure as well as ad hoc configuration. The packet forwarding capability allows for multiple ad hoc networks exchange packets between any two nodes. Packet exchange can be possible by using the HIPERLAN ID (HID) and node ID (NID) pair of addresses. Selective nodes can have store-and-forward capability to assist in multihop routing.

5.7.2. Channel Access

Channel access in HIPERLAN is provided by the CAC sublayer of the DLC layer.

Every station in a cell is required to listen to the medium just like in CSMA used in Ethernet. A station that finds the channel to be idle for 1700 ms can transmit right away. This is similar to the DCF of IEEE 802.11 MAC. Prioritization for channel access is implemented through a three-phase access mechanism called *Elimination Yield Non-Preemptive Multiple Access* (EY-NPMA). The first phase is for access prioritization. One of the five priority levels is assigned to each contending station in terms of the equivalent number of slots of 168 bits each. A station could also be assigned one of the two priorities (Low/High) within each of the five slotted priorities. The allocation of priorities is based on packet lifetime (*time after which the packet loses its usability*). The packet life is usually measured in terms of normalized residual lifetime (NRL). If p_0 is the highest priority and p_4 the lowest priority then the NRL bounds are given by $p_0 \leq 2^0 \times 10\,\text{ms} \leq p_1 \leq 2^1 \times 10\,\text{ms} \leq p_2 \leq 2^2 \times 10\,\text{ms} \leq p_3 \leq 2^3 \times 10\,\text{ms} \leq p_4$. In other words, the highest priority is given to a packet with NRL of less than 10 ms and the lowest to a packet with NRL greater than 80 ms. These priorities can be imagined as being the slot delays for starting to sense the channel. An MT with a higher priority packet will start sensing earlier than a station with a lower priority packet. So, when a lower priority station starts sensing, the higher priority one may already have gotten into the next phase, which is contention. In other words, as soon as sensing starts, all stations are eliminated, except the ones with highest current priorities that start sensing simultaneously. When the priority timer reaches zero, the contention timer starts.

5.7.2.1. Contention.

In this phase, that can continue for up to 12 slots of 212 bits each, the surviving stations transmit a random burst and then sense the channel at the end of the burst. If a station finds an idle channel, it will be the survivor of the contention phase. Apparently, the station/s with the longest burst will be the winner. This is opposite of the backoff concept, in which the station with the lowest backoff would be the first to find the station idle. The channel sensing in this case is not allowed 100% of the time, but depends on a permission probability.

5.7.2.2. Yield.

Figure 5-10 and Figure 5-11 show the prioritization and contention phases of three stations. In Figure 5-10, the low-priority station B is eliminated when its delayed sensing results the other two (A and C) sending data burst on channel, thus making it non-idle. However, in Figure 5-11, station C is shown to have generated a longer random burst than station B, thus being the sole winner. It is possible that more than one station might generate the same length of burst in the contention mode. In such case, another random wait of up to 9 slots of 168 bits each will form the final and *yield* phase of

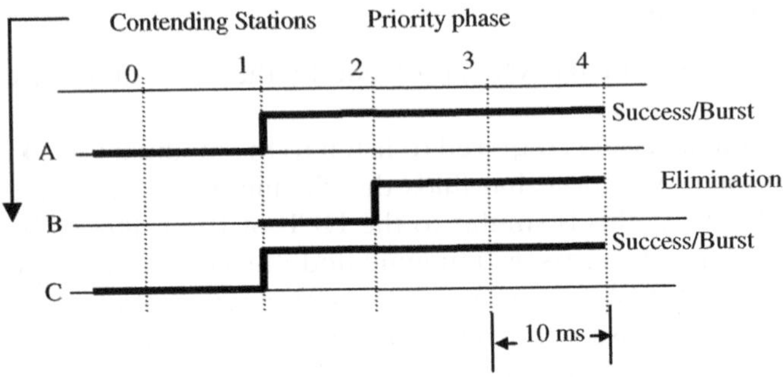

Figure 5-10. One of the three stations eliminated in priority phase.

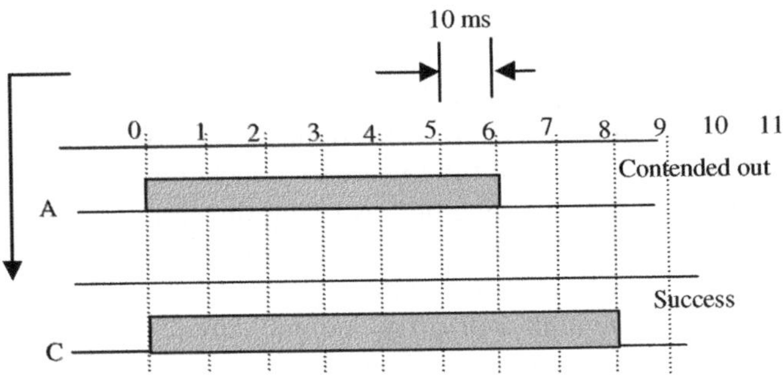

Figure 5-11. Station C survives contention due to longer random burst.

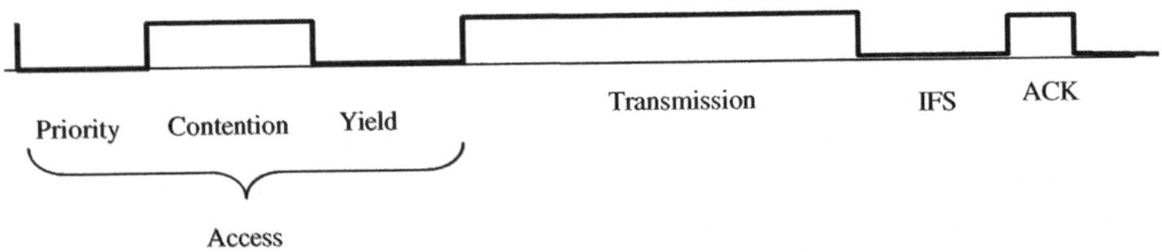

Figure 5-12. Transmissions are ACKed in HIPERLAN.

channel access. After the final random wait for the *yield* phase, a station is ready to transmit data. Since all transmissions in HIPERLAN are ACKed, a timing diagram for packet transmission will appear as shown in Figure 5-12.

5.7.3. Multiple Access

The multiple access in HIPERLAN 1 is part of channel access (i.e., PDMA) and a contention, transmission, acknowledgement paradigm is followed. Collision may still occur, albeit with a very low probability. Backoff timers and

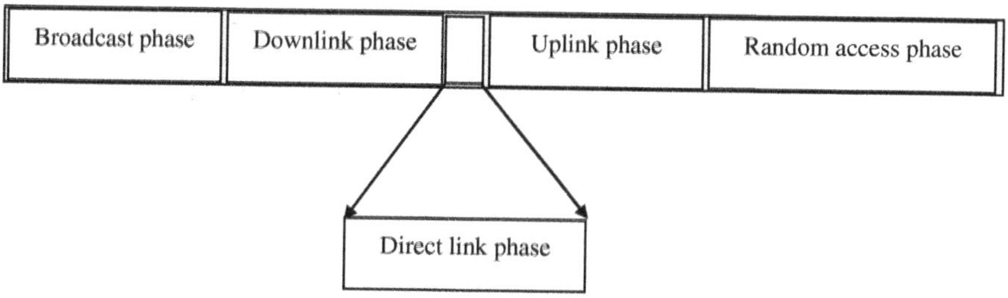

Figure 5-13. HIPERLAN 2 MAC frame.

probability of permission in contention and yield are used for further reduction in collision.

5.8. HIPERLAN 2

HIPERLAN 2 has been designed as an access network for an ATM backbone. Therefore, its MAC protocols reflect this capability.

5.8.1. Channel Access

Slotted ALOHA is used in a part of MAC frame reserved for this purpose. Stations contend for time slots in the other parts of the frames. These times slots are allocated in downlink, uplink, and direct link phases of the frame. It is known that the maximum throughput of slotted ALOHA is less than 40%. This is due to collisions and leaving empty slots (no one contending for them). The result of an access attempt is announced in another part of the MAC frame called the *broadcast phase*. See multiple access for details. See, for example, [12][13][14].

5.8.2. Multiple Access

The HIPERLAN 2 MAC principle is based on the paradigm of cellular networks. It uses a 2ms TDMA frame[4], with the MTs contending for time slots and AP/CC (central controller) allocating them. A time slot has duration of 400ns. The slot allocation is centrally managed by AP (CC in case of ad hoc topology) and is entirely dynamic. The MAC frame consists of 5 phases, as shown in Figure 5-13.

5.8.3. Broadcast Phase

The broadcast phase is from AP/CC to MTs and it contains the status information of current slot allocation. MTs trying to reserve a slot in the previous

[4] In practice, 2ms is so short that ARQ can be used, even for voice packets.

frame know the result in the broadcast phase of current frame. Its duration is 20 μs fixed. Three channels, broadcast channel (BCH), frame control channel (FCH), and access channel (ACH), convey complete organization of the rest of the MAC frame to all stations; in particular, FCH provides transmission resource allocation information to MTs that have accessed the channel successfully. ACH gives feedback about the status of the newly registering channels.

5.8.4. Downlink Phase

During this phase, data are transmitted from AP to MTs in the slots announced during the broadcast phase. All MTs listen to the broadcast phase to check if there are data for them. If their ID appears in the recipients' list, they synchronize to the slot in which they are supposed to receive data during the downlink phase. Its duration depends on the number of MTs in a cell.

5.8.5. Uplink Phase

In this phase, slots are used by MTs to send data to the AP. Its duration is 12 μs fixed.

5.8.6. Direct Link

Direct link transcends the restrictions of transmitting in uplink and receiving in downlink phases. In this small field, transmission resources can be allocated to the transmitting and receiving MTs simultaneously, such that one station can transmit and the other can receive from a single slot.

5.8.7. Random Access Phases

In this phase the MTs use Slotted ALOHA to contend for channel resources. Prioritization is allowed to have multimedia capability.

5.9. USER AND DATA PRIVACY

HIPERLAN 1 has no mechanism for user authentication. For data privacy encryption is optional. However, the random sequence generation algorithm is secret [15] as reported in [16]. For data encryption, a common set of shared keys *HIPERLAN key-set* is employed. The algorithm uses as input the secret key and an initialization vector (IV). The IV is transmitted with every MPDU.

HIPERLAN 2 provides many choices of security, including authentication and inter-AP security information exchange [17]. These include no authentication for open systems, pre-shared key authentication (using HMAC-MD5 algorithm—see Chapter 8) and three RSA algorithms with keys of 512, 768, and 1024 bits. The procedures for exchange of authentication keys for pre-shared key authentication and public keys for RSA algorithms are not rigorously specified. Either AP or MT can authenticate, thus extending the authentication mechanism to ad hoc configuration. Data encryption is done by using DES standard, and optionally 3DES.

5.10. POWER MANAGEMENT

Power save mode is provided in HIPERLAN by defining a type of station that can participate in this mode. These stations, the so-called *p-savers*, are provided power saving capability with help from complementary stations called *p-supporters*. The p-savers communicate their schedule for the sleep and awake times to the p-supporters. If a packet arrives during the scheduled sleep time, the p-supporters store the packet until the scheduled awake time [18].

Another power-saving feature of HIPERLAN is the low-rate header transmission. Frame header determines whether a frame is destined to the current recipient. A low-rate header consumes less energy than a higher-rate data part of the frame. However, the data-part is not received if a station knows (from the header) that it is not the intended recipient.

5.11. MULTIMEDIA SERVICES

Since HIPERLAN 2 uses TDMA/TDD, it is ideally suited to providing multimedia. In fact, the original development of HIPERLAN 2 was to provide a WLAN interface for the ATM backbone and Quality of Service (QoS) provisioning was one of the main attractions of ATM. Consequently, a mobile station in HIPERLAN 2 can specify QoS parameters used in ATM network (throughput, delay, loss, jitter, etc.) Enhancements have also been suggested [19].

One of the biggest strengths of HIPERLAN 2 arises from the absence of collisions and that it has a high *actual throughput*. HIPERLAN/2 54 Mbps standard provides an actual throughput above 40 Mbps. Compare this with the 20 Mbps or below of IEEE 802.11a (54 Mbps maximum). This certainly results in better quality of audio-visual applications [20], even with simple priority queueing. In addition to ATM, HIPERLAN 2 is projected to provide interfaces to IP as well as 3G cellular networks [21]. Since the standard is poised to provide interface to a variety of networks (including the IEEE 1394 standard), a major amount of flexibility is embedded in the standard. This includes MAC frame scheduling to the choice of error-control (EC) mechanisms. EC, being a major fraction of the HIPERLAN 2 DLC, provides a choice among selective acknowledged, unacknowledged, and repetitive transmissions. In case a retransmission mechanism is chosen, further flexibility is provided in choos-

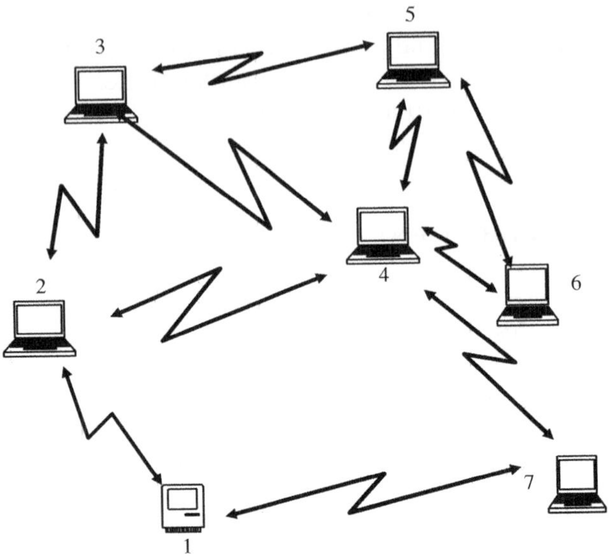

Figure 5-14. Mobile terminals 2 and 4 act as forwarders.

ing the EC parameters, for example, number of retransmissions, window size, and so on [22].

5.12. ROUTING

When HIPERLAN is configured as infrastructure network, the access point (AP) also functions as central controller (CC). Thus it provides the routing and bridging function as well as resource management. When HIPERLAN is realized in ad hoc form, the function of the CC is distributed in nature and any station can act as CC. In that case, if a station is registered with more than one CC, it can also act as frame-relaying station. Routing is a major strength of HIPERLAN. All MTs can act as forwarding stations (*forwarders*) in an ad hoc HIPERLAN. MTs use a control packet called *Hello* packet to share neighbors' information with every MT around them [23]. The forwarders use the information in the Hello packets to set up the routing table.

Other routing mechanisms can be defined at the top of the HIERLAN 2 DLC, such as an inter-operability routing mechanism, proposed in [24] for dual mode HIPERLAN 2[5]. Also, [25] proposes an extension of the relaying capability by requiring stations to use a relay station within a single network instead of raising power levels when away from the destination [26]. Figure 5-14 shows an ad hoc network with MT number 2 and 4 acting as forwarders.

[5] The dual node HIPERLAN 2 is defined at 60 GHz to support highly congested urban areas.

5.13. MOBILITY SUPPORT

Since HIPERLAN is designed for low mobility, the only mobility support provided in the standard is the capability of signal measurement. An MT keeps measuring the signal to noise ratio from more than one AP simultaneously. When connection with the existing AP becomes weak, it can request the association to be shifted to a new AP with clearer signal. Data may be lost during handoff.

5.14. MAC FRAME

As shown in Figure 5-13, the HIPERLAN 2 MAC frame consists of phases. Within these phases logical channels are defined over transport channels [27]. Figure 5-15 [28] shows various transport and logical channels. Here's a brief description of each of the transport channels.

BCH Contains cell information.

FCH Contains frame composition information.

ACH Contains the results of contention from the previous frame.

LCH Contains 54-byte MAC PDUs with 48 bytes of user data (=ATM cell payload).

SCH Contains 9-byte signaling data.

RCH Allow contention for access.

Figure 5-15 showcases the transport channels.

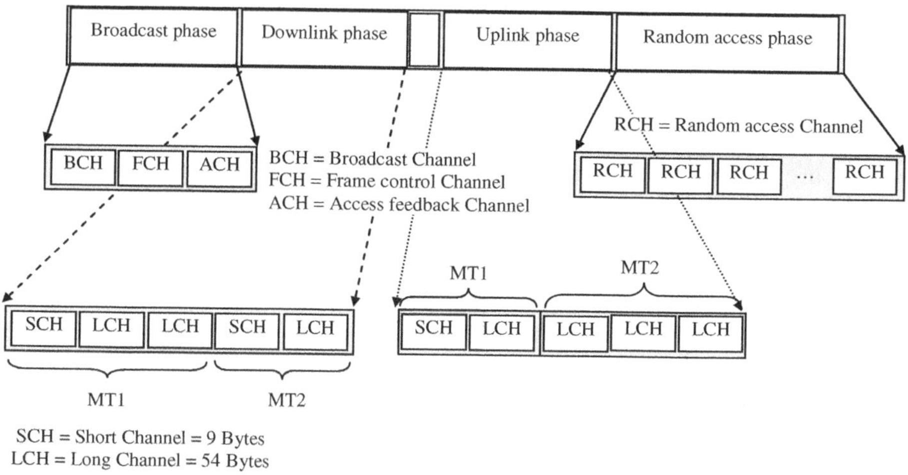

Figure 5-15. HIPERLAN2 MAC frame channels.

Logical channels are defined within the transport channels (see [27] for a mapping). Here's a brief definition of each:

SBCH = Slow broadcast channel. Carries a variety of information from AP accessible to all MTs from handover, MAC ID assignment for non-associated MT, encryption seed, convergence layer information, etc. It is downlink only.

DCCH = Dedicated control channel. Carries messages between an MT and AP using radio link control (RLC) sublayer signals. These messages pertain to DLC connection control and association control functions. It is bi-directional.

UDCH = User data channel. Carries convergence layer data between an MT and AP. It is bi-directional.

LCCH = Link control channel. Carries information between the error control (EC) functions of MT and AP. It is bi-directional.

ASCH = Association control channel. Carries information for association and re-association request from the MTs to AP. It is uplink only and is used only during the handover process.

Since HIPERLAN 2 has a connection-oriented mode of operation, information among various functions of the standards is exchanged in the form of messages instead of frames. These messages are communicated using the above logical channels. Each of these channels is not, however, dedicated to be used only for a particular message. A large number of messages could be exchanged using a small number of logical channels, depending on various phases of connection (set up, data exchange, handover, encryption, etc.).

5.15. TELECONFERENCING CAPABILITY

HIPERLAN 2 supports multicasting and broadcasting that can be used for teleconferencing capability. The multicast PDUs are not acknowledged and are transmitted only once. However, the broadcast PDUs, though not acknowledged, are transmitted with temporal diversity (multiple instances transmitted per MAC frame).

Temporal diversity enhances the chances of reliable reception. In addition to the group multicast function, the standard also defines what is called the *N*unicast mechanism*, which is using N unicast addresses to make one multicast group. Using this mechanism, retransmissions can be employed as if they were unicast in nature.

5.16. DATA LINK CONTROL (DLC) LAYER

The HIPERLAN standard has a complex DLC layer that consists of channel access control (CAC), medium access control (MAC), error control (EC), and radio link control (RLC) functions, and their sub-functions. At the top of the DLC, a convergence layer provides interfaces to ATM, IP, IEEE 1394, and 3G UMTS systems. There is no LLC as such, but DLC contains functions of an LLC and many more.

A DLC connection is referred to as DLC user connection (DUC), which is a unique combination of MAC ID and DLC connection ID. It is through this connection that peers of CAC, MAC, EC, and RLC exchange user and control data.

REFERENCES

[1] J. Yeh, and Chen S. C., 'Support of Multimedia Services with the IEEE 802-11 MAC Protocol,' IEEE International Conference on Communications, Vol. 1, pp. 600–604, 2002.

[2] A. Banchs, Radimirsch, M., and Perez, X., 'Assured and Expedited Forwarding Extensions for IEEE 802.11 Wireless LAN,' Tenth IEEE International Workshop on Quality of Service, pp. 237–246, 2002.

[3] Sunghyun, Choi, 'IEEE 802.11e MAC-level FEC performance evaluation and enhancement', *IEEE Globecom'02*, November 2002.

[4] Daqing Gu and Zhang, Jinyung 'Evaluation of EDCF Mechanism for QoS in IEEE 802.11 Wireless Networks', *Mitsubishi Electric Research Laboratory (merl)*, Technical Report TR-2003-51, May 2003, Available from http://www.merl.com/reports/docs/TR2003-51.pdf

[5] IEEE 802.11e draft D/4.1, Part 11: Wireless Medium Access Control (MAC) and physical layer (PHY) specifications: Medium Access Control (MAC) enhancements for Quality of service (QoS), Feb. 2003.

[6] Sunhyun Choi, del Prado, Javier, Shankar, Sai, and Mangold, Stefan, 'IEEE 802.11e contention-based channel access (EDCF) performance evaluation', *IEEE ICC'03, 2003*. Available from http://path.berkeley.edu/dsrc/reading/03-ICC-EDCF.pdf

[7] Stefan Mangold, Choi, Sunghyun, MayOle Klein, Peter, Hiertz, Guido, and Stibor, Lothar, 'IEEE 802.11e Wireless LAN for Quality of Service', *European Wireless 2002*, Florence, Italy, February 2002. Available from http://www.ing.unipi.it/ew2002/proceedings/H2006.pdf

[8] Arunesh Mishra, Shin, Minho, and Arbaugh, William, 'An empirical analysis of the IEEE 802.11 MAC layer handoff process', *ACM SIGCOMM Computer Communication Review*, v.33 n.2, April 2003.

[9] IEEE 802, 'Practice for multivendor access point interoperability via an inter-access point protocol across distribution systems supporting IEEE 802.11 operation', *IEEE Draft 802.11f/D3*, January 2002.

[10] G. Anastasi and Mingozzi, E., 'Stability and performance analysis of HIPERLAN', *IEEE INFOCOM 1998*, Available from http://www.ieee-infocom.org/1998/papers/02a_1.pdf

[11] Jorg Habethat, Dutar, Romain, and Wiegert, Jens, 'Performance evaluation of HIPERLAN/2 multihop ad hoc networks', *European Wireless* 2002, Available from http://www.ing.unipi.it/ew2002/proceedings/H2005.pdf

[12] Emil Ebdom and Henriksson, Henrik, 'Design comparison between HiperLan 2 and IEEE 802.11a services', *Master's thesis*, Department of Science and Technology, Linkoping University, Sweden. Available from http://www.ep.liu.se/exjobb/itn/2001/as/001/exjobb.pdf

[13] Edgar Bolinth, Lappetelainen, Antti, Ojala, Jussi, Pauli, Mathias, Kramiling, Andreas, Journe, Thierry, Lacroix, Dominique, Bohnke, Ralf, Wegmann, Bernhard, and Schulz, Egon, 'QoS enhancements for HIPERLAN/2', Available from http://jungla.dit.upm.es/~ist-mind/publications/H2-QoS-Enh-v08.pdf

[14] ETS 300 652. High Performance Radio Local Area Network (HIPERLAN) Type 1; Functional specification. ETSI, 1996.

[15] TR 101 054. Rules for the management of the HIPERLAN Standard Encryption Algorithm (HSEA). ETSI, 1997.

[16] Sami Uskela, 'Security in Wireless Local Area Networks', Available from http://www.tml.hut.fi/Opinnot/Tik-110.501/1997/wireless_lan.html#TR054

[17] Marco Casole, 'WLAN security: status, problems and perspectives', *European Wireless*, 2002. Available from http://www.ing.unipi.it/ew2002/proceedings/sec002.pdf

[18] Raja Jurdak, Videira Lopez, Cristina, and Baldi, Pierre, 'A survey, classification and comparative analysis of medium access control protocols for ad hoc networks', *IEEE Communications Society Surveys*, Jan. 2004, Available from http://www.comsoc.org/livepubs/surveys/public/2004/jan/jurdak.html

[19] Edgar Bolinth, Lappeteläinen, Antti, Ojala, Jussi, Pauli, Mathias, Krämling, Andreas, Journé, Thierry, Lacroix, Dominique, Böhnke, Ralf, Wegmann, Bernhard, and Schulz, Egon, 'QoS enhancements for HIPERLAN/2', http://jungla.dit.upm.es/~ist-mind/publications/H2-QoS-Enh-v08.pdf

[20] Jim Geier, 'HIPERLAN/2: An efficient high-speed LAN', http://www.wi-fiplanet.com/tutorials/article.php/2109571

[21] Stephen McCann and Flygare, Helena, 'HIPERLAN/2 public access interworking with 3G cellular systems', *European Wireless 2003*. Available from http://www.ing.unipi.it/ew2002/proceedings/H2003.pdf

[22] Goran Malmgren, Khun-Jush, Jamshid, Schramm, Peter, and Torsner, Johan, 'HiperLan Type 2—An emerging world wide wireless LAN standard', *International Symposium on Services and Local Access*, 2002. Available from http://www.issls-council.org/proc00/papers/6_3.pdf

[23] Zhengping Zuo, 'In-building wireless LANs', http://www.cse.ohio-state.edu/~jain/cis788-99/ftp/wireless_lans/

[24] Athanasios Vaios, Oikonomoo, Konstantinos, and Stavrakakis, Ioannis, 'A centralized routing scheme supporting ad hoc networking in dual mode HiperLan/2'. Available from http://www.intracom.gr/downloads/pdf/news/publications/2003/aveiro2.pdf

[25] D. Hollos and Karl, H., 'A protocol extension to HiperLan/2 to support single-relay networks', *Proceedings of Ist German Workshop on Ad Hoc Networks*, pp. 91–108, Ulm, Germany, March 2002.

[26] Jost Weinmiller, Schläger, Mortin, Festag, Andreas, and Wolisz, Adam, 'Performance Study of Access Control in Wireless LANs: IEEE 802.11 DFWMAC and ETSI RES 10 HIPERLAN', *Mobile Networks and Applications (Special Issue on Channel Access)*, Vol. 2, No. 1, pp. 55–76, June 1997.

[27] Martin Johnsson, 'HiperLAN/2—The broadband radio transmission technology operating in the 5 GHz frequency band'. Available from http://www.hiperlan2.com/presdocs/site/whitepaper.pdf

[28] Romain Rollet, Rosier, Corinne, Bonville, Herve, and Mangin, Christophe, 'Field trial results at DLC layer of a HiperLAN/2 prototype', *IEEE Vehicular Technology Conference Spring 2003*, Available from http://www.mitsubishi-electric-itce.fr/English/scripts/Publications%20pdf/2003/rollet_vtc03spring.pdf

CHAPTER 6

MOBILITY AND INTERNET PROTOCOLS

The simplicity and flexibility of the Internet Protocol (IP) have been its main driving attributes. The IPv4, once thought to be soon out of phase, is still the dominant internetworking protocol in the world. As a connectionless protocol, it defines a simple header with 32-bit address spaces for source and destination IP layers. The header error checksum gives some protection to the header part of the packets. The fragmentation function allows the use of a mechanism to adapt to smaller maximum permitted packet sizes in the intermediate networks in a heterogeneous end-to-end connection. Time to live (TTL) can be used to get an estimate of how many hops a packet has traveled. There is no routing protocol that is a part of IP-itself. An IP packet can be encapsulated by another IP packet or a lower layer packet. IP can encapsulate any protocol data unit (PDU) as long as it has a protocol number for identification. In summary, two instances of IP-based hardware or software could be perfectly interoperable with numerous differences in their implementation of how packets are routed, how they are given IP addresses, and even what they carry. Nevertheless, address distribution (and, hence, routing hierarchy) in IP has traditionally been related to the domain name service (DNS). DNS is supposed to give network addresses a human taxonomy—of stable geographic locations. Due to this, IP addresses have been allocated permanently to geographically fixed networks. The concept of subnet masking has made locations ever so permanent.

If one concludes from the above paragraphs that mobility is not built-in in IP, one needs to be reminded that neither is routing. Source routing may be

Wireless and Mobile Data Networks, by Aftab Ahmad

an exception, but it is not the general choice of routing on the Internet. IP is, in fact, flexible enough a protocol for implementing mobility functions. However, one big difference between routing and mobility management is that routing can be entirely distributed and may require less cooperation among network administrations than mobility. Perhaps it is due to this reason that IPv6 has been designed to have more generic functions to assist in mobility than IPv4. In this chapter, we will look at some of the options for managing the mobility of hosts using IP and related protocols. We will start by defining types of mobility and go on to discuss session initiation protocol (SIP) to manage mobility at higher layers, followed by enhancements in IPv4. Finally we will have a discussion on IPv6 mobility management functions.

6.1. MOBILITY IN INTERNET APPLICATIONS

IP was designed for mainframes and minicomputers. It works the same way with workstations and desktops. In fact, as the performance gap among these machines has narrowed, IP has not been affected much by this, except perhaps by their proliferation. The popularity of notebooks, palmtops, and personal digital assistants has changed the scenario. The new voice-over-IP (VoIP) applications are further evidence that IP mobility is critical to current and future applications. A mobile user with a device requiring an IP connection does not have to be wireless, however. In fact, as pointed out in [1], there are three kinds of mobility when it comes to IP in cellular infrastructures. These are: (1) reconnectivity when a user may move to another domain and plug in to the network to establish a new IP connection, (2) portability (*macromobility*), when a user may do the same thing as in reconnectivity but may be wired or wireless *and* keep a permanent presence at a 'home network', and (3) mobility (*micromobility*), when the user is wireless and mobile *while using the connection continuously*. The three types offer different problems and could, in fact, be required by a single user all the time. Here are general possible solutions.

6.1.1. Reconnectivity

Reconnectivity does not require any change in the classical IPv4, as a new IP address can be allocated either manually or automatically using dynamic host configuration protocol (DHCP), as shown in Figure 6-1.

6.1.2. Portability

At least four major additions to the classical implementations of IPv4 are needed to provide portability. These are: (1) to have another node receive IP datagram in the home network (mobile *home agent*), (2) to have a way for the mobile home agent to keep track of the mobile host's whereabouts (*registra-*

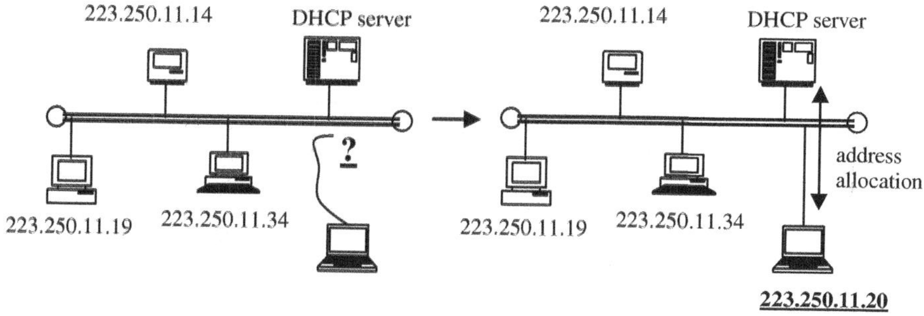

Figure 6-1. Reconnectivity can be as simple as connecting to a wall socket.

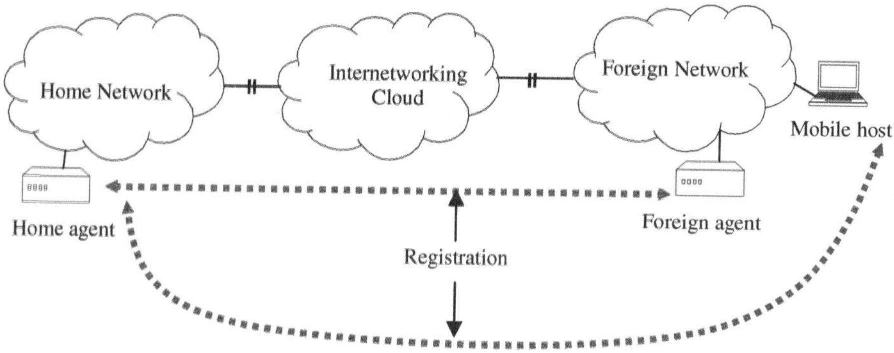

Figure 6-2. Components of mobile IP (macromobility, could be wired or wireless). Shown wired connections.

tion), (3) if a mobile host moves from one domain to another, to have a node (*foreign* agent) at the next network inform the mobile home agent about the next network, and (4) to have a way of allowing two-way datagram exchange from the new location to a *correspondent* host (CH) transparently. Figure 6-2 shows the interaction of these four components.

6.1.3. Micromobility

There are many ways to handle micromobility, and each mechanism will have its requirements. We will use a hypothetical scenario. Continuous mobility and connectivity may require a way of having an IP address to be registered across more than one node that can deliver the IP datagram to the mobile host. Additionally, a mechanism is required for keeping current location (*registration*) among the nodes that serve a given IP. An example is shown in Figure 6-3, where a group of base stations have the IP address of a mobile host listed as *active* or *passive*. In this example, the IP address-allocation is done by higher level 'address server'. As soon as the IP address is allocated using some dynamic mechanism such as DHCP, it is published to all the serving base sta-

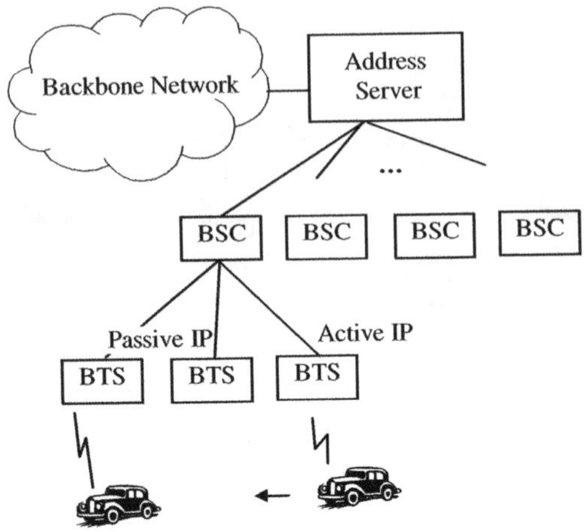

Figure 6-3. Example of micromobility handling.

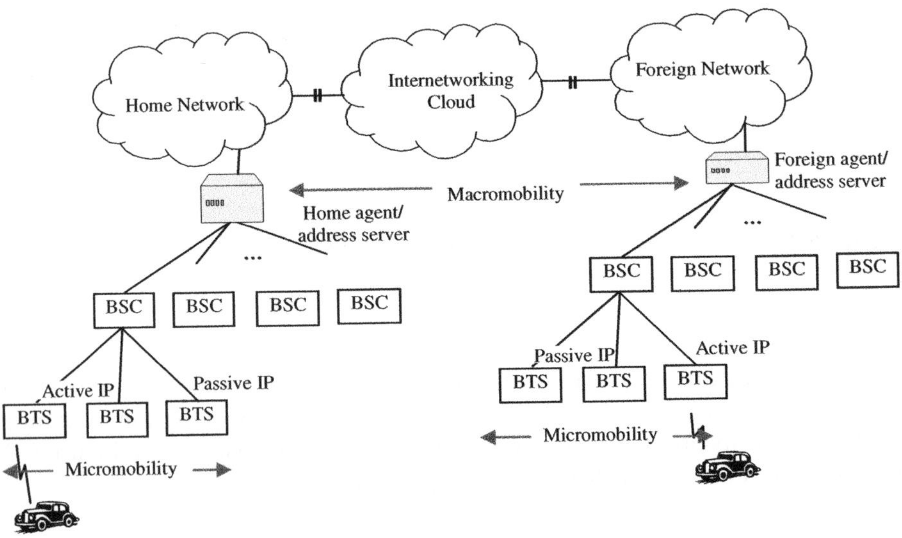

Figure 6-4. Network-wide mobility management.

tions under this address server. The base station, which started in the address-acquisition process, has the IP address of the mobile host as active while all other base stations have it as passive. If the mobile host (MH) moves from one base station to another, the datagrams from this MH are treated like any other datagrams of the active IPs. A *handoff* process is executed for incoming datagrams.

Figure 6-4 shows a realistic scenario in which micromobility management is carried out in the address server area and macromobility management is performed in the interdomain handoff.

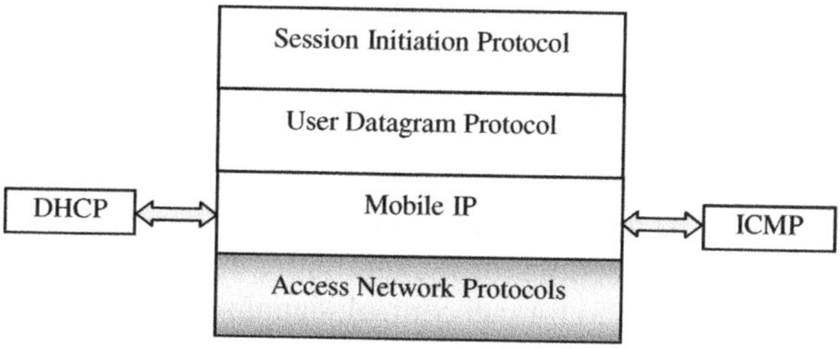

Figure 6-5. Key Internet protocols for managing mobility.

6.2. INTERNET PROTOCOLS FOR MOBILITY

As apparent from the discussion on three forms of mobility, there is no critical need of enhancements for reconnectivity. For macromobility, the infrastructure is best met by mobile IP [2]. For micromobility, a cellular network-like infrastructure is needed. This would ideally be on the same lines that wireless ATM mobility was supposed to be handled, to avoid frequent hard handoffs. Several protocols have been proposed (See the 'protocol explosion' on p. 10 of Ref. [1].) In the following pages, we will discuss three protocols that are critical for providing end-to-end continuous service to a wide area IP-based mobile wireless network. These are the session initiation protocol [3][4], mobile IP, and cellular IP [31]. Figure 6-5 shows a relation among various protocols that act together to provide multimedia services to mobile hosts in cellular network. In the end, we will have a section on mobility-related functions in IPv6, since leading companies (such as, Ericsson and Nokia) believe that, due to increasing peer-to-peer applications, (as opposed to client-server applications), IPv6 has to be deployed for cellular IP networks.

6.3. SESSION INITIATION PROTOCOL (SIP)

The architectural difference between SIP and IP is that SIP is an application-level signaling protocol whereas IP is a network layer data (plus signaling) protocol. Functionally, SIP provides only signaling capability between two applications. In 3GPP, SIP is used to establish and terminate telephone calls.

6.3.1 SIP versus H.323 and HTTP

In many ways, SIP is an alternative for ITU's H.323 architecture. However, the SIP protocol and H.323 architecture don't share anything beyond goals.

SIP is a text-based session management protocol that allows two applications (the caller and the 'callee') to establish understandings in the form of uniform resource identifiers (SIP URI)s. It is extensible, flexible and widely interoperable (e.g., with service discovery protocol (SDP) and real-time streaming protocol (RTSP)). It does not require specialized hardware and is implemented in software entirely. H.323 is binary coded multiplayer architecture that is rather inflexible (three versions to include more functions). It contains two protocols (H.245 and H.225) for call control, signaling, and authentication. SIP shares much with HTTP, including message coding mechanism and programmability. SIP has been designed keeping the HTTP flexibility in mind and uses its request/response model. The regular 'bake-offs' help incorporate extensibility as it arises.

6.3.2. SIP Provisions

SIP is a signaling protocol for the Internet applications. As such, it does not provide services, but provides the infrastructure in the form of primitives to design services. These primitives can be used for exchanging a number of messages in the original protocol and its extensions. Fundamental session management capabilities provided by SIP include allowing applications and users to establish, move and terminate sessions in client-server paradigm. For networks with mobile users, SIP can be used to locate end systems of communications to help in managing user mobility. A number of security related functions are also included in the protocol. These include encryption, denial of service (DOS) prevention mechanism, privacy, and authentication. Additionally, it provides for negotiating media (data coding mechanism, etc.) parameters and modifying them during a session. It does not concern itself with the actual information to be exchanged in a session. Figure 6-6 shows a typical usage scenario.

Table 6.1 lists SIP entities. Each entity performs tasks that consist of a series of *SIP transactions*. SIP transactions constitute the SIP requests and responses. All SIP entities except the proxy server contain transaction layer. The *transaction user* is the layer above transaction layer. The job of the transaction user layer is to generate application layer responses to request retransmissions where applicable, and timeouts etc.

6.3.3. SIP Request Types

The purpose of request messages is to invoke action. The request messages map into what SIP can perform. Clients send a request message and it is routed to servers through a proxy server that acts as a client to the destination server and a server to the invoker of request. Here's a list of the six SIP requests and their meanings, as defined in RFC 3261.

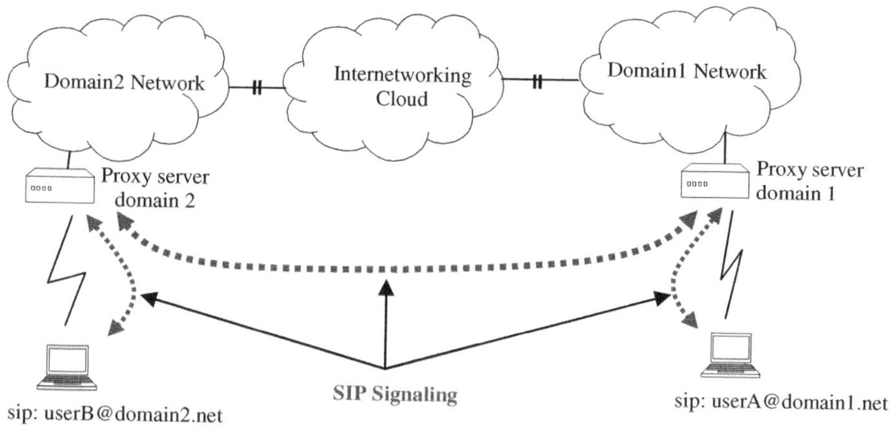

Figure 6-6. SIP is a signaling protocol, in which users can be addressed using e-mail like addressing.

TABLE 6.1. SIP Defined Network Elements

SIP Network Element	Function
User agent (UA)	A UA can act as a user agent client or server. User agent client (UAC): Generates Request transactions. User agent server (UAS): Generates Responses to the requests. Back to back user agent (B2BUA): A concatenation of UAC and UAS. Redirect server: A UAS that redirects UAC to the changed user location.
Proxy (server)	A proxy (server) routes a request from a UAC to an appropriate UAS. It can modify the request (hence, proxy) acting as a client. Stateful proxy: A proxy that maintains the state machines of client and server during session. Stateless proxy: A proxy that does not maintain the state machines of client and server during session.
Registrar	A server that handles location service for a domain. The request type REGISTER contains this information for the registrar.
Transaction users	The layer above the transaction layer of SIP. It includes the protocol functions (called cores) of UAS, UAC and proxy server.

1. REGISTER. This request is sent from a client to registrar to register in a domain. In mobile networks, it will be needed every time a user moves to a new domain.
2. INVITE. Sent for requesting the setting up of a new session with another SIP entity. 'reINVITE's may be used to make changes in the on-going sessions.
3. ACK. The last signaling message in a successful session set up sent from the requesting client.
4. CANCEL. Sent by the client to cancel the last request.
5. BYE. Terminating a session.
6. OPTIONS. Message sent to a proxy server asking what capabilities are available at this server.

6.3.4. SIP Response Types

The responses carry the status of requests if any additional data is needed. Each response carries a three digit number *xyz* that describes the response type. The *x* can have a value 1 though 6 and it determines the response class. The remaining two digits determine the exact type of response. Table 6.2 gives a list of response classes.

6.3.5. SIP Operation

Figure 6.7 shows an instance of SIP operation. It has two phases, a signaling phase during which session is set up and terminated, and a data exchange

TABLE 6.2. SIP Response Classes

Status Code	Class	Meaning
1*yz*	Provisional	The request is being processed, e.g., Trying (100), Ringing (180).
2*yz*	Success	The request is successfully received and understood, e.g., Ok (200).
3*yz*	Redirect	Continue the request with further action, e.g., try another recipient, e.g., Moved temporarily (302), Moved permanently (301).
4*yz*	Client error	The request is bad or can't be met at this time, e.g., Unauthorized (401) or Busy (486).
5*yz*	Server error	The server failed to comply with a legitimate request, e.g., Server internal error (500).
6*yz*	Global failure	The request can't be fulfilled anywhere among all servers, e.g., Busy everywhere (600).

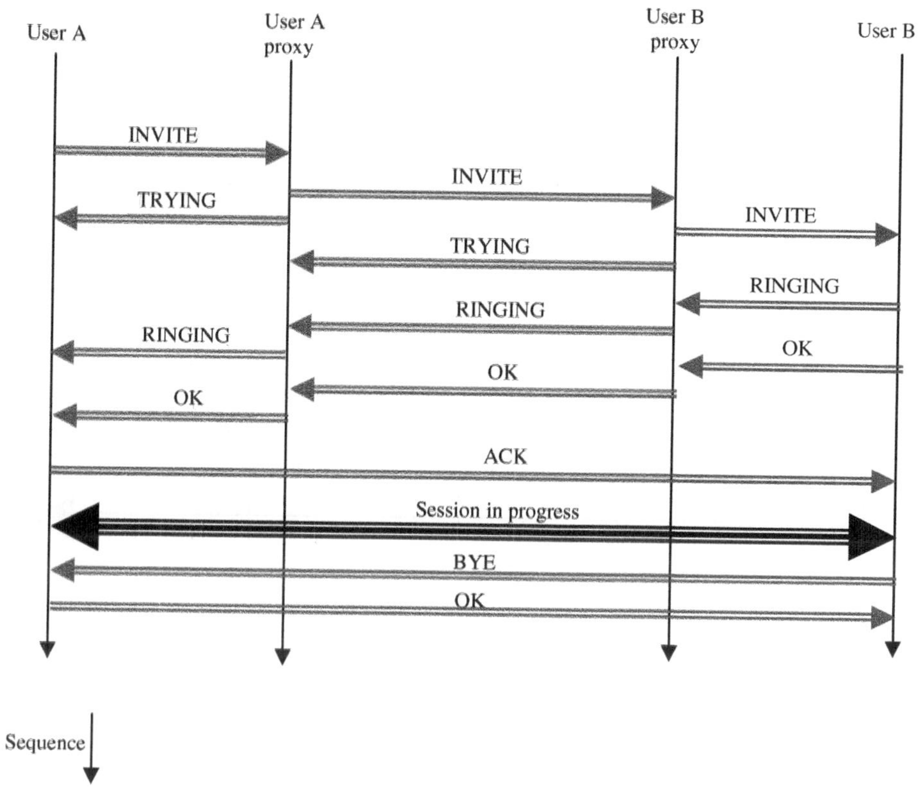

Figure 6-7. Message sequence for media session instance using proxies (RFC3261).

phase, during which the two users exchange data, for example, interactive instant messaging, phone call, and so on.[1] The network path for signaling does not have to be the same as for data exchange [5]. User A in domain 1 (sip: userA@domain1.com) requests a connection with the User B at domain 2 (sip: userB@domain2.com). It sends the INVITE message to its proxy server (proxy A). The proxy A forwards this message to proxy B (with a provisional message TRYING back to User A), the proxy server for domain B forwards it to User B, and so on. Note that each SIP user has an address made unique by the http or e-mail-like notation.

6.3.6. SIP and Cellular Networks

SIP can handle terminal and personal mobility with the help of REGISTER and reINVITE request types. REGISTER can be used to register the mobile

[1] Data exchange phase is among the applications, and SIP does not provide any service during this phase.

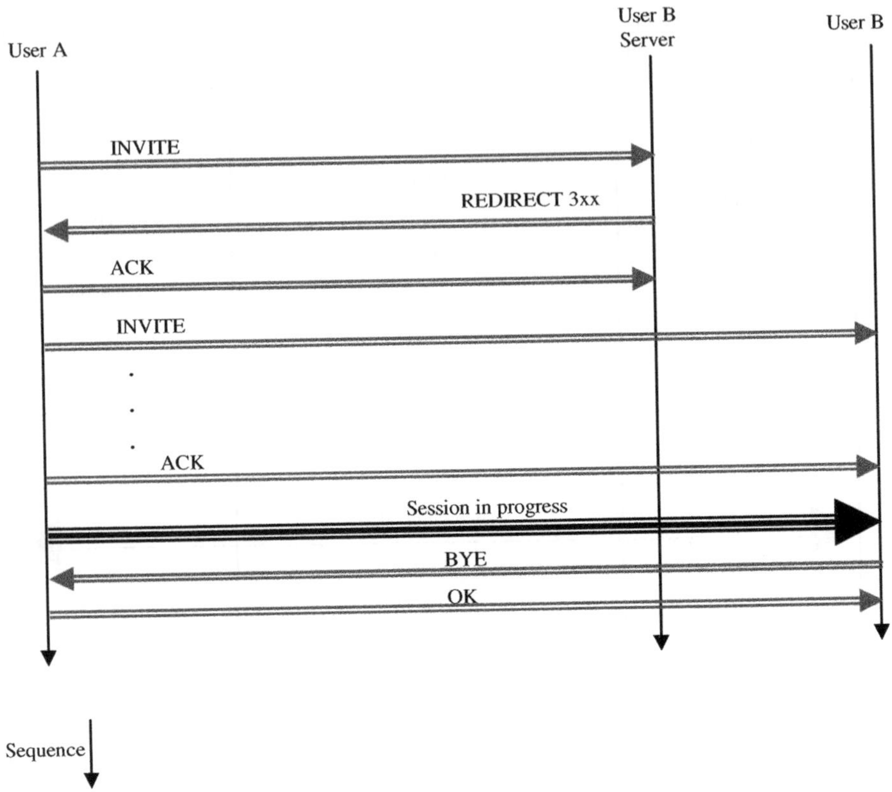

Figure 6-8. REDIRECT used to set up connection with user B, who moved out.

host (MH) in home SIP server database to show the new domain. Once the home server knows the current location and a corresponding foreign SIP server of the mobile, it can forward all requests to the mobile host through its new server. The MH, after receiving redirected INIVITE, can use reINVITE to directly request a session with the calling user. See Figure 6-8 for example of demonstration of this concept. In fact, using user authentication, SIP can provide personal services to mobile users wherever they can reach a proxy server and authenticate themselves. Once the REGISTER request message can be used for address binding to new locations and foreign severs, mobile IP tunneling and triangular routing can be used with SIP acting as the signaling protocol for mobile IP.

As pointed out in [6][7], SIP can handle mobility and heterogeneous access network interfaces in a variety of ways. The two leading interoperability projects for 3G systems (3GPP and 3GPP2) are expected to use SIP for mobility management in the IP part and all-IP versions of the networks.

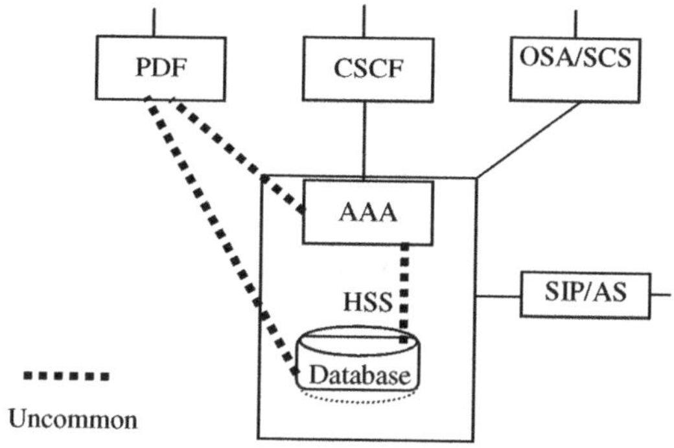

Figure 6-9. HSS with SIP/AS in proposed harmonized reference model for 3G IMS part of the systems.

6.3.7. SIP and 3GPP, 3GPP2

3GPP network consists of three sub-networks to provide multimedia [8]. These are the traditional GSM circuit-switched domain, GPRS-based packet-switched domain (IP or non-IP) and the IP multimedia domain. Mobility management in the IP multimedia domain is managed by SIP as of Release 3 of the 3GPP. The 3GPP2 network consists of two sub-networks [9]. These are the traditional one, based on CDMA, and the IP multimedia subsystem (IMS). IMS uses SIP for application-level mobility management. The SIP application server (SIP A/S), connected to the home subscriber system (HSS), provides a quick application and service-development environment, as shown in Figure 6-9 ([9]) for a proposal of harmonized architecture for 3GPP and 3GPP2.

6.4. MOBILE IP

Mobile IP is more of a commonsense use of IP for mobility.[2] As mentioned earlier, IP standard stops short of specifying every procedure being an essential part of the protocol (e.g., any routing protocol can be used at IP routers). This has given the protocol one of its main strengths, that is, flexibility of how it is deployed. Mobile IP is more of a deployment mechanism for IP under the established, that is, hierarchical, geographically fixed, routing and domain name service (DNS) systems. Note that it is possible, at least in theory, to redeploy IP (even IPv4) in order to get rid of 'geo-sensitive' routing and make it a completely or partially a mobile protocol. Of course, even if this can be done, the cost is not conceivably affordable, due to the enormous heterogeneity embedded in the Internet infrastructure.

[2] Mobile IP without the suffix 'v6' is generally considered Mobile IPv4. We stick to this convention.

Mobile IP is best suited to macromobility and travel-related mobility (reconnectivity) for fixed wireline access while traveling. However, this applies equally well to macromobility in wireless cellular networks. In fact, the travel-related mobility (reconnectivity) issue is better resolved by DHCP allocating IP addresses in the foreign network and the mobile host configured to get the IP address automatically from a DHCP server. In the following, we will first describe the components of the mobile IP followed by the general example. Then we will discuss its limitations and usage in 3G wireless cellular networks.

6.4.1. Mobile IP Components

There are two types of components of a mobile IP system: network elements, and procedures and logical elements. Among the network elements are mobile host (MH), mobile home agent (HA), mobile foreign agent (FA), and correspondent host (CH). Among the procedures and logical elements are registration, agent discovery, tunneling, care-of address (CoA), home address, and mobility detection and deregistration. We will look at each of them briefly.

6.4.1.1. Mobile Host (MH). A mobile host (MH) is an Internet device, such as a notebook or a PDA, that could potentially be moving from one domain to another and should thus be equipped with mobile IP.

6.4.1.2. Home Address. If the MH has a permanent address in the home domain, it is called home address.

6.4.1.3. Correspondent Host (CH). Corresponding host is another Internet host device with which MH is communicating during an instance of a Mobile IP connection. There could be a higher-layer connection, or it may be a connectionless exchange between the MH and CH.

6.4.1.4. Mobile Home Agent (HA). A device (e.g., a router) with which the MH is registered along with an IP address in its home domain. As the MH moves around, it will keep HA informed about its new locations. Therefore, the HA always knows where a MH is and can reach it. A domain could have more than one HA.

6.4.1.5. Mobile Foreign Agent (FA). A device (e.g., a router) that registers MHs visiting its domain. It is quite possible that the MH uses the IP address of the FA for itself while it is visiting a foreign domain.

6.4.1.6. Mobility Agent (MA). Mobility agent (MA) is the term used to describe either the HA or the FA.

6.4.1.7. Mobility Detection. There should be a mechanism for the MH to figure out whether it is in foreign, new foreign, or the home domain. In other words, the mobility-detection function is part of Mobile IP. The mobility-detection mechanism could be implemented in many way, for example, by making all registrations with a timeout and then renewed, by making it necessary for the HA and the FA to keep announcing their presence, or for the MH to keep sending some kind of keep-alive messages. There are two algorithms specified for mobility detection, one to be used under all circumstances and the other to be used in some cases. The first of these algorithms uses a lifetime from the agent advertisement message and the second compares the network prefixes of the received agent advertisement. The first method can be used under all circumstances, but it has no way of detecting mobility before the end of lifetime. Therefore, if a MH moves out of the range of the current MA soon after updating its registration, all transmissions until the next update will be lost. The second algorithm is not useful if the prefix length of the new MA is not known to the MH. The method is recommended to be used only if the prefix lengths are known.

6.4.2. Agent Discovery

Agent discovery is a type of mobility-detection mechanism. This process consists of the MAs broadcasting messages (Agent Advertisement) of their presence using one of the ICMP message types called *Router Discovery*. If an MH does not receive any agent advertisement message, it could simply send an Agent Solicitation message to invoke an agent advertisement message [10]. Registration or de-registration follows once a change (i.e., mobility) is detected or the existing registration is timed out. It is important that the TTL field in the IP header for looking for an MA should be set, so that the packet does not leave the domain within which registration is sought. For example, a TTL = 1 should be set if the 'All Mobility Agents' is the destination. If registration is to be done with a known MA, then the TTL should be set to reach that agent and not beyond.

6.4.3. Registration

The visiting MH registers the new location with its HA. However, in order to register a foreign location, the HA needs a foreign agent. Therefore, the MH must find a FA in the new location and then register it with its HA. The registered FA also registers this MH as a visiting MH in its domain and allocates an address (care-of address, see below) to it to be used only as long as the MH is visiting this domain.

6.4.4. De-registration

The process of registration to own HA is called de-registration because it also means termination of registration in any other domain.

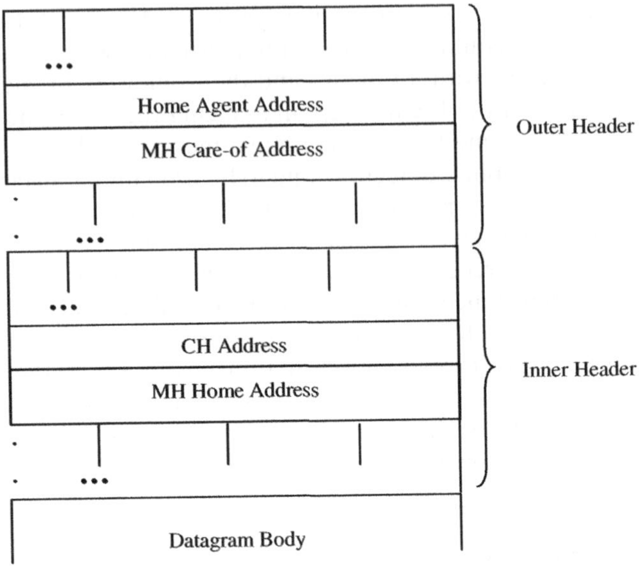

Figure 6-10. IP in IP encapsulation for tunneling.

6.4.5. Care-of Address (CoA)

Once in a foreign domain, the mobile host needs a new IP address to which incoming datagrams would be forwarded. This is called the CoA. The CoA could be either an address allocated to the MH only, called *co-located* CoA (e.g., using DHCP), or it could be the address of the FA. In the latter case, all datagrams addressed to the visiting MH must be received by the FA first. The FA needs a mechanism to know which MH a packet belongs to. This can be done by preserving the IP header with the home address of the MH inside the received datagram header. From the outer header, the packet is routed to the FA and from the inner header, it is forwarded to the MH, as shown in Figure 6-10.

6.4.6. Tunneling

Tunneling is used by HA to send IP datagrams to the MH when the MH is in a foreign domain. Tunneling implies use of double protocol headers, in this case, an IP over an IP. Here is the usage scenario of tunneling. When the MH moves from one domain to another, the corresponding host (CH) is to be kept unaffected by this mobility. The CH keeps sending packets to MH at the original home address. These packets are received by the HA. These packets have the CH address as the source IP address and the MH address as the destination IP address. The HA forwards these packets to MH to the care-of address while preserving the original header. So, the HA adds another IP header (*encapsulation*), showing its own address as the source address and the CoA

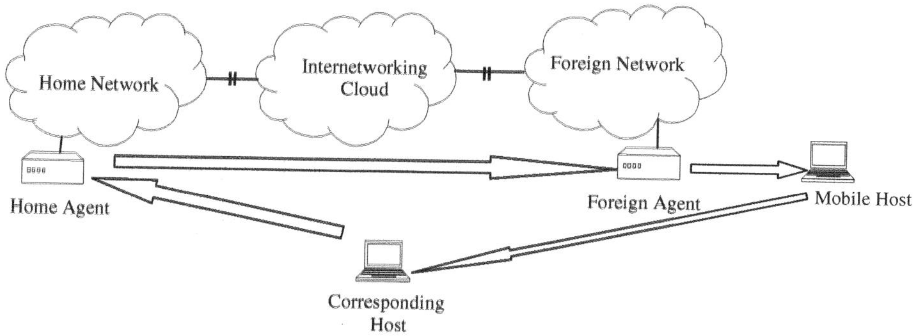

Figure 6-11. HA uses tunneling to forward packets from the CH to the MH in a foreign network.

address of the MH as the destination address. This can be visualized as making a tunnel of the outer IP headers between the HA and FA through which the inner IP packets for the MH are forwarded. Figure 6-11 shows tunneling from the home agent of a mobile host to the current foreign agent. A mobility agent may optionally provide shortened encapsulations instead of using the full IP header. Two such mechanisms are given in [11] and [12].

6.4.7. Mobile IP Usage Scenario

Consider the IP network part of a cellular system with three domains, as shown in Figure 6-12. The MH, registered in domain 1, and exchanging IP packets with the CH in domain 3, moves to domain 3.

First, Figure 6-12(a) shows the two hosts (MH and CH) communicating as if there is no mobile IP. In Figure 6-12(b), the MH moves to domain 3. It may happen that it still can get the outgoing packets to the CH, but it will not be able to receive packets from the CH because the CH is still sending to its home address. Therefore in Figure 6-12(b), the MH uses agent solicitation, followed by registration request (RRQ), for which it uses UDP port 434. The MH gets a CoA, co-located or the FA's IP address. With the co-located address, the MH can now communicate directly with the CH from domain 3. In general, there could be three problems of having this capability in Mobile IP. These are:

1. *Address space.* If all MHs are given their own CoAs, the address space in IPv4 will be further limited.
2. *Privacy.* If the MH wants to hide its mobility from the CH, it cannot be given the new CoA.
3. *Firewall* (not applicable to this particular case). If the MH moves to a domain that is firewalled by the access router of the CH domain, the packets from the new CoA address of the MH may not be able to reach the CH anyway.

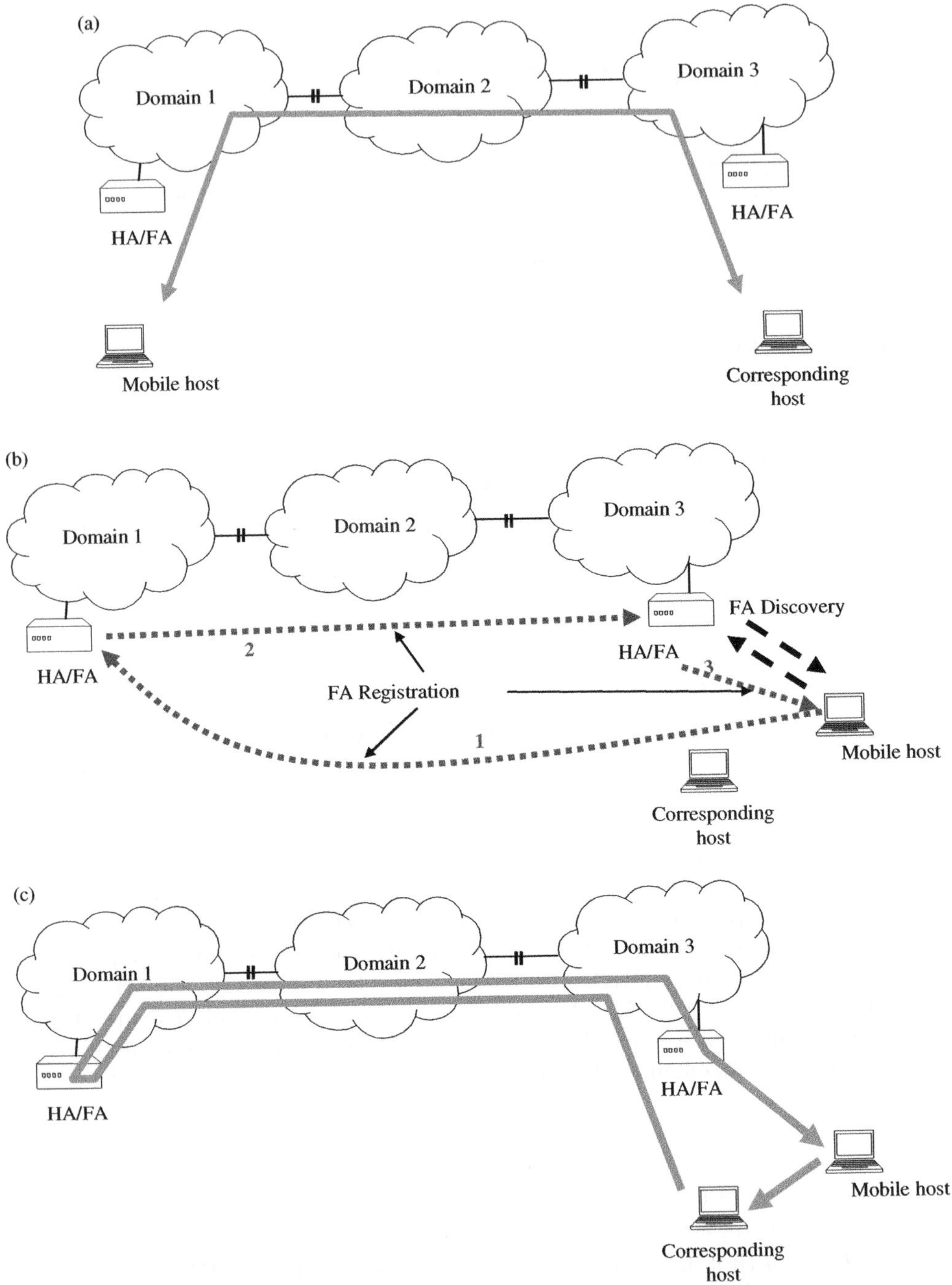

Figure 6-12. (a). MH and CH exchanging packets from their respective domains; (b). In Domain 3, MH discovers FA and registers it with its HA; (c). Use of tunnel can be quite inefficient in this case.

In addition to these, new higher-layer bindings may result in excessive delay or lost data. In our example, we assume that the CH domain can receive packets and the FA address is used for tunneling. Now, as shown in Figure 6-12(c), the MH is sending data to the CH directly through the new IP route, while the CH is sending data to the HA of the MH. These data are transmitted to the FA from the HA using tunneling and are delivered to the MH by the FA by uncovering the tunneling IP header.

6.4.8. Security Measures in Mobile IP

When used in wireless environment, mobile IP can be subject to all classical attacks and vulnerabilities plus the problems inherent to wireless channels and networks. A few security measures are proposed as part of the standard. Prominent among these are nonces and time-stamping of the registration messages so that an intruder may not replay a stolen registration message to authenticate it and act as the MH. Timestamps allow the recipient node (i.e., the MA) to disregard a message with a timestamp older than expected. The nonces (a unique random number to be used only once) can help figure out if the current message has already been used.

6.4.9. Limitations of Mobile IP

The biggest problem with Mobile IP is that it may not be implemented ubiquitously due to the heterogeneous nature of the Internet. However, once and wherever it may be available, it still has many limitations in terms of operation and performance. As already mentioned, mobile IP in a wireless cellular setup is only for macromobility and not fast-paced micromobility, to provide a continuous IP connectivity during handoff. Following is a list of problems and some proposed solutions. Refer to [13] and references therein for details.

1. Tunneling Delay. As seen in the example of Figure 6-12, the tunnel simply adds a round trip delay. In terms of QoS, this could be the difference between a good and bad connection. There are a number of solutions proposed for this, such as the use of a co-located IP address, so that the MH and the CH can communicate directly, assuming there are no other hindrances. This is called *route optimization* (Ref. [6] of [13]) and results in having the HA sending mobility binding updates to CH.

2. Discontinuities in Communications. It might also happen that a mobile host moves from one foreign host to another and the registration process halts packet exchange. If the CH sends any packets during this time, they are either lost or discarded due to TTL expiry. Besides, as pointed out in the original RFC, too many overhead packets for registration drain battery from the MH in a wireless setup. This problem can be resolved by extending the operation

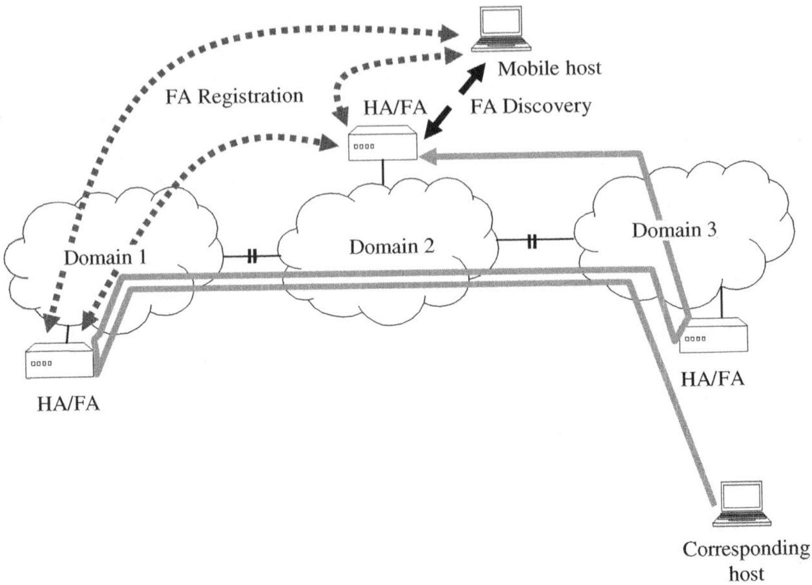

Figure 6-13. MH moved from one FA to another. The previous FA maintains the data tunnel and forwards packets to the MH though a new FA until registration of the new FA is completed at HA.

of the last FA until it gets a signal of registration completion from the HA or the next FA. Until then, from the knowledge of the current FA, the last FA simply forwards all received packets for the MH to the new FA. Once it receives a registration complete message from the new FA, it deletes the MH entry. Figure 6-13 shows this process.

Another solution proposed in reference [8] of [13] is the regional registration. In regional registration, a larger domain has a hierarchy of FAs. These FAs are all available with their IP addresses to be used for tunneling. Alternatively, all tunneling can happen via the GFA, the gateway FA. However, the MH registers only within the FA that is the highest (or lowest) in the hierarchy (e.g., with the gateway MA). Therefore, as long as the MH remains in a larger domain, it does not have to register from one FA to another. Other solutions, as proposed in Ref. [5] of [13], include using layer information to inform the mobile host of the possible movement before it occurs (network-assisted, mobile, and network-controlled or NAMONC handoff) and mobile knows about it. The second proposal suggests allowing the next FA to use layer 2 information to start *pre-registration* before it is informed by the mobile of its mobility (called *network initiated mobile terminated* or NIMOT handoff). These two mechanisms are viable because layer 2 detects mobility before layer 3. Registration and tunneling delay are best handled by IPv6, which will be briefly discussed later in this chapter.

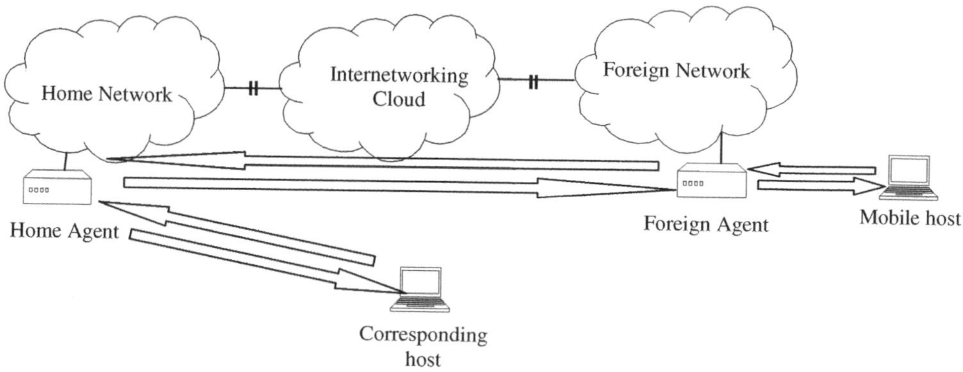

Figure 6-14. Reverse tunneling.

3. Firewall. Private networks are usually firewalled these days. If an MH goes to a foreign network that is not allowed by the domain of the CH, the packets will be blocked. A solution to this problem is reverse tunneling, in which all packets to and from the MH and CH pass through the HA. Figure 6-14 shows reverse tunneling [14]. As seen from the figure, reverse tunneling adds another tunneling delay. However, reverse tunneling is useful also if the MH does not want to lose the higher protocol bindings (e.g., TCP/UDP sockets).

4. Security. All new FAs must be authenticated in order to avoid vulnerabilities due to redirection attacks and host identity stealing. This poses two problems, namely, authentication delay and possibly a lowered level of security implemented in the foreign domain. This problem requires an administrative coordination between the two domains as well as a standard way of authentication that can be implemented Internet-wide.

The protocol requires the support of mobility security associations. A mobility security association is a security context between two nodes that carries information to be used in authentication, such as the encryption algorithm, any associated keys (secret, public, private) and a mechanism for replay detection and protection. It is specified that the mobility security association should be indexed by security parameter index (SPI) and node's IP address. The SPI identifies a security context used and must be exclusive of the range (0–255). This is a complex requirement set and an obvious impediment to universal implementation of mobile IP.

5. Intra-domain Mobility. As mentioned in RFC 3344, mobile IP is not suitable for micromobility. If a mobile host moves among link connections within the same domain, it still needs a handoff procedure to work with a different link layer. The new link layer could be either a WLAN or a 3G cellular access network connection with a new base station. Mobile IP is not intended for this kind of mobility, and it will be addressed separately.

6.4.10. Mobile IP Messages

The protocol needs only a few messages for registration and agent discovery processes. For registration, two new messages have been defined (Registration Request and Registration Reply). These messages use UDP with well-known port 434. For Agent Discovery, the ICMP messages of Router Discovery and Router Solicitation are used. Extensions are defined for Agent Advertisement and Registration messages. There are three extensions for Agent Advertisement, relating to a byte of (1) padding, (2) mobility agent advertisement and (3) prefix-length advertising. These are called *One-byte Padding*, *Mobility Agent Advertisement*, and *Prefix-Lengths*, respectively. The extensions for registration relate to padding (again, One-byte) and authentication. These are called *One-byte Padding*, *Mobile-Home Authentication*, *Mobile-Foreign Authentication*, and *Foreign-Home Authentication*.

6.4.11. Internet Standards for Cellular Networks

IETF RFC numbers 3141 [15], 3131 [16], 3113 [17], and 3002 [18], indicate the extent of activity within IETF related to the application of Mobile IP in cellular networks. Due to the need of close coordination between IETF and 3GPP2 group, RFC 3131 proposes to have a 3GPP2 contact person in the steering group (IESG) to act as 3GPP2 Liaison. Also, an observer from IETF may be appointed for 3GPP2 on a per case basis. A similar recommendation is made for a contact person with 3GPP in RFC 3113. It is also indicated in RFC 3113 that the task group (TSG-SA) in the 3GPP project team has been made for interaction with IETF. The admission that Mobile IP is not for micromobility has helped its cause with the cellular networks. Otherwise, as pointed out in the 2000 IAB Internetworking Workshop [18], the cellular operators were much concerned about the slow handoff of mobile IP. Their other major fear was that the communications between HA and FA might become a bottleneck for the cellular network. This was largely handled by route-optimization procedures, mentioned above.

Another rather major issue concerning collaboration of cellular and IP technologies is in the way IETF and cellular operators 'own' their services. IETF believes in open architectures, which lead to flexibility, private and open innovations, as well as a plethora of proposals and problems. Cellular operators emphasize service and give their users what has been dubbed as 'walled garden'. This does not provide the user any more flexibility than usual to use the service, but takes care of many fears such as security and available throughput. Nevertheless, 3GPP has set an ad hoc group, the 'Mobile IP ad hoc group' within TSG-SA. The 3GPP specifications contain the IP multimedia subsystem (IMS), which is projected to provide IP multimedia services in UMTS. Likewise, the cdma2000-based 3GPP2 project employs mobile IP in phase 2 of the phased all-IP approach under IMS-OMA (open mobile alliance) interoperable VAS (value-added service) (see Figure 7-3, p. 21 of Ref. [19], e.g.). The all-IP is to be culminated in Phase 3. The reader is referred to the chapters 7 and 8 on data in cellular networks for more details.

6.5. MOBILITY MANAGEMENT IN AN ACCESS NETWORK

A cellular access network consists of an air interface between the mobile host (we assume an IP host) and a base station. The base station function is divided into transceiver and control (e.g., BTS and BSC). A single switch (MSC) controls a number of BSCs (base station controllers), and each BSC, in turn, controls a number of base station transceivers (BTSs). Within an IP cellular access network a BSC can still control a number of BTSs, each BTS communicating with a large number of mobile stations (MS)s.[3] Let's look at different phases of operation of an imaginary IP access network. Mobility can be handled by mechanisms similar to ones proposed for wireless ATM (e.g., Ref. [20]). The next subsections address a general phased approach.

6.5.1. Address Allocation

A commonsense picture for controlling mobility consists of a BSC or equivalent control unit allocating IP addresses to MSs through their respective BTSs, by using some kind of DHCP or its enhancement. Once an IP address is allocated to an MS, all BTSs under this BCS are informed about the IP allocation. All BTSs mark this IP address as a *passive IP address*, except the one with whom the MS is actually communicating. This BTS, let's call it the home BTS, marks this IP address as *active IP address* and expects to have IP datagrams to and from its address. Figure 6-15 shows the address allocation procedure. If a MS has a permanent IP address, it registers itself with the BSC on turning ON through a BTS. The end result of registration is the same, that is, it is registered as active with one of the BTS and passive with all others.

6.5.2. Data Communications

The MS that just got an IP address registered with the BSC is free to send and receive IP packets through the connecting BTS. The BSC keeps a record of which BTS the MS is connected with, and simply forwards all packets to it. If the MS does not send or receive packets for a certain time, the address can be de-allocated, depending on the policy of the network administration.

[3] Mobile station is more akin to cellular terminology (also mobile node MN, and mobile terminal MT. Mobile host MH is closer to the Internet terminology. Mostly, we have not differentiated among these terms.

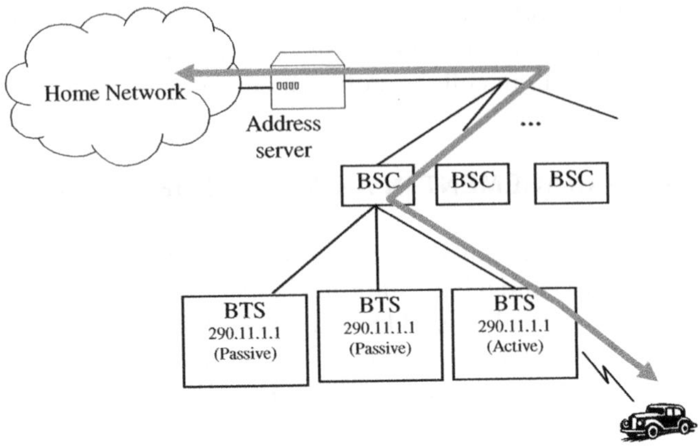

Figure 6-15. Active and passive address allocation.

6.5.3. Mobility

If at some point in time the MS moves closer to another BTS and transmits a packet to it, the BTS, on checking that this MS had a passive IP address, requests a network-controlled handoff procedure. The handoff procedure is quick in this case and consists of the BSC notifying the previous BTS to make this address passive (or use some timeout mechanism). The new BTS automatically makes this address active on receiving the first datagram.

It is quite possible that one or more incoming IP packets were delivered to the previous BTS during this time. This problem can be resolved in many ways, including the simple soft handoff mechanism, in which the MS communicates with both BTSs for some time. Figure 6-15 shows this mechanism of mobility management. If the MS crosses over to another BSC, mobile IP or some such protocol is used for handoff to another domain.

In the above hypothetical mobility-management mechanism, many improvements can be added to make the process and routing more efficient. For example, taking the control another level above (MSC), a much larger area can be covered. Detecting the direction of mobility and informing the future BTSs in advance could result in more efficient handoffs. However, all these additions result in tradeoffs that suit certain situations and don't suit certain others. With this general view of a typical mobility-control procedure, we will describe cellular IP, which was based on principles similar to the above model. For other proposals to handle micromobility, see references, especially HAWAII [21].

6.6. CELLULAR IP

Cellular IP is one of the proposed protocols that provide micromobility management functions. We will, however, drop the prefix *micro-* in further discussion. As we saw in the discussion on mobile IP, there are various phases and network components required for managing mobility, starting from Agent Solicitation to Tunnel management. All these functions cause enough delay and consume enough network resources in control signaling that the use of mobile IP will be prohibitive in an environment in which a large number of mobile hosts are changing domains and would like to keep a continuous connection. Apparently, there is a need to have a mobility-management protocol for this type of environment. Cellular IP, proposed by a group at Columbia University, follows the *passive connectivity* paradigm of cellular voice technology [22][23], where idle mobiles register a lot less frequently than active ones, and their locations are known to be only in a general area. A paging mechanism is used to find the exact location of a mobile host if they receive a call.

6.6.1. Components of a Cellular IP System

The cellular IP (cIP) system [24] consists of a number of components to allow access, paging, and mobility management. The concept behind cIP is the same as mobility management of voice terminals in GSM MAP or IS-41-based core networks employing GSM and one of the many IS-41-related air interfaces, that is, to allow the idle mobile stations (called a passive MH) to have discontinuous (less frequent) transmission. As long as they can be traced to be in a larger *paging area* (see below), they do not have to register every move during passivity.

6.6.1.1. Active and Passive Mobile Hosts. A passive MH is one that does not have a current communication session (to send and/or receive IP datagrams). Once it starts sending or receiving IP packets it becomes an active MH. A timer is used to change the status from active to passive if no transmission occurs during a fixed timeout period.

6.6.1.2. Base Station. A cellular IP (cIP) base station has the capabilities of access point and IP router. It performs all mobility-related tasks for the mobile host (MH).

6.6.1.3. Gateway Router. All mobile hosts in an access network register with the gateway router. The gateway router is connected to several base stations on one side and to the Internet on the other side. The visiting mobile hosts use the gateway IP address as their care-of address. All packets to the MH first reach the gateway, from which they are routed to the MHs through the base stations to their respective IP (home) addresses.

6.6.1.4. Base Station Routing Cache. The base station routing cache is a routing table stored in a base station for each active MH. The entry includes the interface as well as the neighboring node from which the last IP datagram communication occurred. It helps base stations implement a hop-by-hop (neighbor-to-neighbor) routing. The route cache is updated by data packets. The base station routing cache remains valid for a *route timeout* that starts after no communication to or from a MH.

6.6.1.5. Route Update Packet. A route update packet is sent by an MH that wants to keep its route cache alive even if it does not have a data IP packet to send for the time being. It is a short IP packet with no data in it. These packets are sent at regular intervals called *route update time*.

6.6.1.6. Uplink/Downlink Packet. An uplink/downlink packet is a packet to/from the gateway from/to the MH.

6.6.1.7. Semisoft Handoff. A cIP-specific handoff mechanism in which the moving host first establishes a routing cache entry with the new base station, a seismic handoff then initiates the hard handoff process. For some time, the gateway or the lower-level crossover switch (which is a parent base station of both the new and old base station) sends incoming packets to both the new and old base stations.

6.6.1.8. Paging Area. Paging area is the geographical region in which a passive MH does not have to give information about its exact location. If a packet arrives for a passive MH in this area, it is paged, or a paging cache is used to search for it.

6.6.1.9. Paging Update Packet. A paging update packet is a type of control packet similar to the route update packet sent regularly by a mobile to update the optional paging cache at the base station.

6.6.1.10. Paging Cache. A paging cache is like a route cache set by a data packet or a paging update packet. A passive MH could sent paging alerts at regular intervals in the form of paging update packets, to allow the base station the option of making a paging cache. A paging cache has a timeout period longer than the routing cache.

6.6.2. cIP Usage Scenario

Figure 6-16 shows a usage scenario for the cIP. Mobile host *a* (MH *a*) sets up a connection with gateway router *k* (gr *k*), through the interface *x* of base station number 2 (BS2). The routing cache for MH *a* has the entry for the interface and the next hop base station (BS1, which is also a crossover switch between BS2 and BS3).

Figure 6-16(a) shows this communications before the need for mobility-management function. Figure 6-16(b) shows that MH *a* has moved closer to

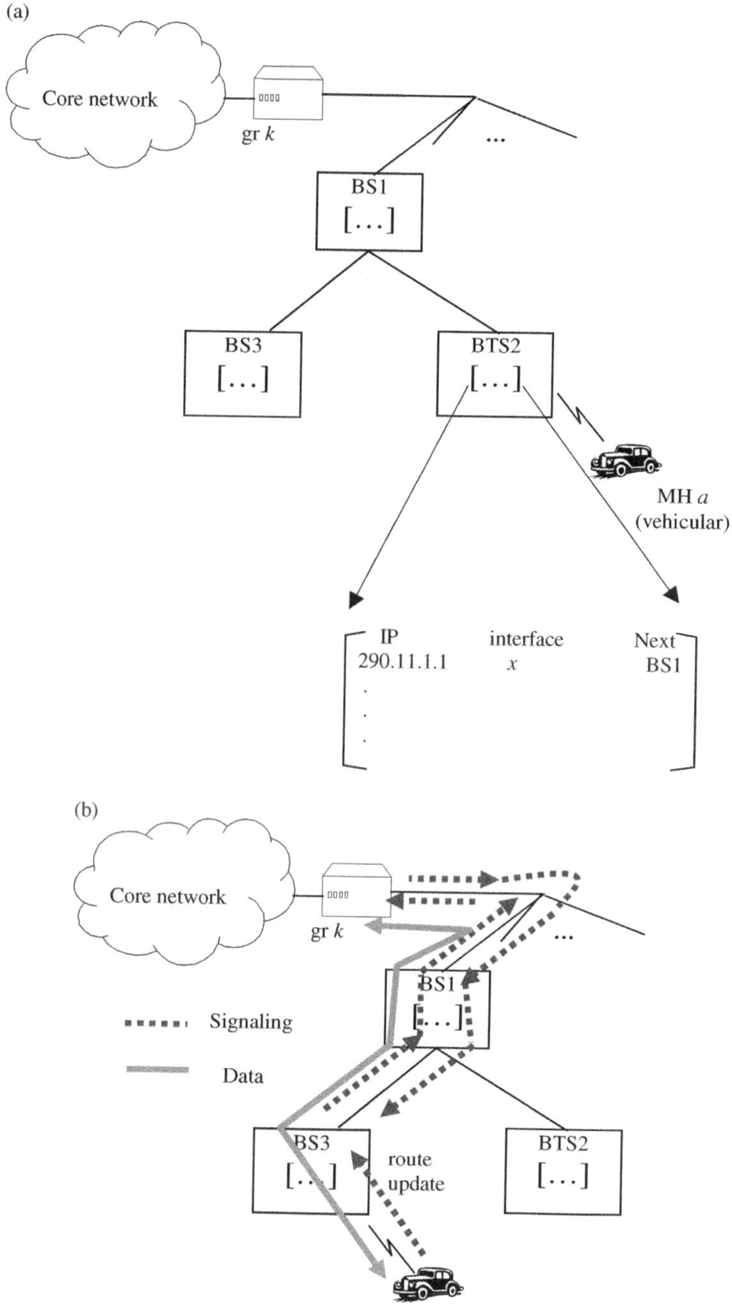

Figure 6-16. (a). cIP, hop-by-hop routing; (b). hard handoff.

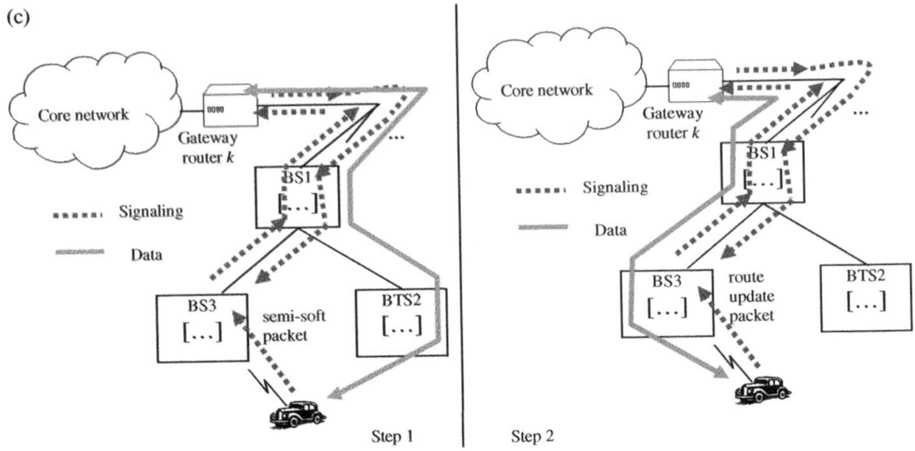

Figure 6-16. (c). semi-soft handoff.

the BS3. There are two choices of handoff, hard handoff and semi-soft handoff, as described below.

6.6.2.1. Hard Handoff. The hard handoff is initiated by the MH *a* by sending a route update packet (Figure 6-16(b)) to BS3. This packet goes to the gr *k* through BS3, thus creating a route for the downlink from gr *k* to MH *a*. By noticing the neighboring hops addresses, the BS3 can use the same route for uplink as well. The next packet to MH *a* is routed through the new route, and possibly through the old route too. A timer keeps the route soft-state active until the next route update or timeout. In this case, a timeout of the soft-state timer will signal the end of route at BS2. A second choice of handoff is the semi-soft hand off described in the following.

6.6.2.2. Semisoft Handoff. In this mechanism, the co-existence of two routes for a short time is exploited by the MH *a* to reduce handoff redundancy. Before the route update packet (read handoff), the moving MH *a* sends a semi-soft packet to the new station. In the semi-soft packet, a request to delay the further incoming packets at the gr *k* or MS1 may be included. The route update will now trigger a handoff, and transmit any delayed packets. Figure 6-16(c) shows this procedure.

6.6.3. cIP and Mobile IP

Depending on where to draw the line between cIP and Mobile IP (MIP), many architectural possibilities exist for an all-IP cellular network. In the following we show two such architectures discussed in [25]. Other references to be checked are [26], [27], and [28] and the references in these papers in addition to links and papers at http://www.ctr.columbia.edu/~andras/cellularip/

In the first scenario, the gateway router of the cIP acts as mobility agent (MA) for inter-domain handoff and mobility management. Thus the MH is loaded with both cIP and MIP for registration at foreign access network and mobility agent (MA). When the MH is in a foreign network, tunneling is used to route packets by the home agent/gateway router to the foreign agent. The foreign agent uses the cIP to deliver these packets to the MH. In the second scenario, both MIP and cIP are improved by employing route optimization in the former and semi-soft handoff in the latter. The HA or the current FA keeps sending the incoming packets to the current as well as the new FA access network.

6.7. IPv6 AND MOBILITY MANAGEMENT

One of the major features of the new generations of cellular networks is a service-oriented design. Thus technologies would be designed to maximize service needs, instead of basing service on the available technology. It is expected that future applications will require more and more peer-to-peer communications, contrasted with the client-server design of the current Internet (read TCP/IP) applications. Due to this, leading companies have emphasized the importance of IPv6 in future cellular environment owing to its stateless feature and expanded address space (see [29] and [30], e.g.). In this section, we will highlight the mobility-related features of the IPv6 that make it a desirable choice for cellular data networks.

6.7.1. Expanded Address Space

One of the limitations of the IPv4 is that the address space is getting thinner. Such is not the case with IPv6, which specifies 128-bit address space. Therefore, care-of addresses can be easily allocated in foreign domains.

6.7.2. Efficient HA Registration

The ICMP for IPv6 has more functions in the Router Advertisement, Neighbor Discovery, Address Autoconfiguration (not in v4), and Router Solicitation messages. Additionally, the Router Advertisement repetition interval can be configured.

6.7.3. Autoconfiguration of IP Addresses

In IPv6, address spaces are allocated for each category of usage. A router in the visited network can help MH calculate care-of address by advertising its network preface and the MH combining it with an interface identifier. This auto-configuration simplifies the address search and allocation in a foreign domain. In fact, there is no need of a foreign agent, as we will see shortly. This mechanism of care-of address is so-called *stateless mechanism*. A *stateful mechanism* would require some server (e.g., a DHCP6 server) to keep a state of address allocation. Both mechanisms are available in IPv6.

6.7.4. Mobility Detection

The same message, from which an MH extracts the router's network prefix, is used for mobility detection. For mobility detection, a mobile host could do one or both of the following: First, it will receive Router Advertisement from all the neighborhood routers and check their network prefixes against that of its *mobility* agent (which will be the same as its own, if no FA is used). If it can find a matching prefix, it concludes that it has not crossed the domain boundary; otherwise, it will conclude that it has moved. Second, in the absence of a Router Advertisement message, it broadcasts a Router Solicitation message. Note that route solicitation is a standard approach and is not added externally. Thus the mobility-detection feature is automatically integrated with IPv6.

6.7.5. Optimized Routing

While IPv4 requires optimized routing to avoid IP tunneling, the IPv6 routing is optimized by default. This eliminates the need for FA to receive IP datagrams for the visiting MH at its address. In fact, there is no need of a foreign agent in IPv6. If the visiting MH needs to know its HA, it can use IPv6-specific *anycast* message. In IPv4, a directed multicast which results in more latency, would be needed for this purpose. Route optimization requires the use of updating IP binding with the higher layer protocols. Three messages, called Destination Options, for exchanging binding information are included as part of the standard. These are (1) Binding Update (BU), (2) Binding Ack (BA), and (3) Binding Request (BR). BU and BA use IPsec authentication. An exchange of BU and BA with the home agent registers the MH back to home if it had moved back to its home domain.

6.7.5.1. Higher Layer Bindings. When a mobile host has set up direct two-way routes with the correspondent, it can use the CoA as the source address in IPv6 header. If this were done in a non-optimized IPv4 header, it would break all bindings with the higher-layer protocols (e.g., sockets with TCP/UDP). However, the IPv6 has an option for Home Address that can be used by the roaming host to convey the home address for higher-layer binding. The correspondent node, using Home Address option, replaces the source address by the home address, thus making mobility transparent to the higher layers.

6.7.6. Security

IPsec is an integral part of IPv6. This automatically takes care of the need of authentication, authorization, and encryption. IPsec can be used for IPv4, but it is not an integral part of IPv4. For IPsec, IPv4 uses a registration procedure based on UDP, adding complexity and latency to the protocol.

6.7.7. Micromobility

IPv6's feature of bypassing FA works for micromobility and multi-access cellular networks. Multi-access networks are the ones that allow many radio access networks of choice for the MH. It is common to have MHs with two or more access interfaces, for example, WLAN, Bluetooth, cell phone, and so on. For such access networks, IPv6 could provide a quick handoff procedure due to the autoconfiguration capability and Home Address Option. A mobile device could keep switching among different access networks with the upper layers keeping the same IP to Transport layer bindings.

6.7.8. Network Support for Application-Level Mobile IPv6

As shown in Nokia's white paper [30], using IPv6 at the application level may be the best way to integrate all types of access networks. It requires the implementation of a home agent, IPv6 support for the MH, user plane support for IPv6, and the implementation of IPsec, as IPv6 uses only IPsec for security. Figure 6-17 shows Nokia's protocol plane vision for employing IPv6 with GPRS backbone and WCDMA access.

6.7.9. Internet and Cellular Networking

The cooperation between IETF and cellular phone operators has been rather late in coming to wireless data technology. The two work differently in developing standards, the former by contributions from anyone to develop open standards, and the latter by keeping in view market pressures, customer expectations, and competition with other providers. The alliance of IETF and cellular operators has, however, shown promise already by an increasing number of customers using their cellular connections for short-burst-based com-

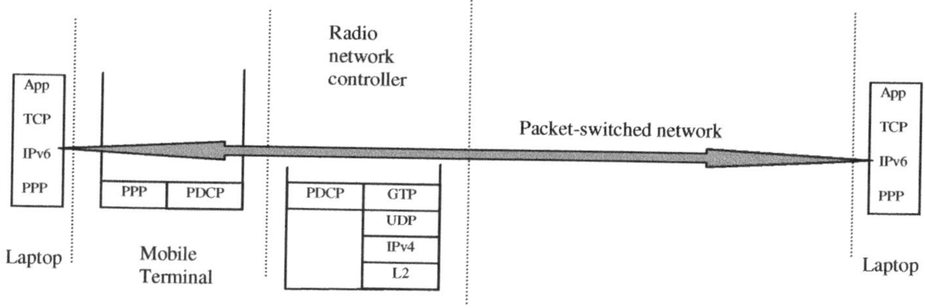

Figure 6-17. Nokia's vision of using IPv6 as end-to-end service for 3G.

munications. Streaming, as well as interactive applications (e.g., push-to-talk, games), are expected to garner a respectable share of the future market. In order to have a cellular network with all-IP connectivity, a number of protocols from the IETF standards will be deployed. One arguably negative result of this is an explosion in the various forums and standards associations and cooperative projects recently materialized for effecting this deployment.

We have not discussed several important protocols, for example, MPLS, which could be potentially a part of future cellular networks. There are two reasons for this, namely (1) the scope of the book and (2) protocols' relation to the cellular or wireless nature of the network. We have restricted ourselves to only those protocols that make a major contribution to the cellular networks in this sense. The future scenario may change and an altogether new picture might surface before these technologies are realized in practice. New books will cover those aspects as they arise!

REFERENCES

[1] Andrej Mihailovic, 'IP mobility', *Center for Telecommunications Research*, King's College London. Available as IP_Mob_Helsinki_Part1.pdf.

[2] Charles Perkins, 'IP Mobility Support for IPv4', *IETF RFC 3344*, Aug. 2002.

[3] J. Rosenberg, H. Schulzrinne, G. Camarillo, A. Johnston, J. Peterson, R. Sparks, M. Handley, and E. Schooler, 'SIP: Session Initiation Protocol', *IETF RFC 3261*, Jun. 2002.

[4] A. B. Roach, 3265 'Session Initiation Protocol (SIP)-Specific Event Notification', *IETF RFC 3265*, Jun. 2002.

[5] Dorgham Sisalem and Kuthan, Jiri, 'Understanding SIP', available from http://www.fokus.gmd.de/mobis/siptutorial/

[6] Ashutosh Dutta, Ling, Yibei, Chen, Wai, Chennikara, Jesmine, and Schulzrinne, Henning, 'Multimedia SIP sessions in a mobile heterogeneous access environment', *3G Wireless*, 2002.

[7] Ashutosh Dutta, Altintas, Onur, Chen, Wai, and Schulzrinne, Henning, Mobility Approaches for All IP Wireless Networks, SCI 2002, Orlando, FL.

[8] Adam Roach, 'SIP for 3G', *Ericsson*, slides available from http://www.cs.columbia.edu/sip/talks/SIP3GPP.ppt

[9] Dean Willis, 'SIP and 3G Wireless' *DynamicSoft*, Slides available from http://www.dynamicsoft.com/news/ presentations/SIPn3GWireless.pdf

[10] Paresh Jain and Kelkar, Rakesh, 'Mobile IP: Enabling mobility for the 3G wireless Internet', white paper available from http://www.tcs.com/0_whitepapers/htdocs/Mobile_IP.pdf

[11] S. Hanks, Li, T., Farinacci, D., and Traina, P., 'Generic Routing Encapsulation (GRE)', RFC 1701, October 1994.

[12] Charles Perkins, 'Minimal Encapsulation within IP', RFC 2004, October 1996.

[13] Ashutosh Dutta, Chen, Wai, Altintas, Onur, and Schulzrinne, Henning, 'Mobility approaches for all-IP wireless networks', International Conference on Information

Systems, Analysis and Synthesis 2002 (SCI2004). Florida, USA, July 2002. Also available from http://www1.cs.columbia.edu/~dutta/research/sci2002-mobility.pdf

[14] G. Montenegra (Ed.), 'Reverse Tunneling in Mobile IP' (revised from RFC 2344), *IETF RFC 3024*, Jan. 2001.

[15] T. Hiller, Walsh, P., Chen, X., Munson, M., Dommety, G., Sivalingham, S., Lim, B., McCann, P., Shiino, H., Hirschman, B., Manning, S., Hsu, R., Koo, H., Lipford, M., Calhoun, P., Lo, C., Jaques, E., Campbell, E., Xu, Y., Baba, S., Ayaki, T., Seki, T., and Hameed, A., 'CDMA2000 Wireless Data Requirements for AAA', *IETF RFC* 3141, Jun. 2001.

[16] S. Bradner, Calhoun, P., Cuschieri, H., Dennett, S., Flynn, G., Lipford, M., and McPheters. M., '3GPP2-IETF Standardization Collaboration', *IETF RFC 3131*, Jun. 2001.

[17] K. Rosenbrock, Sanmugam, R., Bradner, S., and Klensin, J., '3GPP-IETF Standardization Collaboration', *IETF RFC 3113*, June 2001.

[18] D. Mitzel, 'Overview of 2000 IAB Wireless Internetworking Workshop', *IETF* RFC 3002, November 2000.

[19] 3GPP2, 'Evolution document', *3GPP2 S.R0038-0*, Version 2.0, February 2004.

[20] Anthony S. Acampora and Naghshineh, Mahmoud, 'An architecture and methodology for mobile-executed handoff in cellular ATM networks', *IEEE Journal on Selected Areas in Communications*, Vol. 12, No. 8, Oct. 1994, pp. 1365–1375.

[21] R. Ramji, La. Porta, T., Thuel, S., and Varadhan, K., 'IP micromobility support using HAWAII', *Internet Draft*, draft-ramjee-micro-mobility-hawaii-00.txt, February 1999 (then work in progress).

[22] Andrew T. Campbell, Gomez, Javier, and Valco, Andras, 'An overview of the cellular IP', *IEEE Wireless Networking and Communications Conference*, Sept. 1999.

[23] A. Katouzian, 'Cellular IP report', *3G Mobile Technologies Conference Publication No. 471, IEE*, pp. 129–132.

[24] Andrew T. Campbell, Gomez, Javier, Kim, Sanghyo, Valkos, Andras G., Chieh Yiwan and Turanvi, Zoltan R., 'Design, implementation and evaluation of cellular IP', *IEEE Personal Communications*, August 2000, pp. 42–49.

[25] Marco Carli, Neri, Alessandro, and Rem Picci, Andea, 'Mobile IP and Cellular IP integration for inter access networks handoffs.

[26] Keara Barret, Carrol, Ray, and va der Meer, Sven, 'Investigating the applicability of mobile IP and cellular IP for roaming in smart environments', *Irish Telecommunications Systems Research Symposium (ITSRS)* 2003. Dublin, Ireland. Available from http://www.tssg.org/papers/20030506_ITSRS_2003/Barret_Carroll_Investigating_Applicability_of_Mobile_IP_and_Celluar_IP.pdf

[27] Xinming He, Jiang, Shun, and Zheng, Xiaojing, 'Secure mobile IP and cellular IP', available from http://netweb.usc.edu/xinming/papers/mobile_ip/859_mobile_ip.pdf

[28] I. Guardini, Fasano, P., and D'Urso, P., The role of the Internet technology in future mobile data systems (WTC/ISS'2000, Birmingham, 7–12 May 2000. Paper available from http://carmen.cselt.it/papers/

[29] Karim El Malki, 'IPv6 and 3G mobile networks', *Core network development Ericsson*, Italian IPv6 Task Force, Milan. Available from http://www.ec.ipv6tf.org/PublicDocuments/IPv6and3G.pdf

[30] Nokia, 'Introducing IPv6 in 2G and 3G mobile networks', white paper. Available from http://grouper.ieee.org/groups/scc32/dsrc/ip/ip_images/3g_wp_allip_mipv6.pdf

[31] Campbell A. T., Gomez, J., and Valko, A. G., 'An Overview of Cellular IP', *Proc. 1st IEEE Wireless Communications and Networking Conference (WCNC'99)*, New Orleans, 21–24 September 1999.

CHAPTER 7

DATA COMMUNICATIONS IN CELLULAR NETWORKS: CDMA2000

Packet data in cellular networks is an emerging technology and is evolving continuously. Until recently, wide area mobility management services for packet-based transmission have been provided by specialized networks designed mainly for business use. These enterprise wireless data networks did not, in general, have open architectures, had low data speeds, cost more, and had high latency not suitable for multimedia. However, for packet-switched data, these networks provided the best solution until the cellular networks evolved out of the second generation. The second generation networks mainly consist of GSM (core and air interface), IS-41 Revisions B and C with IS-54/IS-136 (TDMA/PCS) and IS-95/IS-95A,B (CDMA) as air interfaces, and PDC. While cellular voice networks were evolving, another change of interpretation has occurred in the way network services are recognized. By data, we don't necessarily mean delay-insensitive services, such as e-mail and file transfer. The Internet is not immune to the defects of high-latency protocols due to an increasing integration of interactive and streaming data exchanged over the web. Consequently, the whole scenario of available technologies, and the way we look at them has gone through a transformation. Now, all information is regarded as data, and data networks include voice networks and vice versa. However, when we refer to voice network versus data network, we usually refer to circuit switching versus packet switching. Here another major change occurred in networking. IP has emerged as a universally unifying networking technology, practically removing all competition. Today, even if we want a bandwidth guarantee, we want IP to provide it, instead of designing another

Wireless and Mobile Data Networks, by Aftab Ahmad

technology or protocol. ATM, though not completely out of the picture, has resorted to its original role as a link technology. Data networks based on technologies other than IP must also have IP, or perhaps go out of business. VoIP has finally become a commercially viable technology. In spite of initial hiccups, DiffServ has started influencing standards at the lower layers for end-to-end QoS transportability. Web services are emerging as a collective killer application. Arguably, web services are going to be the 'IP' of the application layer protocols. Perhaps the biggest obstacle in this integration is data security. Standard generalized markup languages, such as XML, have the ability to prepare data structures with multiple parts, each part associated with its own attributes of accessibility and QoS.

The concept and implementation of QoS is getting support from high-speed standards in all networking areas, from fixed wireline, mobile wireline, fixed wireless and mobile wireless standards. The access networks (usually QoS bottlenecks) are faster than ever, the evolution of IP switching into generalized MPLS has made bandwidth guarantees possible on IP links, wavelength division multiplexing (WDM) is promising ever higher bandwidth backbone connections, and the backbone traffic is far from the point where congestion could pose serious situations. Consequently, we don't really need complex definitions of QoS. In fact, the 3G cellular systems provide only four categories of user QoS; all application types are expected to get service under one of these four. In this chapter and the next, we look at the selective cellular systems technologies for packet-switched data. We will start the chapter with a short account on 'enterprise' wireless data networks, and then get right to the heart of one of the two prevailing 3G systems, cdma2000. We will divide the 3G discussion into air interface and core networks. The discussion on their evolution may partially cover part of the earlier generations. It is not our intent to discuss 'before 3G' (be3g) technologies. Also, we focus on packet switching only.

In the discussion on cdma2000, we will look at the PHY and MAC for the revision D of cdma2000 as specified in 3GPP2, followed by the all-IP network architecture model. Before cdma2000, we will briefly mention the five air interfaces originally considered by the ITU, forming the basis of the 3G cellular systems worldwide. Of these five, two will be given remaining coverage of the topic under the banners of cdma2000 (this chapter) and WCDMA (next chapter).

7.1. BUSINESS WIRELESS DATA NETWORKS

Due to the incapability of the cellular systems to provide packet-switched data, enterprise wireless data networks have been serving the business community. These networks are not high speed, but they provide a TCP/IP interface. The overall coverage is quite comprehensive and, in the United States, almost the whole country gets signal from one or the other network. In the following, we will briefly describe the characteristics of some of these networks.

7.1.1. Cellular Digital Packet Data (CDPD) Network

The CDPD is an open standard that began service as strictly an overlay network over the American Mobile Phone System (AMPS) in the 800 MHz band. It provides an IP connection at 19.2 kbps raw bit rate in full-duplex mode. At the physical layer, CDPD employs Gaussian Minimum Shift Keying (GMSK) with Reed-Solomon code. For security, RSA's algorithm RC4 is used for data encryption. The CDPD network covers about 80 metro areas and many companies are service providers. CDPD has been negatively affected by packet-switched services in cellular networks. Cingular and AT&T are predicted to shut down their CDPD networks and migrate to the 2.5G network to be phased out in 2005 [1].

7.1.2. ARDIS

Motorola and IBM started the ARDIS network and the company to initially provide support for IBM service engineers in the early 1980s. It provided data rates of 4.8, 9.6, and 19.2 kbps with packet interfaces to the X.25 network. Later, TCP/IP was added as well. The network has over 15000 base stations covering much of the United States [2]. ARDIS is a proprietary (semi-open), half-duplex service. It uses 4'ary FSK modulation and trellis coding for error control. The error-control procedure is assisted further by link-by-link CRC. The link protocol used is radio data link access procedures LAP (RD-LAP). At the MAC layer, it uses slotted digital sense multiple access. The authentication of uniquely assigned IDs provides some security.

7.1.3. RAM Data Networks

RAM data network uses Mobitex architecture, originally developed by Ericsson. Like CDPD, it uses GMSK at the physical layer. The bit rate provided is 8 kbps. At the MAC layer, the RAM network uses inhibit sense multiple access (ISMA). ISMA is a type of CSMA with one bit used for busy/idle channel condition. RAM network uses a pretty complete cellular concept with frequency reuse and cell splitting. It employs the hamming code for error correction.

There are some other wireless data networks that provide half- and full-duplex services. RadioMail, Metricom Richochet (data systems), SkyTel (SkyTel Corp.), EMBARC (Motorola), and MobileComm (Bell South) (message systems) are some examples. SMR (Nextel) is another overlay networks (besides CDPD), in the 800 MHz range, based on Motorola MIRS technology. It uses TDMA with six conversations per channel. SMR has deployed many base stations in each area. Comparison tables are available for feature of these systems at http://www.cs.berkeley.edu/~randy/Courses/CS294.S96/xMobileData.pdf

Also check [3] for some more details.

7.2. CELLULAR DATA NETWORKS

As mentioned in Chapter 2 on network architectures, cellular systems have a wireless air interface and a fixed, wireline, or wireless backbone. The network architecture for the backbone network, called the *core network*, has heavily borrowed from the PSTN signaling technology. As a matter of fact, most all systems until the second generations have SS7 as core network with an added part to maintain registration databases and handle mobility. The air interfaces usually have the equivalent of the bottom three layers of the OSI-RM. These provide signal conditioning and RF interface at the PHY, logical channels at the link layer, and message communications at the equivalent of network layer. SS7 is for a circuit-switched system, such as PSTN.

From the network design point of view, the main difference between circuit-switched and packet-switched networks is the signaling system. In circuit-switched networks, signaling systems provide mechanisms for circuit establishment, supervision, and termination. With advanced systems like SS7, many other functions provided by the signaling system make the network a type of *intelligent* network. Arguably, the main cost of a circuit-switched network will be in signaling system. Due to these reasons, circuit-switched networks have a complex protocol plane for the control part of the network. The packet-switched networks do not necessarily need a complex control plane. However, other weaknesses, such as recovery mechanisms for lost and overflowed packets and need for authentication and authorization, may end up having a packet-switched network with a cost comparable to the circuit-switched network. However, no signaling system can equate the simplicity and ubiquity of the IP protocol.

7.2.1. Cooperation Explosion

The PSTN, cellular networks, the Internet, paging systems, wireless local area networks, wireless personal area networks, and wireless fixed and mobile metropolitan area networks have generally been advancing in their own directions. That is, until recently. The unifying IP has brought the vision of creating a cutoff plane in the protocol stacks of all these networks. Figure 7-1 shows this integration effect of IP.

One of the effects of this 'IP unity' is that with a cellular network providing IP services, the access network becomes universal, and it has to accommodate all other networks as access networks. A second major effect is an explosion in the cooperative groups, agencies, organizations, and forums. Figure 7-2 shows some such groups influencing cellular networks of the future.

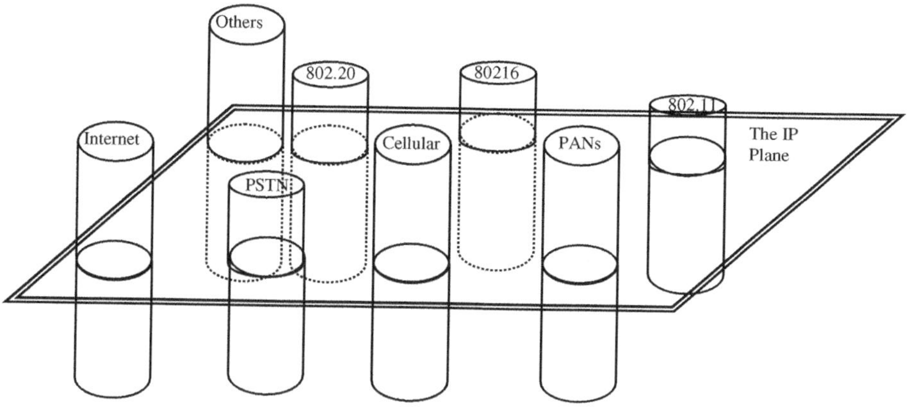

Figure 7-1. The "IP unity" of communications networks. All networks have or are destined to an IP interface.

7.2.2. 3G Air Interfaces

As part of IMT-2000, five air-interface standards were specified. These were:

1. *Wideband code division multiple access* (W-CDMA). Also called Direct Spread. It was defined under the umbrella name IMT-DS. A more specific name is UMTS terrestrial radio access—frequency division duplexing (UTRA-FDD). Duplexing relates to the division of system bandwidth in the uplink and downlink. An uplink channel, or reverse channel, is defined in the direction from mobile terminals station (a cell phone) to the base station. Downlink channels are also called *forward channels*.[1] WCDMA was originally proposed in [6]. W-CDMA is the 3G CDMA air interface for GSM, GPRS, and EDGE systems. EDGE is sometimes called 2.75G, GPRS as 2.5G and GSM as a 2G system.
2. *cdma2000*. Also called *multicarrier cdma*, broadband cdmaOne, umbrellas name IMT-MC. cdma2000 is the evolution of IS-95 to 3G. The intermediate states are IS-95A (2.5G) and IS-95B (2.75G). It is based on the North American cdma system. It has evolutionary stages in terms of services as well as rates. In terms of rates, the names are 1xRTT (Radio transmission technology) and 3xRTT. At this time, the service evolutions of 1xRTT are 1xEV-DO (evolution data only) and 1xEV-DV (data and voice).
3. *Time division—CDMA* (TD-CDMA). Also, with the umbrella name of IMT-TC and called *time code*. More specific name is UTRA-time

[1] We are being insensitive to the actual terminology used by the standards for uplink/reverse channels and downlink/forward channel because of its unimportance for a general understanding.

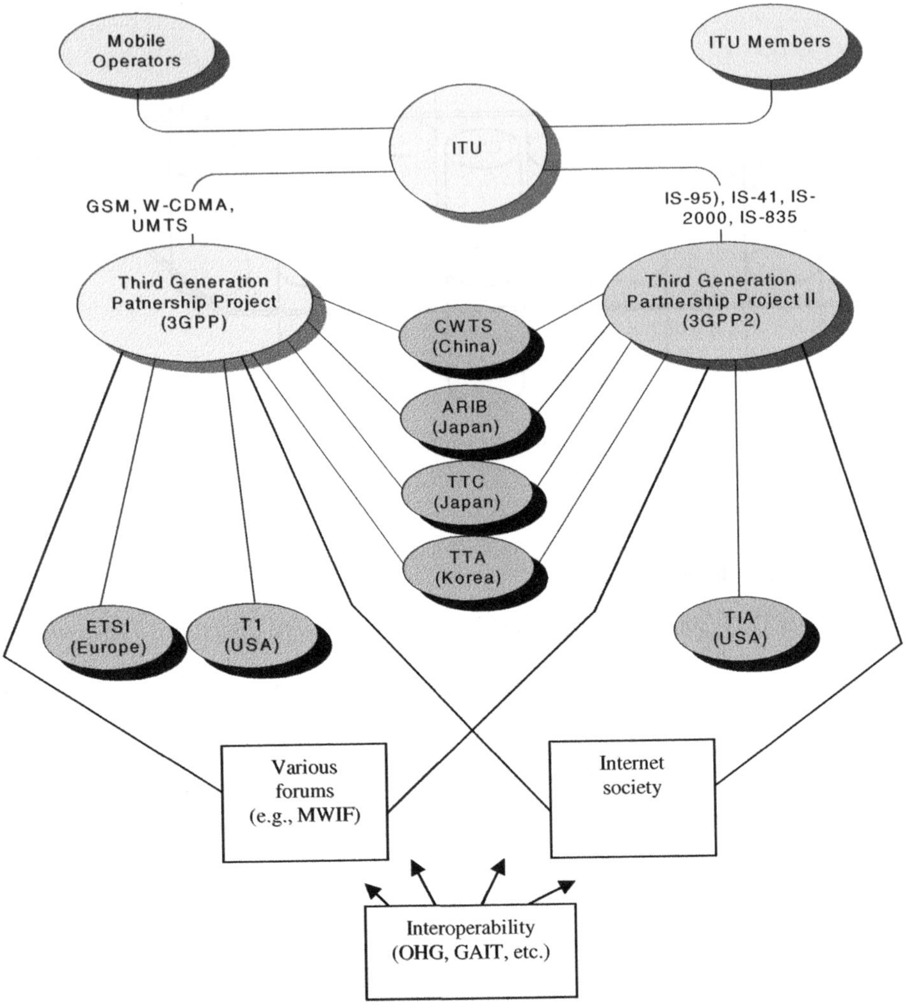

Figure 7-2. The cooperation explosion (modified from [4]).

division duplexing (UTRA-TDD). In relation to Chinese proposal, it is often referred to as TD-synchronous CDMA (TD-SCDMA). This is an entirely new initiative, mainly pushed forward by China. However, it has some significant benefits (once realizable). These include asymmetric transmission most suitable to HTTP and small size handset for cell phones. It was originally proposed in [8].

4. and 5. *Single carrier, UWC-136*, also the umbrella name IMT-SC, and FDMA/TDMA systems, DECT, under the umbrella name of IMT-FT were also part of the IMT UTRA. For a discussion see [4] and [5].

7.2.3. UMTS Terrestrial Radio Access (UTRA)

The UMTS access network consists of the mobile station (MS), node B or basestation, radio network controller (RNC), and the serving GPRS support node (SGSN). It uses W-CDMA as the physical layer multiple access and spreading scheme. For channel access, slotted ALOHA is employed. Three types of packet transmissions are possible in the uplink (from mobile station to base station). There are: (1) single short packet on the access channel, (2) single long packet with successful access mechanism, and (3) multiple packets with successful access mechanism. We will discuss WCDMA in the next chapter.

7.3. RELEASE D FOR CDMA2000 BASED ACCESS

Figure 7-3 shows the air interface layers of the cdma2000 standard as defined in the 3GPP2 Release D [7]. The physical layer for the Rev. D, defined in [9], specifies the operational requirements for the cdma2000 mobile station and base station, such as the transceiver frequency and power characteristics, modulation, reverse channels, error-control coding, spreading, performance measures, and so on. The MAC sublayer, defined in [10], specifies functions, procedures, and performance for the medium access protocols. The specification includes multiplexing and QoS service, logical channels, MAC layer addressing, and so forth. The higher layers depend on the applications services and how they are managed and authenticated. In the following, we will

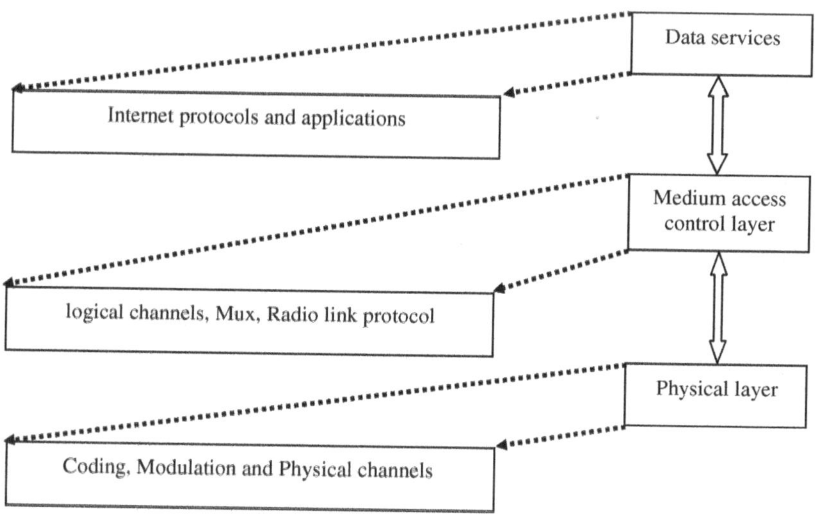

Figure 7-3. cdma2000 air interface layers, as defined in 3GPP2.

summarize some of the physical and medium access control layers characteristics relating to packet data transmission. The terms forwards link and reverse link are used in IS-95 standards.[2]

As pointed out in Ref. [11] there are only a few changes in Rev. D from Rev. C, such as hybrid ARQ (HARQ) at the physical layer, reducing the retransmission delay, some new reverse link channels, for example, reverse packet data channel (R-PDCH), reverse packet data control channel (R-PDCCH), reverse request channel (R-REQCH), and reverse secondary pilot channel (R-SPICH). Also added are some new channels in the forward direction, for example, forward grant channel (F-GCH), forward indicator control channel (F-ICCH), and forward acknowledgement channel (F-ACKCH). Other enhancements are done for broadcast and multicast services, and to increase the length of the equipment identification number to 52, which was 32-bit equipment serial number (ESN). Revision C is the first specification that is termed as 1xEV-DV (1x evolution data and voice) [12], which introduced the packet data channel in the forward direction and acknowledgement channels in the reverse direction for packet data with speeds in excess of 3 Mbps. Revision D provides the packet data enhancement in the reverse direction complementing the forward channel enhancements of Revision C. Additionally, other service enhancements add more value to the standard and are summarized in the following, as per Ref. [11].

7.3.1. Fast Call Setup (FCS)

The fast call setup is introduced to meet the rising demand in push-to-talk over cellular (PoC) service. In earlier versions of cdma, PoC was not possible due to a long call setup (paging, access and allocation) procedural delay. In Rev. D. much of the setup delay can be eliminated for PoC. This is possible due to reduced slotted mode, in which slot size is reduced or a no-slot mode is supported. Second, the access procedure can be by-passed by allocating the reverse channel with the paging message. Third, a fast service resume procedure can substantially cut down the service negotiation delay.

7.3.2. Mobile Equipment Identifier (MEID)

The MEID is a 56-bit equipment ID projected to replace the 32-bit electronic serial number (ESN). MEID will provide all functions of ESN, such as generating long code mask for spread spectrum, identify equipment, authenticate subscriber, and allocate international mobile subscriber ID (IMSI) for roaming. The ESN length warrants a change, as the total number of IDs (2^{32}) is diminishing quickly. Addition of another 14 bits increases the total number by over 16000 times.

[2] cdmaOne is a trade mark of the CDMA development group (cdg.org), and cdma2000 is a registered trade mark of Telecommunications Industry Association, USA (TIA-USA).

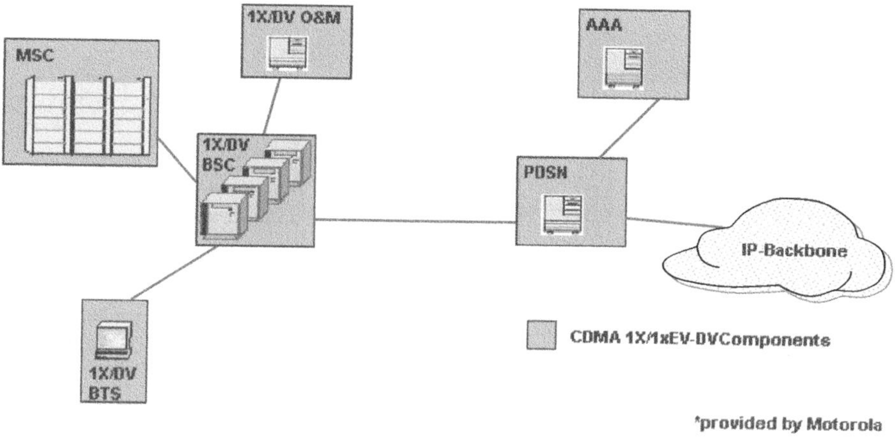

Figure 7-4. cdma2000 1xEV-DV-related components [13].

7.3.3. Broadcast and Multicast Services (BCMCS)

Enhancements in the forward supplemental and fundamental channels (F-FCH, F-SCH) allow for three ways in which these channels could be used for BCMCS. In type 1 of such channels, the F-SCH is shared among the idle MSs to provide a data rate of up to 307 kbps. An upper-layer BCMC message carries the attributes of BCMCS to the mobiles. Transmission efficiency can be enhanced by using optional Reed-Solomon outer code. Type 2 BCMCS provides less data rates (9.2, 14.4 kbps) with power control function for optimal use of power. In this type, the MSs use F-FCH for data and control channels for power and individual control functions. Each mobile uses its own long code mask. Soft handoff is possible with this type of multicasting service. In the third type, F-SCH is shared by the mobiles in traffic state, as against idle state of type 1.

BCMCS adds the capability of point-to-multipoint communications to the cdma2000 system, and can be used for any multicast and broadcast type application, such as streaming, live conferences and pay-per-view multimedia.

7.4. CDMA2000 STANDARD

Figures 7-4, 7-5 and 7-6, taken from [13], show some of the network components for the cdma2000 network in evolution.[3]

[3] IS-41 Revision D was the first ANSI-41 publication, Revision E promises international roaming, (1XRTT = Single-carrier, radio transmission technology). Downward compatible with IS-95. Theoretical above 600, practical, above 300, commercial products about 150 kbps, 1xEV is an improvement over 1XRTT. Uses 16QAM and Turbo coding to reach up to 2.4 Mbps, 1xEV-DV (3.1 Mbps, downlink peak rates) backward compatible with 1XRTT but not with 1xEV-DO.

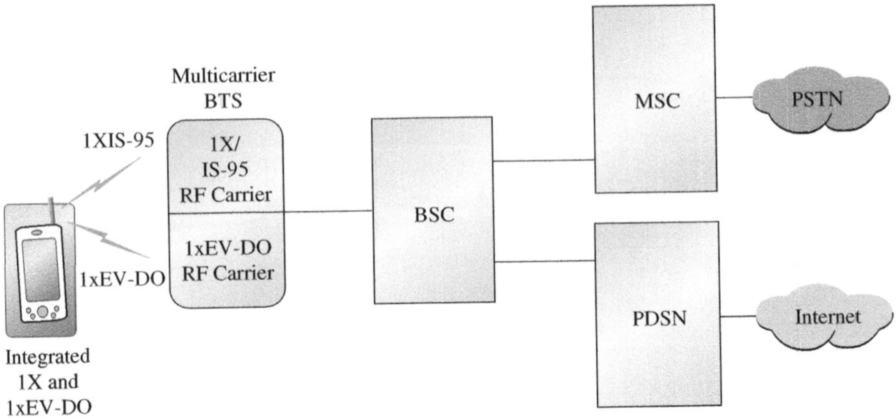

Figure 7-5. cdma2000 enhancements in 1xEV-DO [13].

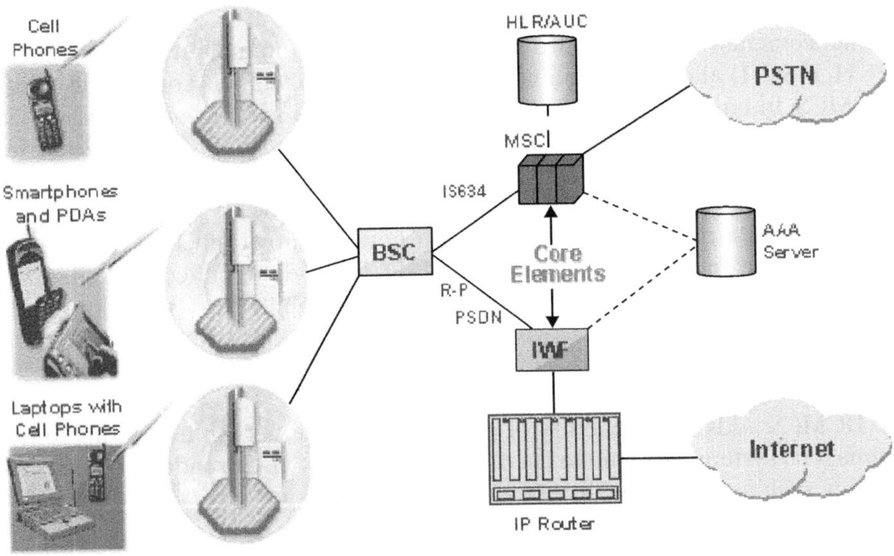

Figure 7-6. cdma2000 network diagram [13].

The first figure (Figure 7-4) shows the components that need upgrading or need to be added due to the high-speed packet data standard of cdma2000, called 1xEV-DV. Figure 7-5 shows the components added or affected due to the earlier high-speed standard evolution 1xEV-DO. The third figure (Figure 7-6) shows cdma2000 network components.

7.4.1. CDMA Timescale

The CDMA mobile station and base station use the global positioning system (GPS) for time reference. Since there is a fixed difference between the GPS and the coordinated universal time (UTC), both can be assumed to work with CDMA.

7.4.2. Physical Layer (PHY)

The PHY of cdma2000 is designed for most of the allocated cellular bands worldwide [14]. Table 7.1 summarizes some attributes of the physical layer. The first column 'PHY attribute' lists the physical layer function, procedure, or channel for the physical transmission of data. The second column lists options and various values for PHY attribute. The third column 'Comments' gives any additional information if considered necessary.

7.4.2.1. Radio Configuration (RC). The standard defines seven radio configurations in the reverse direction, depending on a combination of factors such as spreading rate, data rates supported, forward error correction, and modulation characteristics. The data rates for these radio configurations (RC) range from 1200 bps to 1.845 Mbps. Of main interest for packet transmissions is the RC7, which defines packet data and related channels. A mobile station that implements reverse packet data channel is required to also have the following channels: reverse acknowledgement channel (R-ACKCH) for acknowledged transmissions, reverse channel quality indicator channel (R-CQICH) for reporting channel quality, reverse packet data control channel (R-PDCCH) for various control functions related to higher layer packet data, and the reverse request channel (R-REQCH).

There are ten RCs in the forward direction, providing data rates from 1200 bps to 3.0912 Mbps. RC10 defines forward packet data channels, requiring a minimum of two to be implemented (if supported by an MS). If a mobile station supports forward packet data channel, Revision D requires it to have reverse fundamental channel (R-FCH), reverse dedicated control channel (R-DCCH), or both with RC 3. Also, reverse acknowledgement channel (R-ACKCH) and reverse channel quality indicator (R-CQI-CH) channels are recommended.

7.4.2.2. Access Channel. The access channel provides slotted ALOHA channel access slots to the mobile stations. This is the same as in earlier versions.

7.4.2.3. Reverse Packet Data Channel (R-PDCH)—10 ms (19.2 kbps–1.84 Mbps). The R-PDCH carries higher layer packet data from a mobile station to base station. It provides 11 different data rates from 174 to 18434 bits per

TABLE 7.1. PHY Attributes of cdma2000

PHY attribute	Standard options (ms)	Comments
Reverse frame duration	5, 10, 20, 40, 80	
Reverse channels	access channel (R-ACH), enhanced access channel (R-EACH), reverse common control channel (R-CCCH), reverse packet data control channel (R-PDCCH), reverse request channel (R-REQCH), reverse dedicated control channel (R-DCCH) reverse acknowledgement channel (R-ACKCH), reverse channel quality indicator channel (R-CQICH), reverse fundamental channel (R-FCH)	RC*k* = Radio configuration *k*. for RC3, RC4 For RC1, RC2, RC3, RC4
	reverse supplemental code channel (R-SCCH)	For RC1, RC2
	reverse supplemental channel (R-SCH)	RC3, RC4
	reverse packet data channel (R-PDCH)	RC7
Forwards channels	Forward pilot channel, transmit diversity pilot channel, auxiliary pilot channel, auxiliary transmit diversity pilot channel, sync channel, paging channel, broadcast control channel, quick paging channel, common power control channel, common assignment channel, forward packet data control	For RC1, RC2

	channel, forward common control channel, forward rate control channel, forward grant channel, forward acknowledgement channel, forward dedicated control channel, forward fundamental channel, forward supplemental code channel	
	forwards supplemental channel	For RC3, RC4, RC5
	forward packet data channel	RC10
Channel spreading	Walsh, quasi-orthogonal, followed by quadrature pairs of PN-codes	Chip rate 1.2288
Modulation types	BPSK, QPSK, 8-PSK, 16-QAM.	8-PSK and 16-QAM later additions
Error control mechanisms	Both forward error correction (FEC) and CRC specified.	Combination of FEC and CRC, dubbed as hybrid ARQ (HARQ) used for reverse packet data channel.
Packet data specific function	Channels (F-PDCH, R-PDCH, F-GCH, R-CQICH, F/R-PDCCH, F-ACKCH, R-REQCH), packet data channel control function (PDCHCF)	PDCHCF terminates packet data channels and implements reliability functions. It is divided into forward (F-) and reverse (R-) parts.

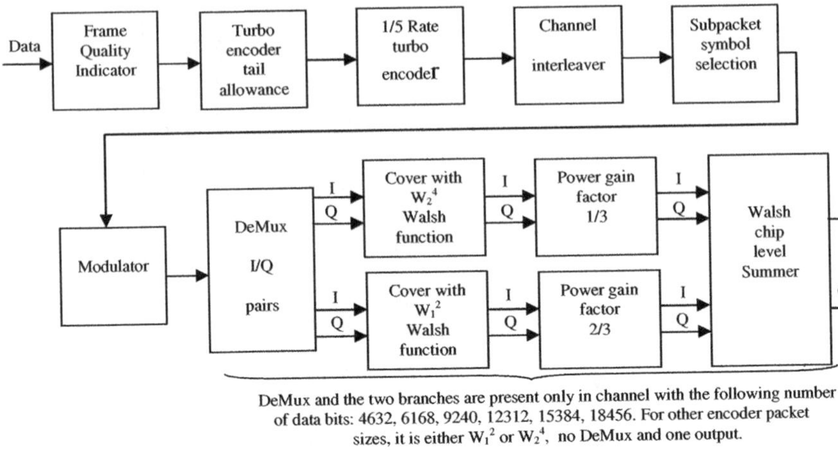

Figure 7-7. Reverse packet data channel structure for 1xEV-DV Rev. D.

frame of 10 ms. The data in this frame employ a turbo coding of rate 1/5. The modulation used depends on the channel rate and can be one of the BPSK, QPSK or 8-PSK. Following modulation, orthogonal spreading is employed using Walsh functions of length 2, or 4 or both.[4] For 1xEV, the Walsh function sequence is repeated every N/1.2288 µs, where N is 2 or 4.[5] Figure 7-7 shows a generic schematic for the Reverse Packet Data Channel (R-PDCH). Table 7.2 summarizes values for various processes (boxes) in Figure 7-7 for packet sizes shown.

7.4.2.4. Transmission. The outputs streams L and K of Figure 7-7 form part of the reverse channel stream.[6] The power-leveled streams from all the channels are summed in the box next to Figure 7-7 (not shown). Here's a list of other channels: reverse pilot channel, reverse secondary pilot channel, reverse dedicated control channel, reverse channel quality indicator channel, reverse supplemental channel 2, enhanced access channel, reverse common control channel, reverse packet data control channel, reverse request channel, reverse acknowledgement channel, reverse channel quality indicator channel, reverse fundamental channel and reverse supplemental channel 1. The outputs of the summed I and Q channels are fed to the complex multipliers. The other input to the multipliers comes from the PN-code that is generated from long code mask specific to a mobile station. Each of the (I and Q) streams has two

[4] Walsh functions are rows of special type of square matrices, each row being orthogonal to all other rows. The effect of orthogonality is that there are no cross-products of the signal energy (that would be a waste of signal power). In technical terms, the dot-product (also called inner product, or scalar product) of two orthogonal functions in zero. In terms of *vectors*, orthogonality implies a 90° angle, as *cosine*(90°) = 0.

[5] For spreading rate of 3 (i.e., 3X, not discussed), it will be one-third.

[6] The encoder packet sizes 192, 408 and 792 have only one stream, the K stream.

TABLE 7.2. cdma2000 PHY in the Reverse Direction

Function	Purpose	Data Bits	Value
Frame quality indicator	CRC error control	192	12
		All others	16
Turbo encoder tail allowance	Make encoder packet size 1/5th of encoder output.	All	6
Turbo encoder	Forward error correction.	All	1/5
Channel interleaver	Interleave channel bits to combat burst errors.	All	Block interleaving with 5 sub-blocks. Dimensions of interleaving matrix depend on channel rates.
Subpacket symbol selection	For modulation.	All	Subpacket durations multiples of 10 ms, up to 3.
Modulator	Digital passband modulation.	192, 408, 792	BPSK
		1560, 3096, 4632, 6168, 9240, 12312, 15384	QPSK
		18456	8-PSK
DeMux I/Q pairs	Separate the encoder into two streams for two Walsh covers.	4632, 6168, 9240, 12312, 15384, 18456	First 1/3 bits sent to the top output and remaining 2/3 to the lower output.
Cover with Walsh function	Generate bipolar sequence of Walsh chips for each modulation symbol.	192, 408, 792, 1560, 4632, 6168, 9240, 12312, 15384	W_2^4
		3096	W_1^2
		4632, 6168, 9240, 12312, 15384, 18456	W_2^4 and W_1^2
Power gain factor	Remove power discrepancies between 1/3 and 2/3 Walsh cover	4632, 6168, 9240, 12312, 15384, 18456	Just reverse the power transmissions so as to have uniform power levels.
Summer	Adds the two Walsh cover streams into one.	4632, 6168, 9240, 12312, 15384, 18456	Not applicable

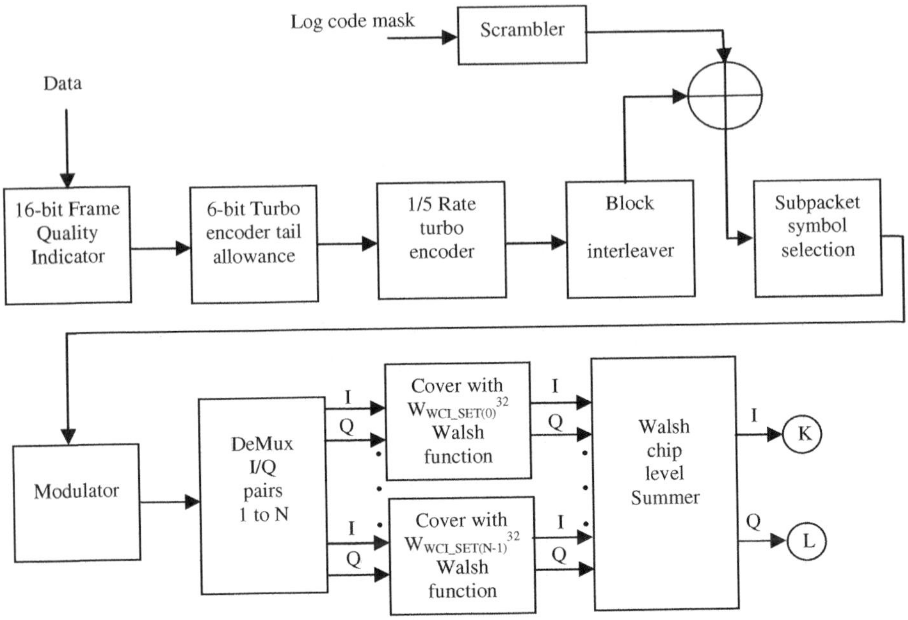

Figure 7-8. Forward packet data channel structure for 1xEV-DV Rev. D.

multipliers, one for the real part and the other for the imaginary (orthogonal) part. Next, the real parts of the two streams (I and Q) are combined as difference to be fed to the in-phase carrier signal (cos $2\pi f_c t$) for transmission as a continuous wave modulated signal.[7] Similarly, the combined difference of the imaginary parts of the I and Q streams is fed to the quadrature-phase carrier signal (sin $2\pi f_c t$) for transmission as a continuous wave modulated signal. Baseband filtering is used before carrier modulation.

7.4.2.5. Forward Packet Data Channel. Figure 7-8 shows a schematic of the operations taking place in the forward packet data channel. Before actual transmission as a continuous wave signal, other functions similar to the ones discussed under 'Transmission' above are performed. The channel consists of 1.25 ms, 2.5 ms, or 5 ms frames. Various data rates are provided for each duration, as shown in Table 7.3

7.5. CDMA2000 MEDIUM ACCESS CONTROL

The Release D MAC sublayer for cdma2000 as part of 3GPP2 is specified in [15]. The MAC sublayer is one of the two sublayers specified as part of layer

[7] If R_I is the real part of the I stream and R_Q, the real part of the Q stream, the combined output is $(R_I–R_Q)$.

TABLE 7.3. Data Rates for Various Forward Packet Data Channel Durations

Channel Duration (ms)	Data Rates (Mbps)
1.25	3.0912, 2.4768, 1.8624, 1.248, 0.6336 or 0.3264
2.5	1.5456, 1.2384, 0.9312, 0.624, 0.624, 0.3168, 0.1632
5.0	0.7728, 0.6192, 0.4656, 0.312, 0.1584, 0.0816

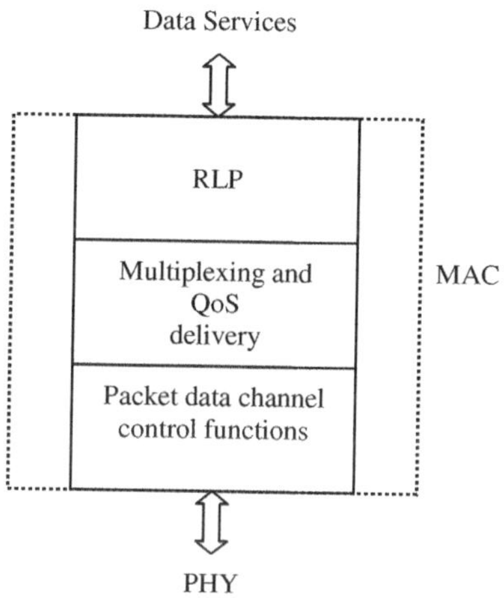

Figure 7-9. cdma2000 MAC sublayer.

2 of the cdma2000, the other one being the link access control (LAC) sublayer. LAC provides many services to the upper layers. These services include signaling, voice, and data application services. The standard provides a multimedia service environment in which high-speed voice and data integration is possible with packet-switched data. QoS management provides a means of service differentiation.

Packet data services are provided through a MAC architecture, shown in Figure 7-9. It consists of a connection oriented radio link protocol (RLP), which uses negative acknowledgements, a multiplexing and QoS delivery entity and packet data channels.

The RLP (radio link protocol) provides best effort delivery with reasonable reliability. In IS-95B, radio link protocol was added to include ARQ to enhance the reliability. The multiplexing and QoS delivery function provides mechanisms for prioritizing access and enforcing the negotiated QoS. In the mobile station, the packet data channel control functions (PDCHCF) map the various PHY packet data channels to the logical packet data channels. For example, the reverse PDCHCF (RPDCHCF) interfaces PHY through the following PHY channels: reverse channels R-REQCH (request channel), R-PDCCH (packet data control channel), R-PDCH (packet data channel), and forward channels F-GCH (grant channel), F-ACKCH (ACK channel) and F-

RCCH (rate control channel). The RPDCHCF interfaces with the multiplex sublayer through the reverse packet data (logical) channel (r-pdch). The FDPCHCF interfaces with the physical layer through the physical layer reverse channels R-CQiCH (channel quality indicator channel) and R-ACKCH (ACK channel) and the forward channels F-PDCCHs (packet data control channels) and F-PDCHs (packet data channels). With the multiplex layer, FDPCHCF connects via the forward packet data channel (f-pdch).

The multiplex sublayer connects to the voice services and RLP using the forward and reverse logical traffic channels (f/r-dtch). Packet service uses the RLP reliability function for negative acknowledgements. In the following, we will summarize functions of MAC sublayer entities.

7.5.1. Mux and QoS (MaQ) Sublayer

As we know from the discussion on cdma2000 PHY, the standard defines several radio configurations (RCs). The MaQ sublayer divides all the configurations into two modes of operation, Mode A and Mode B. A radio configuration of less than or equal to 2 corresponds to Mode A operation and a radio configuration above 2 is considered Mode B of operation. MaQ organizes data in *MuxPDUs*. Each MuxPDU consists of one or more *data blacks*. The data block is the unit of information exchanged between the MaQ and logical channel (or the connected service). One or more MuxPDUs constitute one PHY SDU, depending on the multiplex option employed. In this way, for a given PHY SDU duration, various data rates can be defined. For example, eight multiplex options are defined in FCH and DCCH for two rate sets (4 each). In this way, each channel uses specified multiplex options, each option defining the size of PHY SDU in terms of MuxPDUs, the size MuxPDU in terms of data blocks, and the size of data block as well. A mobile station uses Mode A or/and B, depending on the radio configurations it supports. Communication with the physical layer occurs through the exchange of primitives with the physical layer channel. Primitives constitute the service interface between the physical channel and the MaQ. Similarly, interface with the PDCHCF is defined in terms of a set of primitives to be exchanged between the two sublayers.

7.5.2. Access Channel Procedures

An access attempt is defined as the process of sending a layer 2 PDU on the reverse link and receiving an acknowledgement. Access attempts consist of access sub-attempts and access probes. A probe consists of 4 to 26 R-ACH frames; a frame having duration of 20 ms. The probe has two parts, a preamble and the message capsule. The minimum length of the preamble is one frame. The probe also defines the R-ACH slot. One sub-attempt consists a probe sequence and may include a maximum of 16 probes. The first probe is transmitted with some initial power level, which is increased in subsequent

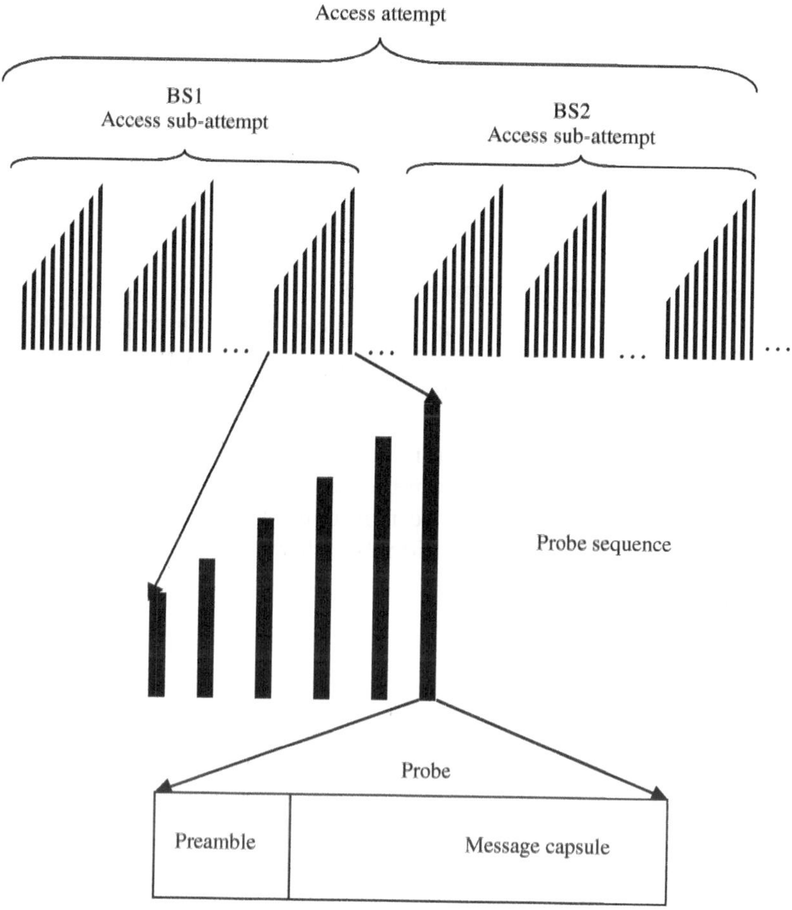

Figure 7-10. Relation among access attempt, access sub-attempt, probe sequence, and probe.

probes. Figure 7-10 shows a relation among probes, sub-attempts and attempts. A pseudorandom generator is used to determine a backoff delay for every access probe sequence, thus randomizing the timing of start of each attempt. Two access modes, basic and reservation are supported to form new and handoff connections.

7.5.3. Packet Data Channel Control Functions (PDCHCF)

The PDCHCF are required to provide packet-switched data services. Voice services can function properly in a relatively lower channel quality situation. However, reliability of packet data must be ascertained above a certain minimum level better than voice data. Since packet data could use retransmissions, the PDCHCF implements retransmission and timing functions for the packet data channels.

7.5.3.1. Forward PDCHCF (FPDCHCF). The PHY channels associated with the forward packet data channel terminate in FPDCHCF. These include F-PDCH, F-PDCCH, R-ACKCH, R-CQICH, and so on. FPDCHCF is used for reliability functions. It performs ARQ of the packets with errors from the base station to mobile station. It provides several other functions for performance enhancement, such as multiple ARQ channels to have more than one retransmission outstanding simultaneously. FPDCHCF controls timings of all transmissions on the forward packet data channels.

7.5.3.2. Reverse PDCHCF (RPDCHCF). This control function (CF) is required if the mobile station supports packet data channels to the base station. These channels include R-PDCH (for packets sent by error control encoder in the form of sub-packets), R-PDCCH (that carries various attributes of the subpackets, such as size), R-REQCH (for traffic control related information at the mobile station, for example, buffer and power thresholds), F-AKCH (for ACKs and NAKs from the base station to the mobile station), F-GCH (for the encoder packet size may consists of multiple sub-packets), F-RCCH (for rate control information, by allowing to increase, decrease, or maintain the authorized traffic).

7.6. ALL-IP ARCHITECTURE

The all-IP architecture is projected to be completed in three phases. Revision 3.0 of 3GPP2 [16] defines an all-IP architecture that divides the networking function into four planes. Before a discussion on these planes, we will briefly describe some of the components of the all-IP network.

7.6.1. Networking Elements

Following is a partial list of functional elements employed for an all-IP network. In some cases, we prefer to describe the traditional elements (e.g., HLR) and the corresponding all-IP element as relevance to the original (e.g. HLRe).

7.6.1.1. Access Gateway (AGW). The AGW provides an interface between radio access network (RAN) and the core network (CN).

7.6.1.2. Authentication Center (AC). The AC stores information to authenticate the legitimacy of a user and equipment.

7.6.1.3. Base Station (BS). The BS provides the radio contact to an MS with the cellular network. It consists of a base station transceiver system (BTS) and base station controller (BSC). A BSC controls and manages one or more BTSs. Each BTS provides transceiver function to a number of MSs.

7.6.1.4. Call Session Control Function (CSCF). The CSCF is the session manager for the user services. It performs a variety of tasks related to session setup, resource allocation and authentication and service provision to the MS in its home network. The CSCF is defined only for the multimedia domain. A proxy CSCF (P-CSCF) acts directly with the MS as well as CSCF for interaction. Sometimes, a CSCF may be divided into two parts for better service placement and load distribution. In such case, an interrogation CSCF (I-CSCF) will provide an interface for the network for queries to a server CSCF (S-CSCF). The MS in this case will interact with the S-CSCF.

7.6.1.5. Databases. The database stores information about subscriber profile, the EIR (see next) data, the dynamic subscriber profile, and so forth.

7.6.1.6. Equipment Identity Register (EIR). The EIR is defined to store equipment identity that can be used for equipment authentication.

7.6.1.7. Home Agent (HA). The HA provides the functions of a mobile IP home agent.

7.6.1.8. Home Location Register (HLR). The HLR stores permanent information of a subscriber and MS. It also keeps track of the mobile user by storing its current location. The IP network defines HLR emulation (HLRe) that uses IP interfaces for signaling.

7.6.1.9. Interworking Function (IWF). The IWF provides an interface between different networks. For example, an IWF is needed between a mobile switching center (MSC) and IP network.

7.6.1.10. Media Gateway (MGW). The MGW provides interface between (circuit-switched) earlier generation network, and packet-switched network.

7.6.1.11. Media Resource Function Processor (MRFP). The MRFP provides services such as, multimedia conference bridges, announcement playback etc. It supports the legacy MS domain as well as multimedia.

7.6.1.12. Mobile Station (MS). MS is the subscriber's unit that has a direct communications with the CDMA access network. Among other terms used in this book a mobile host (MH) could be either an MS (PDA or a notebook), or a device (notebook) connected through an MS (cell phone). The same applies to a mobile node (MN). MS contains the user identification module (UIM) that contains user specific information. Without UIM, the MS is called mobile equipment (ME). A ME could be one of the MTk ($k = 0$, no external interface, $k = 1$, ISDN external interface and $k = 2$ means non-ISDN external interface). Other terms, for example, terminal adapter (TA) and terminal equipment (TE) has the same meanings as in ISDN terminology. Throughout this book, we use various terms for the MS. It is usually clear from context.

7.6.1.13. Mobility Management (MM). The mobility manager (MM) receives register requests from a mobile station and authenticates the request with the AAA server to provide multimedia domain services, such as page response and handoff.

7.6.1.14. Mobile Switching Center (MSC). The MSC is the main switching unit in an IS-41 network. It interconnects base station controller (BSC) systems. It uses home location register (HLR), visitor location register (VLR), authentication center (AC), and message center (MC) for various databases and short message services (SMS). The gateway MSC (GMSC) also provides an interface to the PSTN.

7.6.1.15. Message Center (MC). The MC provides a medium for temporary storage of short messages for short message service (SMS).

7.6.1.16. OSA-Service Capability Server (OSA-SCS). The OSA-SCS provides resources to the access network that applications need while being executed. The OSA-application programming interface (OSA-API) is one of the changes in cellular networks that have resulted from All-IP architecture. OSA-API is the OSA-SCS interface with the application servers. It also grants authorization for applicable resources in conjunction with AAA and position server. It also exists for the legacy network and relays server application information between application servers and MS domain support network. See the next chapter for more on OSA.

7.6.1.17. Packet Control Function (PCF). PCF manages the flow of packets between the base station and packet data serving node (PDSN). The PDSN provides an interface between base stations and packet data networks (PDNs). It establishes a link layer connection with the MS for packet-switched connections. A PDN is an IP network. It could be any other network that provides packet switching.

7.6.1.18. Policy Decision Function (PDF). PDF makes policy decisions about resource management to support QoS. PDF is responsible for resource decisions within its own core network.

7.6.1.20. Visitor Location Register (VLR). The VLR stores information (Registers) for a visiting MS and coordinates this information with its HLR. The all-IP NAM defines VLRe, the VLR emulation that has IP interfaces.

7.6.2. Planar Architecture

The all-IP network architecture mode (NAM) defined the following planes:

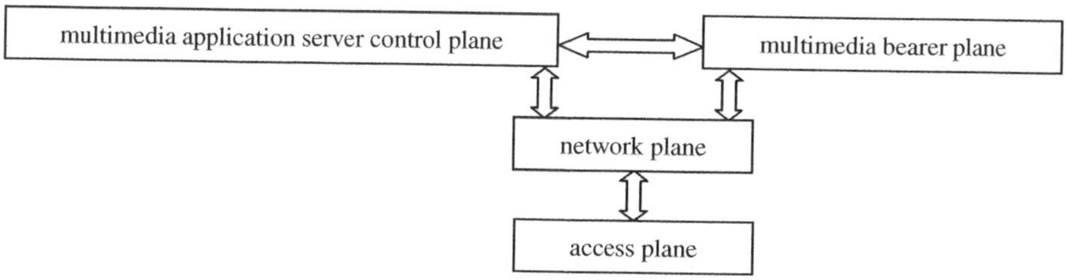

Figure 7-11. A planar view of the 3GPP2 functional network.

TABLE 7.4. Elements of the Access Plane in an All-IP NAM

Element	Interfaces/to	Function
MS	47 (U_m)/BTS in RAN.	Terminal equipment (TE) part, link layer, air interface, signaling, and so on.
RAN	47/MS, 20/AAA, 31/AGW(data), 35/AGW(control).	BSC, RNC+PCF, BTS, MM
AGW	Other than RAN, ?/other access networks.	Access gateway to interconnect MS with core IP.
AAA	gg/AAA(local)-to-AAA(home network)	Authentication, authorization and accounting.

1. Access plane;
2. Network plane;
3. Multimedia bearer plane; and
4. Multimedia application server control plane.

Figure 7-11 shows the relation among various planes.

7.6.2.1. Access Plane. The elements of the access plane with their functions are shown in Table 7.4.

Figure 7-12 shows access plane functional diagram. The other access network can be any of the existing and future wireless data networks, for example, IEEE 802.11, IEEE 802.16, IEEE 802.20, and so on.

7.6.2.2. Network Plane. The network plane is responsible for end-to-end IP connectivity between a mobile host and other IP network nodes. Table 7.5 shows the elements in the network plane. This plane interacts with the access plane through DiffServ, RSVP or AGW/PDF and with other IP networks (for example, multimedia bearer) through generic interfaces.

Figure 7-13 shows the network plane.

TABLE 7.5. 3GPP2 All-IP Rev. 3.0. Network Plane Elements

Element	Interface	Function
MS	47(U_m)/BTS	IP capable, Mobile IP, QoS client
HA	22/AAA, 41,43/AGW	Registration and forwarding
PDF	23/AAA, 28(G_o)/AGE	Enforce resource management to meet QoS within its own core.
AAA	21/AGW, 23/PDF	For IP layer AAA

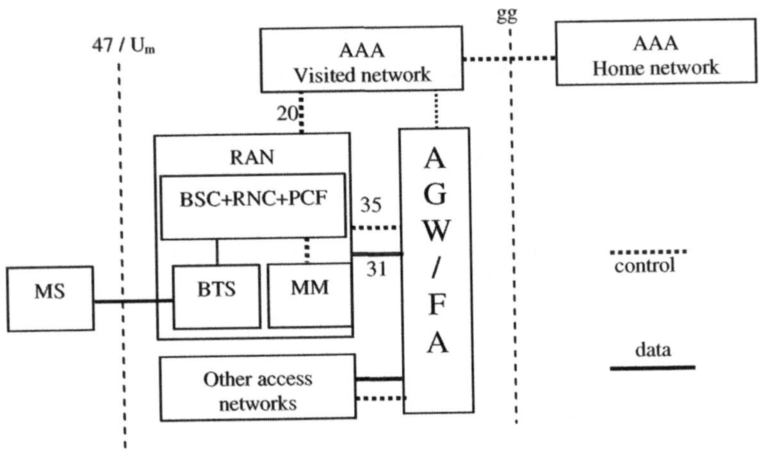

Figure 7-12. Access plane functional elements and interfaces in 3GPP2 All-IP Rev. 3.0.

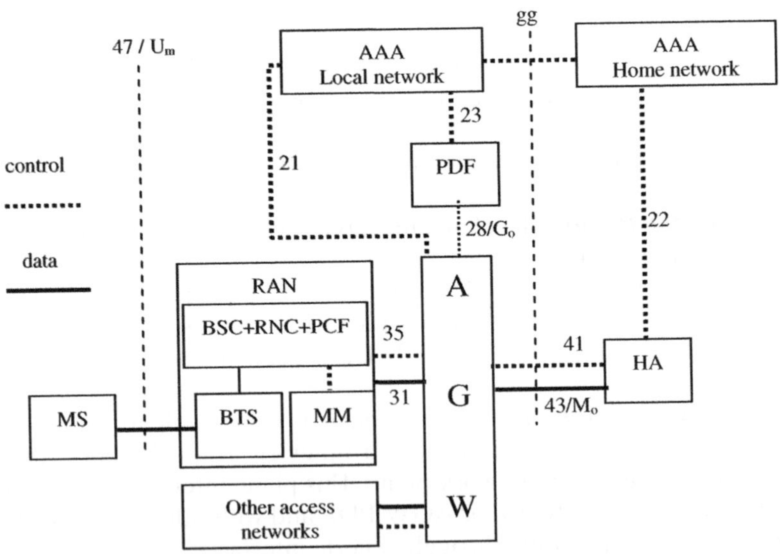

Figure 7-13. Network plane functional elements and interfaces in 3GPP2 All-IP Rev. 3.0.

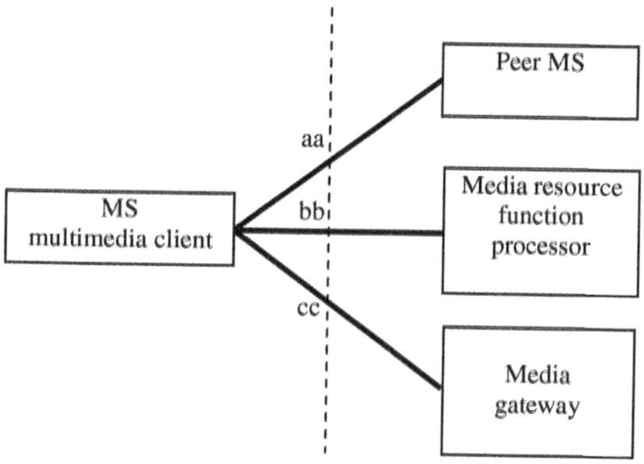

Figure 7-14. 3GPP2 Rev. 3.0, multimedia bearer plane.

TABLE 7.6. 3GPP2. Components of Multimedia Bearer Plane

Element	Interface	Function
MS	47(U_m)/BTS	Multimedia client or application
MRFP	bb	Multimedia conference bridges
MGW	aa	Interface between legacy and IP
Internet	Depend on the peer entity.	Peer entity on the Internet

7.6.2.3. Multimedia Bearer Plane. The multimedia bearer plane is above the network plane as shown in Figure 7-14. It provides multimedia bearer services to the MS, including IP address of the addressee if needed, QoS provision, and data delivery. The multimedia bearer plane has the components listed in Table 7.6.

7.6.2.4. Multimedia Application Server Control Plane. This plane implements the call control services and applications of the multimedia domain of the IP network. The functions of this plane can be implemented on various access infrastructures. The applications in this plane could have peer-to-peer communications or communicate via the multimedia bearer plane. The components of the multimedia application server control plane are shown in Table 7.7 and Figure 7-15.

7.7. SUMMARY

With the recent cooperation between cellular operators, their respective standardization agencies, and the IETF, it is apparent that the future of cellular networking is in IP. However, due to imperfections of current IP service pro-

TABLE 7.7. 3GPP2 Rev. 3, Multimedia Application Server Control Plane

Component	Interfaces	Function
MS	The application service interfaces. See below.	Multimedia client and/or application.
OSA-service capability server	10/Position server, 8 (OSA-API)/ applications server, type A, 11(Sh)/AAA, 12(ISC)/CSCF	Resource authentication and allocation.
Call session control function	12(ISC)/OSA-SCS and Application servers type B, 16(Cx)/AAA	
AAA (for applications)	15/Position server, 16(Cx)/ CSCF, 11(Sh)/OSA-SCS and Application type B, 4/Database	Authentication for resource allocation
Applications servers	8(OSA-API)/OSA-SCS, 11(Sh)/ AAA, 12(ISC)/CSCF	Provide network interface and resources to applications.
Database	4/AAA	

visioning, especially accounting and QoS enforcement procedures, there is going to be no sudden change from the traditional cellular network to all-IP networks. In this chapter we saw several advances in the North American cdma system toward IP-based bearer services. These changes have come first due to high speeds in the form of cdma2000 1xEV-DO evolution, then packet-bearing channels in the two revisions of 1xEV-DV, and finally in terms of an all-IP network architecture model (IP-NAM). The IP-NAM is assumed to be using session initiation protocol (SIP) and mobile IP, discussed in the previous chapter. In the next chapter, we will look at the other major standard for 3G cellular systems, that is, wideband cdma (WCDMA). Again, we will look from the packet data transmission point-of-view only.

REFERENCES

[1] Derek Keron and Bin, Hu, 'US wireless data market: enterprise segment', *Kerton Group,* http://www.kerton.com/papers/WirelessDataAbstract.pdf

[2] ARDIS company, 'ARDIS DataTAC 4000: Software developers reference guide', Revision 2, Jan. 1997.

[3] Jeff Morris, 'Guide to wireless data networks', http://www.sierrawireless.com/news/docs/wirenet.doc

[4] Brough Turner and Orange, Marc, '3G tutorial', *nmscommunications:* nmscommunications.com/3Gtutorial, presented at *Fall VON 2002.*

[5] Tero Ojenpera and Prasad, Ramjee, 'An overview of air interface: Multiple access for IMT-2000/UMTS', *IEEE Communications Magazine*, Sep. 1998, pp. 82–95.

[6] T Ojanpera et al, 'Design of a 3rd generation multirate CDMA system with multiiser detection—MUD-CDMA', *Proc. ISSSTA'96,* Mainz Germany Vol 1, 1996, pp. 334–338.

[7] 3GPP2, *Introduction to cdma2000 spread spectrum systems: Revision D*, 3GPP2 C.S0001-D, Version 1.0, February 2004.

[8] A. Kein and Baier, P.W., 'Linear unbiased data estimation in mobile radio systems applying CDMA', *IEEE JSAC* Vol 11, 1993, pp. 1058–1066.

[9] 3GPP2, *Physical layer standard for cdma2000 spread spectrum systems: Revision D*, 3GPP2 C.S0002-D, Version 1.0, February 2004.

[10] 3GPP2, *Medium access control standard for cdma2000 spread spectrum systems: Revision D*, 3GPP2 C.S0003-D, Version 1.0, February 2004.

[11] R. Thomas Derryberry, Hsu, Alan, and Tamminen, Walt, 'Overview of cdma2000 revision D', www.cdg.org/resources/white_papers/files/Overview_of_cdma2000_Revision_D.pdf

[12] Motorola, 'Technical overview of 1xEV-DV', *White paper, CDMA Development Group* (CDG), Version G1.4, Sep. 2002.

[13] CDG, '3G evolution', *CDMA development group*, www.cdg.org 3GPP2 C.S0002-D, Version 1.0, February 2004.

[14] 3GPP2, *Band class specification for cdma2000 spread spectrum systems: Revision 0*, 3GPP2 C.S0057, Version 1.0, February 2004.

[15] 3GPP2, 'Medium access control standard for cdma2000 spread spectrum systems—Release D', *3GPP2 C.S0003-D, Version 1.0*, February 2004.

[16] 3GPP2, *IP network architecture model for cdma2000 spread spectrum systems'*, 3GPP2 S.R0037-0 v3.0, Aug. 2003.

[17] 3GPP2, Network Reference Model for cdma2000 spread spectrum systems, 3GPP2 S.R0005-B, Apr. 2001.

[18] 3GPP2, '3GPP2 system capability guide: release B', *3GPP2 S.R0003-A* , available from http://www.3gpp2.org/Public_html/specs/S.R0003-A_SCG_Release_B_v1.0.pdf

CHAPTER 8

DATA COMMUNICATIONS IN CELLULAR NETWORKS: W-CDMA

The Group Special Mobile (GSM), later to be rechristened as Global System for Mobile communications, brought the first widely used system in countries now using ETSI standards. With eight 577-μs slots per frame using a 200 kHz carrier, a number of roaming, portability, security, and messaging services, GSM has never looked back. The earlier slow data rates capped to 9.6 kbps were improved by high-speed circuit-switched data (HSCSD) and packet-switched general packet radio service (GPRS). Internet connectivity for VoIP was offered through H.323. HSCSD, by virtue of allowing more than one channel (slot) per connection and data compression techniques [1] increased speeds to V.34 or above levels.[1] GPRS added packet data structure by adding functions, channels, and serving nodes. However, GPRS data rates are not significantly higher for a wideband service (generally less than V.34). GPRS capability of allowing packet data terminals, defining an array of packet data channels, adding packet data protocols, such as GPRS tunnel protocol (GTP) and, most importantly, two GPRS service node (GSNs), are the major contributions of GPRS toward packet data services over cellular networks.

The following concepts are central to the understanding of W-CDMA networks.[2] The serving GSN (SGSN) functions as the MSC-equivalent for packet

[1] Increase in data rates in such cases is not directly proportional to voice bit rate and number of slots because data need extra protection from channel error, which translates into bandwidth consumption.

[2] W-CDMA is the same thing as WCDMA, unless otherwise specified.

data, while leaving Internet and public data network interfaces to gateway GSN (GGSN). A promotion in data speeds of GPRS was achieved through EDGE (enhanced data rate for GSM evolution). By providing more efficient modulation schemes, channel equalization, and link quality control mechanism, EDGE promises data rates in excess of 400 kbps. It also took a stride of compatibility toward North American TDMA system (IS-136) by adding 8-PSK. All in all, the GPRS network set the stage for 3G packet-based networks. It was therefore natural that IMT-2000 suite of air interfaces was based on the core network of GPRS—with some additions. In this chapter, we will look at 3G network employing the wideband CDMA (W-CDMA) as the air interface. The study of WCDMA is different from cdma2000 due to the fact that the former is an entirely newly designed system, while the latter is an evolution of North American cdmaOne. While the cdma2000 has evolved from IS-95 through IS-95A, IS-95B to IS-2000, W-CDMA was the first step toward a CDMA-based network for ETSI standards to be implemented over GSM infrastructure. As part of the 3GPP project, W-CDMA includes both the FDD and TDD air interfaces. We consider only the FDD W-CDMA. In GSM and its extensions, the air interface and core networks (GSM-MAP) are not regarded as two disjoint networks as is the case of North American standards (ANSI-41 core and multiple air interfaces). However, owing to the impact of the deployment of the air interface W-CDMA, the universal mobile telecommunications system (UMTS) network (radio access plus core network) is sometimes regarded as the W-CDMA network. We use the two terms interchangeably. In reality, W-CDMA is used only at the physical layer (PHY) in the UMTS terrestrial radio access network (UTRAN). In fact, we also use "3GPP" [2] to imply UMTS and W-CDMA systems, just like we used the term 3GPP2 for cdma2000. Since we consider only packet data transmission, the circuit-switched parts and signaling network do not qualify for a discussion, except wherever convenient.

8.1. COMPONENTS OF THE UMTS NETWORK

Figure 8-1 shows various networks, systems, and components of a UMTS network in a rather familiar interconnection paradigm.

Most of the components are inherited from the GPRS networks. There are some additions too. The call session control function (CSCF) provides a mechanism for a number of subscriber services relating to privacy, mobility, and resource allocation. As we will see, HLR is, in fact, a part of home network that has a session layer signaling additive for 3G networks. Figure 8-1 shows the most familiar components from a GPRS network. Two important service aspects of the latest standards in W-CDMA network, not shown here, are the IP multimedia service (IMS) and open service access (OSA) application program interface (API). In future releases of the 3GPP standards, the radio access networks will provide a generalized wireless access mechanism for user

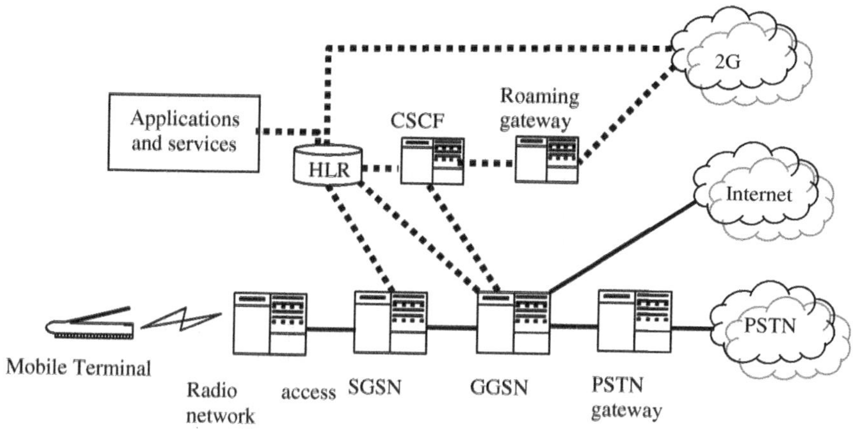

Figure 8-1. IP reference architecture for 3GPP [23].

terminals with conventional (phone) capabilities as well as wireless LAN stations.

In the rest of this chapter we will discuss various aspects of this network.[3] We will broadly divide the topics into two: the network protocol architecture and the network service architecture. The protocol architecture is described under "UMTS network domains" and the network services architecture under "UMTS services". Due to the scope of the chapter, we don't promise to be all-inclusive, rather we give a general idea on a technical level.

The UMTS network consists of domains and strata. Domains are interconnection of functional units and strata are protocol sets among the functional units and applications.

8.2. UMTS NETWORK DOMAINS

A network domain is a set of functional entities. The domains make a hierarchy. If the UMTS Network domain (UND) consists of all network components, then below it are the infrastructure domain and the user equipment domain. Figure 8-2 shows the domain hierarchy.

The interconnection of domains is through a series of *reference points*. The dotted line between the access network domain and user equipment (UE) domain is called the U_u interface and will be considered later under the UMTS terrestrial radio access network (UTRAN).

[3] Due to the complex nature of UMTS network, we could be sketchy at best. Even the 3GPP partners have recognized the fact that there is a need for understanding salient 'features' of the network. In fact, in preparing this chapter we have made frequent use of the 3GPP features documents [3] for Release 99 and [4] for Release 5.

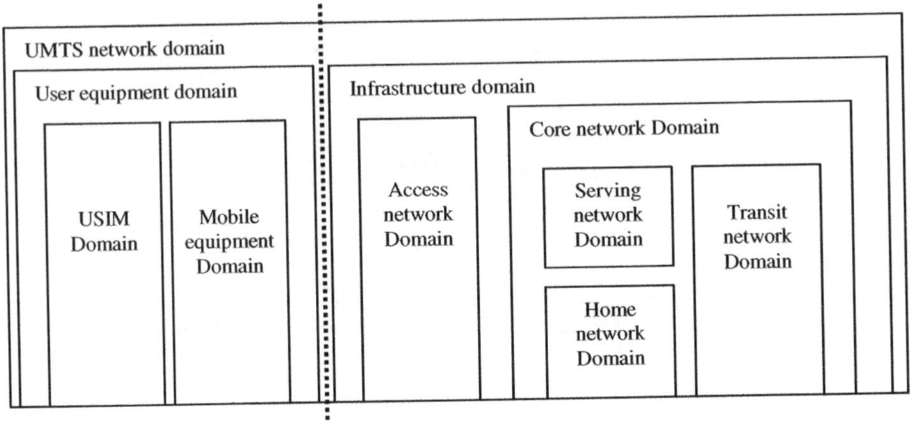

Figure 8-2. Domain hierarchy of UMTS network.

8.2.1. UE Domain

The UE domain consists of end-user equipment and its interconnection services. This is one of the two large domains and contains ME (mobile equipment) and USIM domains. The ME domain contains mobile terminal (MT) with a transceiver device and terminal equipment (TE) with an application. An example of TE is a notebook computer, and an MT could be a cell phone. Using wireless LAN access the ME domain could be installed in a single unit, such as a notebook with WLAN card. The USIM domain consists of a card with IC chip with hard-coded user and equipment identification numbers, encryption keys, and a PIN or password to operate the card. The UE domain can be in a single unit, for instance, a PDA with WLAN/3G access and a USIM slot to operate the cell phone unit of the PDA.

8.2.2. Infrastructure Domain

All network functional components other than the ones in UE are part of the infrastructure domain. The access network and core network domains are contained within this domain. The access network controls radio connection to the network through a set of protocols and procedures. The core network provides the circuit-switched and packet-switched services. The service architecture uses the serving network domain for negotiation and control of application services. It interacts with the home network domain that stores the subscription information on a permanent basis. A visiting network could request some of the subscription information, in order to decide what types of services should be provided to a roamer. The home network also helps user equipment and the person to be authenticated so that he or she can be authorized to use network resources as a legitimate subscriber. The transient network could be one or more autonomous or operator-administered transport service networks providing circuit-switched or packet-switched services.

8.3. STRATA

A stratum is a set of protocols between two or more entities. *Application strata* are end-to-end protocols including user and service provider applications. These strata use services provided by other strata. The *transport strata* provide this service and also service to signaling data. Serving strata provide service to transport strata to route data from source to destination. *Home strata* are a set of protocols and related functions used for handling and storing subscription services. Various strata function on the OSI-RM paradigm of exchanging primitives through service access points (SAPs) or other identifiers.

8.4. RADIO ACCESS NETWORK (RAN)

The UMTS terrestrial RAN (UTRAN) was redefined from earlier versions of GSM and its data network extensions. In Release 99 of the 3GPP, these additions included USIM in the user equipment domain, Node B, and Radio Network Controller (RNC) in access domain and interfaces relating to these entities and relating to packet switching. This is shown in Figure 8-3. RNC and Node B constitute radio network subsystems (RNS).

The air interface (U_u) between the user equipment and Node B, uses W-CDMA at the PHY. RNC performs several packet-switching related functions, such as QoS control with radio resource management function. Figure 8-4 shows the protocol architecture of the radio access network.

Some of the tasks performed by Node B are: radio transmission/reception at the interface with user equipment (interface U_u), forward error correction, rate adaptation in response to channel quality, spreading/de-spreading, modulation, making signal measurement reports, and communicating with to the

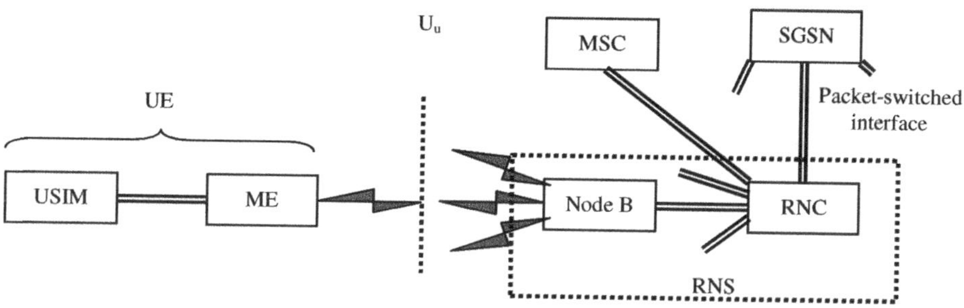

Figure 8-3. USIM, Node B and RNC connections in 3GGP.

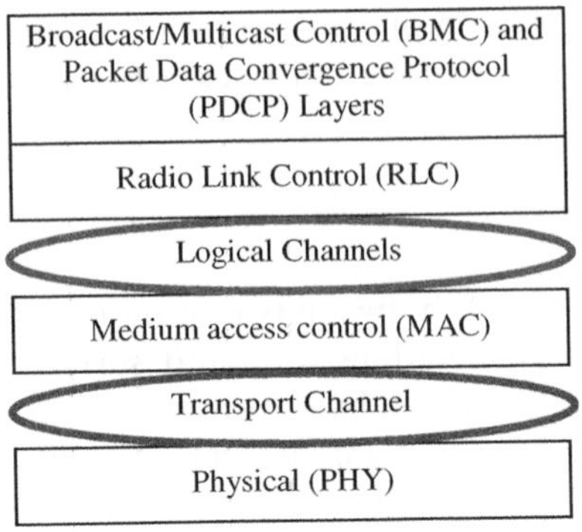

Figure 8-4. Protocol layers packet data in RAN.

RNC [5]. It also participates in power control and soft handover.[4] Protocols are defined between pairs of functional units across their conceptual separation point. These points are termed *interfaces*. For example, the interface between RNC and Node B is called I_{ub}. Here is a brief description of the protocol across U_u, as per Figure 8-4.

8.4.1. Transport and Logical Channels

The transport and logical channels constitute service access points (SAP)s for peer-to-peer communications. The transport channels exist at the interface of PHY and MAC and the logical channels between MAC and RLC. The transport channels represent the information form and organization, while the logical channels represent information type. The *common* transport channels can be used by any user, and thus must carry the user information for identification. The use of common transport channels may result in allocation of *dedicated* transport channels that are specific to users, therefore not requiring identification as part of the information carried. Two broad types of logical channels are the *control* channels and *traffic* channels. The former carry control information (common and dedicated), and the latter carry user data. Transport channels map to the physical channels in one direction and logical channels in the other, as shown in Figure 8-5.

8.4.2. Physical Layer (PHY)

Table 8.1 summarizes some characteristics of the PHY for W-CDMA [6][3] [7][8].

The radio link control (RLC) provides services (reliability) similar to the RLC in 3GPP2, packet data convergence protocol (PDCP) provides transparency to higher layer protocols from the details of the underlying packet

[4] Both of these processes include signal strength measurement.

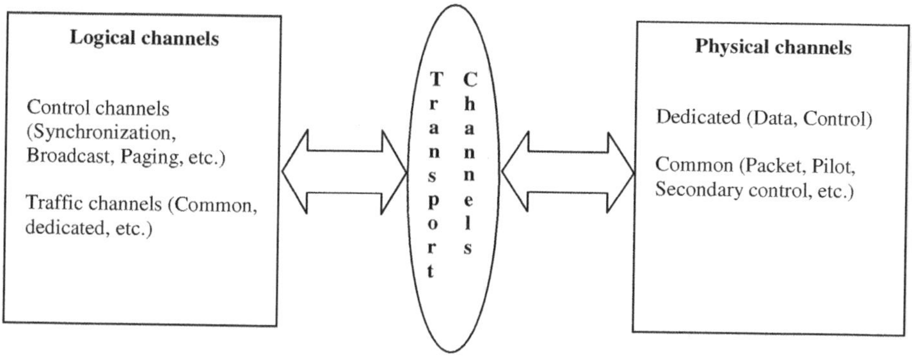

Figure 8-5. Transport channels map to PHY channels in one direction and Logical channels in the other.

switching architecture of the network, broadcast/multicast control protocol converts broadcast and multicast messages from RNC to adapt radio interface. The MAC sublayer provides a range of services through a number of functions. Table 8.2 summarizes the distinguishing features of these protocol layers [9].

8.5. UMTS SERVICES

Table 8.3 shows a list of some of UMTS services and their relation to the earlier generations.

Among the services listed in Table 8.3, we will expand the discussion about open service access (OSA), as it has the potential to create a new application and service development business environment, which could employ the experiences from Internet application development and grow independent of the cellular operators.

8.6. IMPROVEMENTS OVER RELEASE 99

Later versions of 3GPP mainly enhanced its open service 'architecture' (now called 'access') and packet-switched services. Release 4 added multimedia gateway (MGW) and IP multimedia subsystem (IMS). The MGW provided an interworking function, to utilize IP structure for circuit-switched calls, such as VoIP. The IMS made this possible. With Release 5, work on IMS advanced to stage-2. IMS is projected to pave the path to an all-IP 3GPP network. All elements of the core network participate in the IMS and its services architecture is based on the IETF-defined open session capability. Before Release 5, ATM was the only transport across various interfaces in the RAN. IP was

TABLE 8.1. Characteristics of W-CDMA

Attribute	Value	Remarks
Carrier bandwidth	5 MHz	5 MHz includes the guard band. However, it must be paired in FDD for a total of 10 MHz per full-duplex connection. TDD requires one 5 MHz carrier (unpaired).
TMD frame	10 ms	
Chip rate	3.84 Mcps	
User data rates	144 kbps (vehicular), 384 (suburban) 2 Mbps (Indoor)	HSDPA (Release 5) provides up to 10 Mbps.
Modulation scheme	QPSK spreading and BPSK (uplink) and QPSK (downlink) for data.	HSDPA uses 16 QAM in addition to QPSK. GPRS used GMASK and EDGDE used 8-PSK.
Spreading type	Channel: Variable length orthogonal spreading User data: Gold sequence	
Power control signal frequency	1600 Hz Range 80 dB uplink and 30 dB downlink one of the in four step sizes (0.5, 1, 1.5, 2 dB).	Up from 2 Hz in GSM.

Spreading factor	Many, from 4 up to 256 in uplink and up to 512 in downlink.	Actual vale depends on data rate (channel condition).
Time slot size	625 μs (15 slots per fframe)	Include control and packet channels for packet data.
Error control	Turbo convolutional	For BER of 10^{-3} one coding layer (inner) of convolutional type, and for BER of 10^{-6} coding layers (Inner and outer) of both types.
Handover types	Intra-frequency Inter-frequency Inter-system	Packet data can have soft or hard handover Hard handover Hard handover
Output power (Max)	33 dBm, 27 dBm, 24 dBm, 21 dBm	With about 4 dB tolerances.
Multiplexing	2 multiplexing scenarios.	Parallel services in individual packet data channels, as well as multiple services combines combined in a single channel.
Receiver sensitivity (RS)	−121 dBm (uplink) and −117 dBm in downlink (for error rate 10^{-3}).	RS is the minimum receivable signal level.

TABLE 8.2. Functions and Services by UTRAN Protocols above PHY

Protocols	Typical Functions	Typical Services
MAC	Transport format selection for transport channels, Priority handling of UE data, Identifying UE on common transport channels Multiplexing/Demultiplexing of upper layer PDUs to and from transport block sets traffic volume measurements, Transport channel type switching, Ciphering of transparent mode RLC Access Service Class selection for access and pilot channels.	Unacknowledged data transfer between peer MAC entities, Execute radio resource control (RRC) sublayer requests for resource re-allocation and reporting local measurements to RRC.
RLC	Segmentation and re-assembly (SAR), error control, sequence control, flow control, ciphering	SAR for user and control data using ACK mode, Unacknowledged mode and transparent mode, QoS maintenance to upper layer, error notification
PDCP	Header compression/decompression of IP stream, transfer of user data	Data exchange between PDCP users, PDCP sequence control
BMC	Message storage, traffic monitoring, scheduling transmission and delivery of cell broadcast message	Broadcast related services.

TABLE 8.3. Service Features of the UMTS Network

Service Name	Important network component	Purpose	Relation to previous generations
MMS (Multimedia messaging service)	MMS relay/server	Content-rich composite multimedia audio or email message with commonly used media formats and protocols, such as MIDI, JPEG, MPEG, GIF.	Introduced in Release 99
LCS (Location services)	Serving and gateway mobile location centers (SMLC/GMLC) and location measurement unit (LMU).	Providing location services in the user's current vicinity.	Adopted and enhanced from GSM version 98.
CAMEL3 (Customized application for mobile network enhanced logic—Phase 3)	Home environment (HE) in the home network and virtual HE (VHE) in the visited network.	Provide a support mechanism for designing services not part of the standard, for example, multiple subscriber profiles (MSP) allow a user to have more than one profile under one IMSI.	Earlier phases in GMS.
EMS (Enhanced message services)	SMS network with enhancements.	Enhance SMS to include pictures, animations and sounds in a message.	Enhancements from GSM short message service (SMS).
MExE (Mobile station execution environment)	Network nodes (internal and externals) and MS (mobile stations).	Execution environment on MS, for example, MS and network capabilities negotiation.	From GSM 98, enhanced for UMTS.
Multicall	Signaling additive	Simultaneous CS subscriptions, with possibility of multiple 64 bearer channels allocated.	UMTS original. Multiple CS data services can be added to voice call.

TABLE 8.3. *Continued*

Service Name	Important network component	Purpose	Relation to previous generations
OSA (Open service access)	OSA architecture including OSA server, OSI API (application programming interface) and home environment.	Allow third-party application development through open interfaces.	UMTS
Super charger	Signaling additive.	Allow retaining subscriber information in visited network and be used in case of revisiting.	Available for GSM and GPRS signaling infrastructures as well.
Follow me	Signaling additive	A range of follow-me services	Originally, an intelligent network service (from public land mobile network), UMTS has new dedicated specifications.
Interworking with packet data networks	A packet domain-specific DHCP relay agent in SGSN and a client PHCP in TE.	Bearer service for remote wireless access to LANs, ISP.	UMTS specific.
PIAFS (PHS Internet access forum specs)	Japanese personal handy phone MS (PHS-MS) and UMTS-TE.	Bearer service to provide interoperability with PHS phone.	UMTS specific
Universal IC card (UICC) and USIM related services.	TE, UICC, USIM and USIM application toolkit (USAT).	A range of services for protocol specification between terminal, network and UICC/USIM.	Enhanced from GSM subscriber identity module (SIM).

designated as the transport protocol in Release 5. This IP transport is not to be confused with the private IP networks that could be used as part of UMTS network in earlier releases. The IP transport in RAN is not valid on an end-to-end basis. The use of IPv4 is optional, but IPv6 is mandatory. Location service enhancements included optional SAS (stand-alone SMLC). SAS provides global positioning system (GPS) data to RNC to be delivered to UE. It can also act as an alternative to RNC for location calculations. Release 4 event-driven location services were further enhanced by adding the concept of periodicity to them. As mentioned in Table 8.1, a new feature HSDPA (high-speed data packet access) is also added in Release 5. It is based on downlink-shared channel (data only) and provides downlink rates of up to 10 Mbps. For a lower bit rate of uplink, HDSPA targets streaming, interactive, and background services. Just like 3GPP2 (cdma2000), it employs hybrid ARQ (HARQ) for additional error control. It also employs efficient modulation and coding schemes.

The most important development in the latest releases of 3GPP and 3GPP2 is a migration to an all-IP network. At the service level, the path to this migration is apparently via the IMS (Internet multimedia subsystem). Release 6 of the 3GPP is expected to have the next stage of IMS. We will devote much of the remaining chapter on IMS 2 as specified in [10]. Retaining the specification order, we will discuss some of the IMS system concepts, followed by example procedures.

8.7. IMS SYSTEM CONCEPTS

Figure 8-6 shows the IP multimedia architecture for Release 5 as per RFC3574 [11]. It consists of three parts: (1) IM CN (Internet multimedia core network), (2) IP-CAN (IP-connectivity access network), and (3) (IP-multimedia-enabled) terminals. The IMS is more of a complete service platform that includes IP (version 6), session protocols (SIP/SDP), authentication service (e.g., using DIAMETER,) and policy enforcement protocol (e.g., COPs). Here is a brief description of each of the three networking parts.

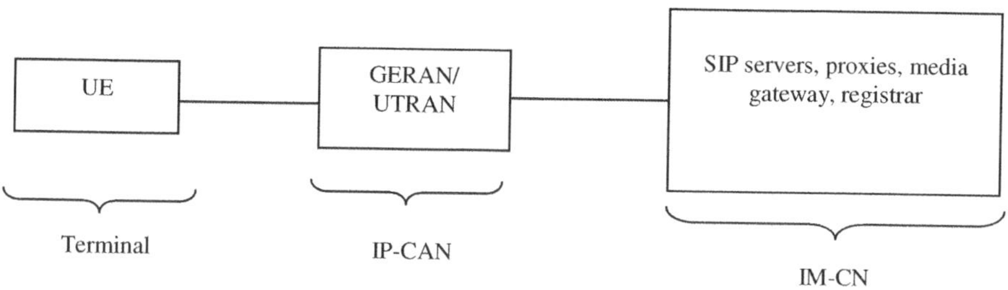

Figure 8-6. Three parts of IMS.

8.7.1. Internet Multimedia Core Network (IM-CN)

All the core network elements that participate in IP multimedia services form what is called the *IP multimedia core network* (IM CN). The multimedia services are based on session initiation protocol (SIP) signaling for session set up, mobility management and session termination as discussed in Chapter 6. It is this part of the 3G systems that has closed the gap between the IETF's open standards policy and cellular operators' rather business-oriented attitude. The 3GPP decided not to "cellularize" the IETF protocols and use them as defined and specified by IETF.[5]

8.7.2. IP Connectivity Access Network (IP-CAN)

The access network that provides connectivity to IM-CN is IP-CAN. It could be GERAN, UTRAN, or both with the GPRS core network. Due to the peer-to-peer nature of IMS services, IPv6 is the only real option in IP-CAN.

8.7.3. Terminals

The terminals in IMS are capable of initiating and responding to Internet session control signaling. When a terminal moves out of the home network, IP-CAN hides mobility and maintains the control and user bearer connections.

8.8. SESSION LAYER ARCHITECTURE

The IMS adds a session layer with an architecture, shown in Figure 8-7 [12]. The IMS network itself does not provide application services; it provides several toolkits to design services [13].

In Release 6, the IMS is projected to harmonize the main three wireless access networks, W-CDMA, cdma2000, and wireless LANs (e.g., IEEE 802.11). The call session control functions (CSCFs) perform the call control functions for the subscriber and callers. There are three types of CSCF, differentiated below.

8.8.1. Interrogation CSCF (I-CSCF)

The I-CSCF handles incoming calls to a home or visiting subscriber. It may direct the incoming call to another server.

[5] This, by itself, is a milestone achieved, which has opened a new world of services and protocols for future cellular networks. In the author's view, this attitude of cellular companies might help both sides, Internet and cellular networks. The Internet is overwhelmed by the abuse of "openness", while the cellular networks are plagued by interoperability issues.

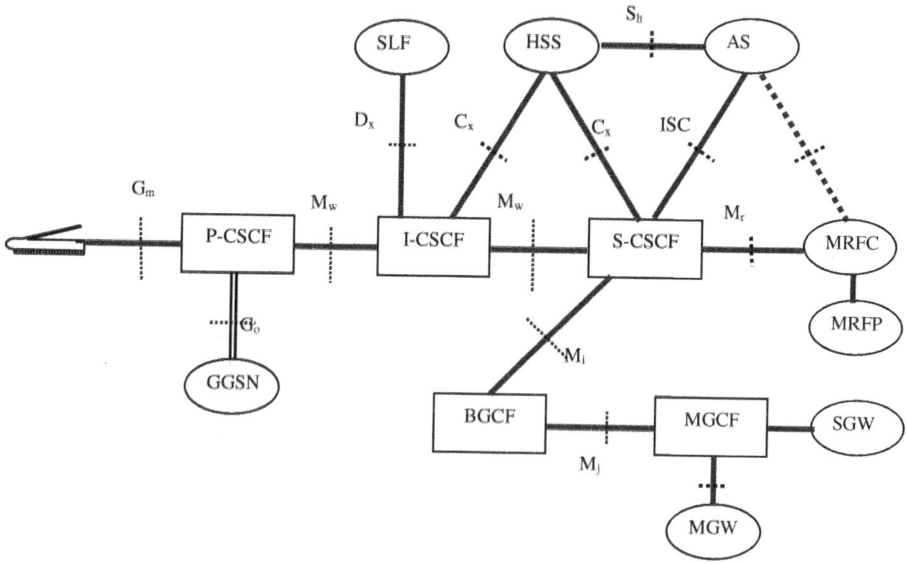

Figure 8-7. Session level architecture [12]. G_m, M_w M_r, ISC, M_i and M_j are across SIP interfaces.

8.8.2. Proxy CSCF (P-CSCF)

The P-CSCF handles all calls initiated within a home network. When a UE has to make a call or invoke a service, it first connects with P-CSCF. The PDF functionality (separate unit in 3GPP2) is contained within P-CSCF in 3GPP.

8.8.3. Server CSCF (S-CSCF)

The S-CSCF performs the subscription session layer signaling functions services for the subscriber, whether the subscriber is at home or in a foreign network.

8.8.4. Home Subscriber Server (HSS)

The HSS performs the database services (e.g., authentication, registration, location) and provides interfaces to the CSCF servers. It is evolved from HLR.

8.8.5. Media Gateways and Associated Control Functions (MGW, MGCF, SGW, BGCF)

These gateways and associated functions provide interworking between IP and their respective bearer channels, for example, PSTN, ISDN, PLMN (public land mobile network).

8.8.6. Media Resource Functions (MRF)

The control (MRFC) and processing (MRFP) entities of MRF provide media stream processing capability. This includes, among others, mixing (e.g., a multimedia conference), announcements (incoming), and media analysis.

The service architecture shown in Figure 8-7 is also what is called *home network*. Home network services can be made available on the network side by constituting a *virtual home environment* (VHE). By emulating a VHE, an operator can provide a subscriber the same personalized services as in a home network, regardless of user's actual location or current equipment.[6] Also, due to the open nature of SIP, more services can be built by third parties without the knowledge of the underlying network. Several toolkits can be employed for new service creation, for example, CAMEL[7], MExE, and open service access (OSA). The OSA harmonizes the service creation among the Internet, 3GPP and 3GPP2 [14].

8.9. OPEN SERVICE ACCESS (OSA)

OSA [15][16] is the third-party service development environment for the latest releases of the 3G networks. It is an application programming interface (API) that terminates at a server called OSA-SCS (OSA service capability server).[8] The capability server provides the OSA interface to the underlying transport network. Thus, all types of services can be integrated into a three-layer architecture [18], as shown in Figure 8-8.

8.9.1. OSA Interfaces

There are several interface classes recognized, three of which are specified in [19]. These are:

1. The interface class, which provides the basic mechanisms to applications so that they can invoke service capabilities. An example of such mechanism is authentication. After being authenticated, an application can access OSA capability servers.
2. The interfaces class, which exists between an authenticated application and particular service capability feature of the OSA architecture. A client might need a particular service feature, for which this interface provides an open mechanism.

[6] As long as the equipment has the service capability. For example, for streaming music, the equipment must have a music player in it.
[7] Not available in 3GPP2.
[8] Specified in OMG IDL as in [17].

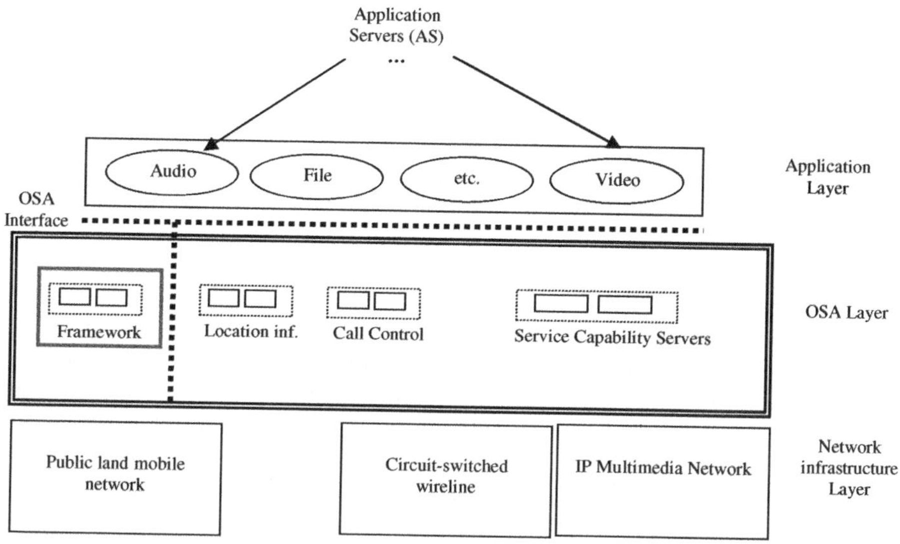

Figure 8-8. OSA adds a service layer upon which applications can be provided by application servers.

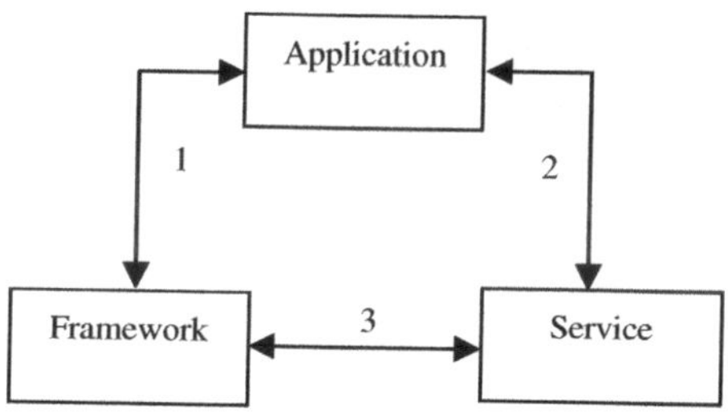

Figure 8-9. OSA-API interface classes.

3. This class of interfaces exists between the service capability function (SCF) and framework (FW) server in order to facilitate a multi-vendor environment. Figure 8-9 shows relation between these interface classes and applications, framework, and service capability functions.

A number of interfaces are defined within these classes. These include the callback interfaces *IpApp* on the application side, *IpClient/IpSvc* interfaces, the client/server interfaces of service capability function (SCF) used by framework (FW), and the *IpFw* interface of the framework (FW) used by service capability functions (SCF).[9] The interfaces could be either dedicated

[9] The callback interfaces are used by service function server to report events.

to a particular session, and hence don't need a session ID, or could be an instance of an interface serving multiple sessions.

8.9.2. OSA Functions

As summarized in [14] the functions provided by OSA can be related to framework, network, and user data.[10] Following is a set of examples in each category.

8.9.2.1. Framework (FW) Functions of OSA. These functions include security tasks of authentication, authorization, capability discovery, integrity management, and service agreement establishment. Multiple authentication mechanisms are supported, including digital signature for non-repudiation. Authorization mechanisms relate to the user getting access privileges to application and the application getting authorization to access service capability functions (SCF). Home environment ascertains eligibility of both before clearing the way to access. In a service discovery phase, an application can request a list of capabilities of SCFs. This is possible because the SCF has to register itself with the framework and let its capabilities be known to FW. Once a service agreement between the application and the home network is agreed, the application signs the online part of the agreement.

8.9.2.2. Network Function of OSA. These functions relate to call and session management, IP multimedia handling, notification, charging, and e-commerce. These call and session functions support circuit-switched calls and data sessions for data collection, redirection capability, and QoS monitoring. IP multimedia handling functions include conference call control and management. Media channel management using RSVP and conference management using part add/delete functions provide resource management. Notification functions can be carried by SMS or WAP-Push for the network application to the terminal applications. E-commerce related functions are provided, to be used in charging and billing. Thus a third party does not have to know the underlying network in order to have the service used and charged. The network functions of the OSA-API cover the whole spectrum.

8.9.2.3. User Data Related Functions of OSA. The complexity of IMS is enhanced over traditional cellular (now including UMTS!) because of the need of lot more user attributes needed for connection authentication, session level mobility management (keeping privacy), and various security threats added due to multimedia (e.g., credit card billing for e-commerce). The home network and virtual home environment manage user information to be used in a personal service environment, regardless of the location. The user data functions provided by OSA deal with issues of managing information about

[10] We follow this presentation on function description.

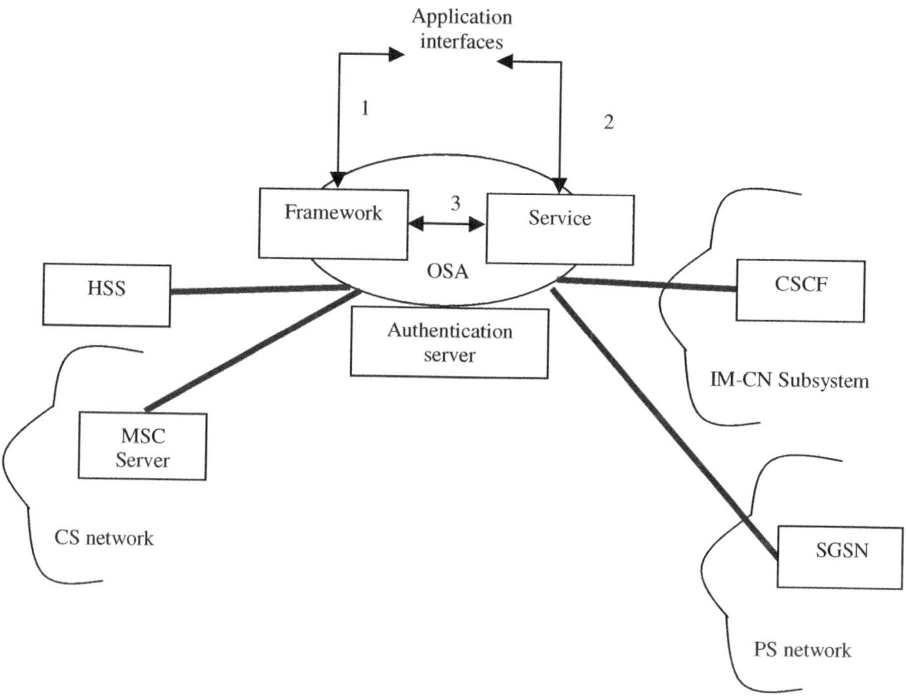

Figure 8-10. OSA service subsystem location in the overall network.

user and subscription profiles. These functions relate to user status (e.g., accessibility, type of terminal used), location (e.g., whether home or visiting and where), user profile (profile attributes for business, home, etc.), terminal capabilities (audio, video, etc.), and the capabilities of the current network if the user is roaming.

By combining the interfaces and functions, the OSA-API forms the service subsystem of the architecture, as shown in Figure 8-10.

8.10. PARLAY

The terms Parlay and OSA-API are sometimes used alternatively and may cause confusion. The Parlay Group (www.parlay.org) is a consortium of companies (from Telecommunications and IT) that has developed the API "Parlay" in order to deliver the services specified by 3GPP OSA-API. Parlay Group and 3GPP have closely cooperated (through a Joint Working Group—JWG) in developing these APIs. Parlay [20] has been adopted for OSA.

8.10.1. Parlay Background. With the move toward all-IP network, it has been a natural trend to base the future networks on services instead of technology-based applications. The cellular companies have traditionally based network design on applications. This has brought up the "killer-application" question for every new technology. With all the glamour that open standards have brought to the Internet, it is obvious that 'open-ness' rids the operator (at least partially) from the responsibility of developing the killer application(s). Before the specification of OSA, cellular operators relied only on protocols. With OSA, a set of open API specifications is now available for third-party vendors to develop applications independent of the underlying network transport. The independence is further achieved in Parlay with the notion of a Parlay/OSA Gateway that is part of Parlay Framework. The Parlay Gateway [21] provides framework interfaces and service interfaces, which is all that a developer needs to know. This has the added benefit that the application developers do not have to worry about security-related issues; it is taken care by the operator at the transport level and HSS at the session level. The list of capabilities offered by the Parlay APIs is long and is quickly being adopted by industry. The Parlay version 4.1 has APIs for all the call management and control functions of OSA. It also has Policy Management and Presence and Availability Management functions. Parlay X is a web services initiative for next-generation network applications.

8.11. IPv4/IPv6 SCENARIOS TOWARDS ALL-IP INFRASTRUCTURE

The introduction of IMS and IMS2 has paved the way for independent IP application development for a given set of session-level signaling services and APIs. Multimedia Gateway (MGW) provides an IP interface for the circuit-switched applications, such as voice service. Session-level mobility management using SIP is also promising for macro-mobility and hiding user presence. This has pretty much released the circuit-switched part of the network from being deployed in future generations, except, maybe, for interoperability reasons (See Figure 8-1.) RFC 3574 discusses some of the possible scenarios for 3GPP's move toward an all-IP network. These scenarios relate to the combinations of IP versions 4 and 6. The two scenario categories of this RFC (GPRS and IMS) will be discussed in the following. Readers are referred to the references and a big stock of documentations and publications available on the Internet using an appropriate search engine.

8.11.1. GPRS Scenarios

Table 8.4 describes various scenarios when a GPRS UE tries to reach Internet service outside of the GPRS network.

8.11.2. IMS Scenarios

One difference between the IMS and GPRS networks is that IMS does not support IPv4. Therefore, an IMS UE is always IPv6. This leaves few

TABLE 8.4. GPRS Scenarios

Scenario No.	UE	Network	Called Terminal	Solution	Comment
1.	IPv4 and IPv6	IPv4 and IPv6	IPv4 and IPv6	Simultaneous IPv4 and IPv6 PDP contexts in GGSN	IPv4 address space shortage may require efficient usage.
2.	IPv6	IPv4	IPv6	IPv6 terminals and Routers allow IPv4 datagrams.	Most likely scenario in the beginning.
3.	IPv4	IPv6	IPv4	Need IPv4 capable GGSN	
4.	IPv6	IPv4 and IPv6	IPv4	IPv6/IPv4 conversion to be done at some point in the fixed or GPRS network.	
5.	IPv4	IPv4 and IPv6	IPv6	Same as above.	

possible scenarios the RFC considers when the called UE is in IPv4 and IPv6+IMS.

If the UE needs to connect with an IPv4 terminal, an IPv6/IPv4 conversion is to be done at a router (or another device) after exiting IMS network when a datagram is going from UE to IPv4 terminal. In the reverse direction, since IMS accepts only IPv6 datagrams, either a conversion must be done before delivering IP packet to IMS or the function should be added at the border device of IMS. The same solution may also be applied for the case when two IMS UEs need to connect through an IPv4 network. Either the IPv4 border gateway routers or the IMS 'gateway router do the conversion.[11]

8.12. 3GPP RELEASE 6 OBJECTIVES

The Release 6 of the 3GPP network is expected to meet several objectives. These are [22]:

1. Phase 2 of IMS with IMS messaging and group management;
2. Wireless LAN interworking for access;
3. Distributed speech recognition (DSR) and other speech-enabled services. These services will increase service accessibility and could result in 'hands-free' multimedia calls;
4. Number portability. Ultimately the same number should be able to move with the subscriber.

There are a number of efforts to align 3GPP and 3GPP2 services and architecture. Harmonization efforts have led to a number of new developments, such as the acceptance of AKA in 3GPP2 security architecture for encryption key distribution, harmonization reference model (HRM), and adoption of OSA by 3GPP2 (IMS and MMD almost aligned as for as service offering is concerned).

8.13. SUMMARY

The W-CDMA air interface, that is part of the IMT-2000 UMTS standard, has paved the way for spread spectrum technologies over the GSM-based network infrastructures, such as GSM, GPRS, and EGPRS (using EDGE). The drive toward an all-IP network architecture has been greatly spurred by the SIP signaling and OSA-based third-party application development. Contrary to previous (and present) practices, the W-CDMA-based cellular network will provide an application development infrastructure (as opposed to delivering specific applications). Harmonization efforts between the cdma2000-based

[11] We use this term in the generic sense, not claiming that such a device is defined.

networks (3GPP2) and W-CDMA-based networks (3GPP) have led to many interoperability successes, such as interoperable IP multimedia structures, encryption standards, and even the emergence of a harmonized reference model.

REFERENCES

[1] Yi-Bing Lin and Chlamtac Imrich, *Wireless and Mobile Network Architecture*, John Wiley and Sons, New York, 2001.

[2] 3GPP, "Third Generation Partnership Project (3GPP): Partnership Project description", *Copenhagen Meeting Presentation*, December 1998.

[3] 3GPP, "Overview of 3GPP release 99: Summary of all Release 99 Features, *ETSI Mobile Competence Centre TSG#23(04)0200.* March 2004.

[4] 3GPP, "Overview of 3GPP Release 5: Summary of all Release 5 Features"*ETSI' Mobile Competence Centre TP-030152*, June 2003.

[5] Tektronix, "UMTS Protocols and Protocol Testing: Primer", www.Tektronix.com/commtest

[6] Seimens, "3G Wireless Standards for Cellular Mobile Services: The Siemens View".

[7] Tero Ojanpera and Prasad, Ramjee, "An Overview of Air Interface Multiple Access for IMT-2000/UMTS", *IEEE Communications Magazine*, September 1998. pp. 82–95.

[8] Ramjee Prasad, *Universal Wireless Personal Communications*, Artech Publishers, Boston, London, 1998.

[9] HUT communications laboratory, "UTRAN Radio Interface protocols", www.comlab.hut.fi/opetus/238/lecture7_RadioInterfaceProtocols.pdf

[10] ARIB/3GPP TSG Services and Systems Aspects, "IP Multimedia Subsystem (IMS); Stage 2 (Release 6)", *3GPP TS 23.228 v6.4.1*, January 2004.

[11] Alain Durand, El-Malki Karim, Murphy Niall Richard, Shieh Hugh, Soininen Jonne, Soliman Hesham, Wasserman Margaret, and Wiljakka Juha, "Transition Scenarios for 3GPP Networks", *IETF RFC 3574*, August 2003.

[12] Mikko Puuskari, "Development of IP Multimedia services and Architecture standards for 3G Networks", *3GPP TSG-SA WG2 (System Architectures) Presentation*, September 2002.

[13] Stephen Haynes, "IP Based Multimedia Services Platform", *ITU-T IMT-2000 and Beyond,* May 2002, Ottawa, Canada.

[14] Jrog Swetina, "Virtual Home Environment) VHE) and Open Services Architecture (OSA)—an overview", *3GPP TSG-T2 #10 (presentation)*, August–September 2000.

[15] 3GPP, "Service Requirements for the Open Services Access", *3GPP TS 22.127.*

[16] 3GPP, "Virtual Home Environment / Open Services Access", *3GPP TS 23.127.*

[17] Maarten Wegdam, Plas Dirk-Jaap, and Unmehopa, Musa, "Validation of the Open Service Access API of UMTS Application Provisioning" PROMS 2001, LNCS 2213, pp. 210–221, 2001, available from www.item.ntnu.no/~thanhvan/doc/ValidationOfOSA.pdf

[18] Lucas Lokstermann, "PARLAY and the 3GPP Open service architecture: TINA ideas and principles, *Presentation: 3GPP-CN5.*

[19] ETSI,"Universal Mobile Telecommunications System (UMTS); Open Service Access (OSA), Application Programming Interface (API); Part 1: Overview" (3GPP TS 29.198-01 version 5.4.0 Release 5), *ETSI TS 129 198-1 V5.4.0,* December 2003.

[20] Ard-Jan Moerdijk and Klostermann Lucas, "Opening the networks with PARLAY/OSA standard and aspects behind the APIs", www.parlay.org/specs/library/Opening_the_networks_with_Parlay_OSA_v3.4.pdf

[21] Zygmunt Lozinsky, "Parlay/OSA—a New Way to Create Wireless Services", *white paper, available from* www.parlay.org/docs/2003_06_01_Parlay_for_IEC_Wireless.pdf

[22] Brough Turner and Orange Marc, "3G tutorial", *NMS Communications*, available from www.nmscommunications.com/Solutions/3GTutorial.html

[23] Rajeswari Harikrishnan and Chapanand Anurat, "The 3GPP and 3GPP2 movements towards an all IP mobile network", May 2002.

CHAPTER 9

SECURITY IN WIRELESS DATA NETWORKS

Ever since the first Internet worm in 1988, work on security of data networks has been following the trail of their success. Much of this work has been advanced in response to security breaches, but a reasonable amount has also progressed in testing various security systems and protocols. A prominent fraction of the work in security has resulted not in adding safety to network data, but in exposing its vulnerabilities (see Ref. [33] for a taxonomy of vulnerabilities). In any system or network, such as the Internet, with millions of parties participating due to a variety of reasons, it sounds unreasonable to make security as a thing secondary to the network design and deployment. Such, however, has been the case, thanks largely to unpredictable scale of the Internet and computer proliferations.

Noticeably, after several years of the first successful attack, followed by numerous attacks of an increasing number of types, networks are still designed with security considered as a bystander instead of it being a part of the network architecture. This is like a town planning with houses that have no outside walls, no or weak roofs, no or easily breakable doors, none of the locks, and no police in the area. Perhaps no government would approve such town planning; networking is a different story[1]. The Internet commerce boom has taken place with a large fraction of global economy being vulnerable to hackers and attackers, but has not resulted in the design of secure networks.

[1] Some recent standards do have security layer or sublayer, e.g. IEEE 802.16 standard for broadband wireless access.

Wireless and Mobile Data Networks, by Aftab Ahmad

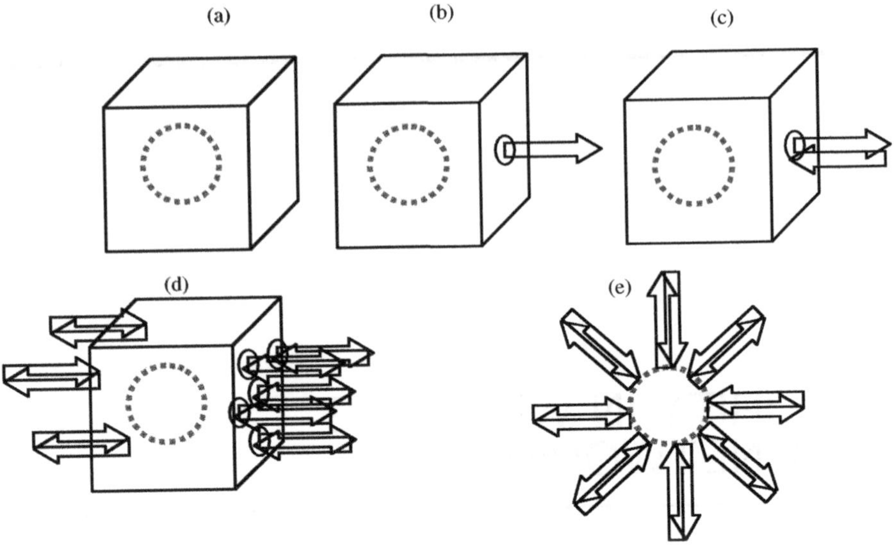

Figure 9-1. Various conceptual levels of security of a data system.

In fact, according to several surveys, the wireless local areas network standard IEEE 802.11 has mostly been deployed without any concern with security.[2]

In this chapter, we will have a look at the technology regarding the security of wireless networks. Our treatment of the topic is technical, however, our approach is more systematic than generally found on this topic. We will avoid much of the common knowledge, usually found easily, about what constitutes vulnerability. Instead, we will use a holistic approach to understanding the security issue and a network-wide treatment. Following this, we will present a security requirement definition along the lines of network architecture, taking the example of the Open System Interconnection reference model (OSI-RM). Following this discussion on *security architecture plane* will be a technical summary of some salient components of security systems available today. We will end the chapter with security measures available for WLANs and cellular networks.

9.1. ASCRIBING SECURITY TO A NETWORK

A network, being an interconnection of communications devices, must allow these devices to send and/or receive data in a form that is secure and understandable by all potential addressees of data. Data storage or processing device that is 100% secure is like a box with no opening for data in or out of the device, as shown in Figure 9-1(a). The data in the box are shown as dotted

[2] The deployment largely took place during or as a result of the e-commerce boom.

circles. There is no way to access it after it is sealed in the box. Figure 9-1(b) shows next lower level of security, in which a 'hole' is made into the box. Now data can either leave the box through this hole or someone can access the data. This hole is like a communication channel that opens data to vulnerabilities. If this channel allows only data out, such as a network management system only reporting network conditions with no way of sending any input, there is the risk of the data getting into the wrong hands. If it allows only data in, there is a chance of the data getting manipulated in a malicious way, such as a virus attack. In other words, having one access channel, even in one way, could open information to security risks. A data system with a two-way channel, such as in Figure 9-1(c), is open to a number of different attacks. A computer on a network is more like Figure 9-1(d), that is, it has a large number of 'holes' in it. Worse yet, a wireless system is like the one shown in Figure 9-1(e), that is, it is, in general, not contained in a secure box, leaving an indefinite number of 'holes' around it.

9.1.1. Why Are Wireless Network Devices a Bigger Challenge?

Figure 9-1 explains in a conceptual way why wireless data networks are more of a challenge to secure than the fixed, wired network devices. If all communications occurs through only one 'hole', such as in Figure 9-1(b) and (c), the security problem can be modeled as having a door with lock on these holes that is opened by certain keys. The key copies are distributed among only those allowed to access the data. A box with n doors, however, needing an equal number of keys, will have a possible $n!$ ways of having one or more compromised entries to such doors. If the probability of damage through one door is p, the probability increases linearly with the number of doors. So, for 100 doors, the probability is 100-fold. The wireless device in a wireless network does not have the box with walls; it is simply open. In other words, the wireless network devices present a two-fold security problem, first, to secure a box and then to secure the doors on this box. In the United States, the government standard FIPS 140-2 defines the minimum requirements for non-classified, but sensitive data. Wireless network devices, by virtue of their exposition, are considered of this type.

9.2. SECURITY NETWORK ARCHITECTURE

A secure system is supposed to provide *access control* (only authentic users or equipment are allowed to use the system), *confidentiality against eavesdropping* (data are accessible to *only* authorized recipients), *integrity* (data are not corrupted by intruder or attacker, or if data are manipulated in any way, the manipulation is detected), protection against *protocol spoofing* (changing the address of data source protocol or sender). Once a connection or device is compromised, there are many ways in which the unprotected data or com-

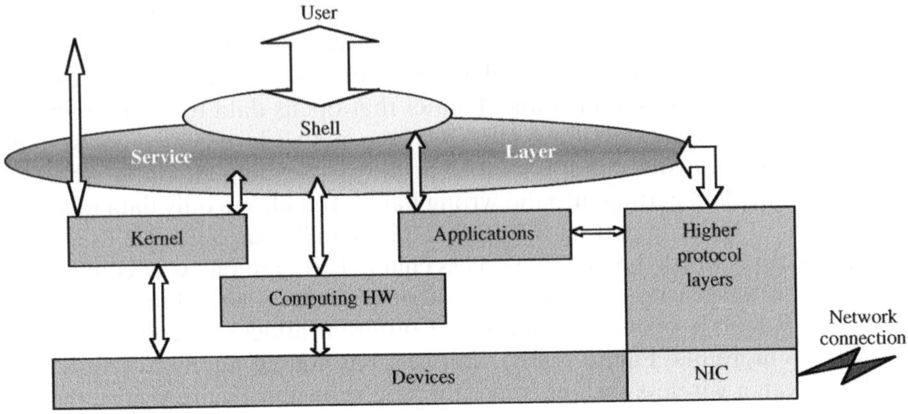

Figure 9-2. Communications systems are typically part of OS.

munications can be put to work. In addition to simple eavesdropping, a more malicious attack is *man in the middle*, in which the attacker is sending information to both the communicating users. The unsuspecting users interpret the information as coming from each other. Protection is also needed against breaking of security system, such as *replays* (gathering the data during non-secure phases and using it to become the authenticated user).

Networks and network devices have grown substantially in complexity in recent years, owing to a multitude of reasons, from advancements in chip technology to e-commerce. A typical computing device, right out of the box, consists of not just an operating system (OS), but also a communications system, which in all possibilities could be more complex than the operating system software. In fact, operating systems today are *network operating systems*, and are typically designed to work with TCP/IP and Internet applications as part of the shipped software. Conceptually, the OS resides at the top of application programs, thus providing the user access to application software (Figure 9-2).[3]

There are a minimum of two ways to the inside of a computer, namely, via the operating shell and through the network connection. The security concerns from the attacks via the OS are typically related with physically accessing the system or using a program with a virus that uses an OS feature to corrupt the data stored in computing hardware (memory and secondary storage). Of course, a virus is dangerous because it does not have a known effect, and could do anything from corrupting existing files, to filling secondary storage, to *masquerading* the processing resources. Access through the operating system is usually controllable due to its visible nature. However, attacks from the net-

[3] Admittedly, OS does a lot more than this, as should be known to anyone reading this book. Of special mention is the OS kernel, which controls devices through a layer of programs called *device drivers*. One of these devices is typically the network interface card, wired or wireless.

working connection are especially dangerous due to many reasons listed below.

1. The apparent reason that the attacker is not visible.
2. The attacker does not have to leave a trace, as in a virus attack.
3. Once compromised, a connection can become a hotbed of malicious activity.
4. The attacker, it is seen, does not stop but could easily make a compromised system a source of similar activities.
5. It is common sense to assume that the attacker who comes in from the network side is more knowledgeable about the working of the communications system than an unauthorized user who attacks from the OS by guessing a password. The 'knowledge' in this case can be a lethal weapon.

9.2.1. Securing a Standalone Device

From the above discussion, it is easy to conclude that a secure operating system and authenticated access mechanisms are pre-requisites to stopping an unwanted attack on a device that is not networked. This is akin to saying that there is only one 'hole' that requires a key to the operating system to render it secure.

9.2.2. Securing a Networked Device

Figure 9-2 shows a single network connection. Modern networks have layers of protocols with typically peer-to-peer logical connections, as shown in Figure 9-3.

A logical connection obeys a set of protocols that perform specific functions on data. The figure shows five layers above the network interface card (NIC—which may consist of a physical as well as a logical connection itself). For example, in an Internet connection with WLAN NIC, the NIC implements the physical and medium access control layers. Above it could be the logical link control (LLC), IP, TCP/UDP and application layer. Another thing shown

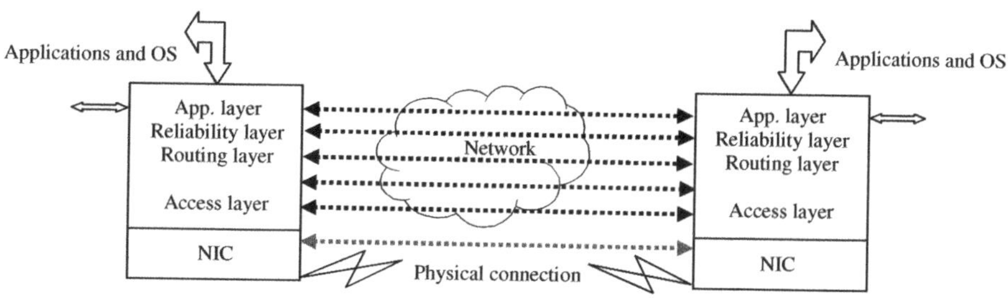

Figure 9-3. A network connection consists of multiple logical connections.

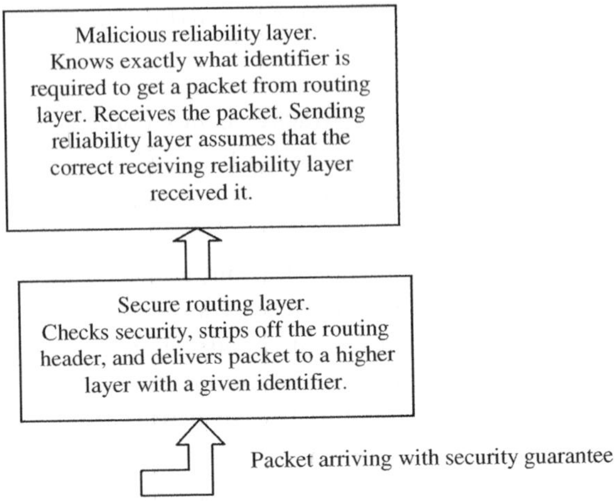

Figure 9-4. Security at one layer doesn't guarantee secure communications.

in Figure 9-3 is the network 'cloud' that covers only the *upper layers*. This is only symbolic, to reflect the characteristics of a network such as defined by the OSI-RM with upper layers that are *end-to-end* as compared with the *lower* layers that are *direct point to point.*

Figure 9-3 is a typical representation of a device carrying or receiving data across a network, be it wired or wireless, Internet or another wide area network. Each of the logical and physical connections could result in creating a security hole in the device. In fact, in most of network application programs, more than one application layer connections are open simultaneously, resulting in a lot more holes than the number of layers. Security has to be provided at each of these holes. Consider the following scenario: If a secure operation at the routing layer is guaranteed, it can only apply to routing. When the routing layer delivers this packet data unit (PDU) to the reliability layer, it is not certain that the receiving reliability layer is the right one or a forged one. A security mechanism between the peer reliability layers is needed to ascertain the authenticity of each, as shown in Figure 9-4.

9.2.3. Securing a Wireless Networked Device

In the open system interconnection reference model (OSI-RM), layers communicate only with their peers through protocol data units (PDUs) and with adjacent layers through primitives. Using the primitives between adjacent layers, a user data block moves from the application programs through the 6 layers in the sending computer to the physical (7th, but layer number 1) layer, from which it is transmitted to the next recipient in the network. Therefore, all layers' logical connections are bundled within the physical layer transmission. This is indicated in Figure 9-5. In a wired network the number of recipi-

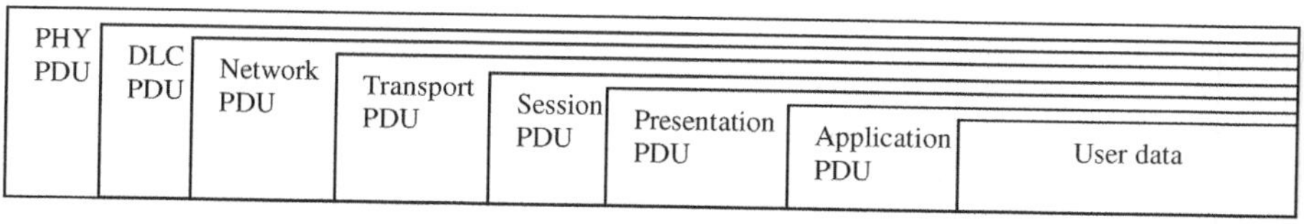

Figure 9-5. Physical layer PDU travels bit by bit, and contains all other layers PDUs in it.

ents of the physical layer signal is less than or equal to the number of devices connected with the physical cable. Therefore, if there are 10 devices and a 7-layer protocol is used, then there are a total of potential 70 security holes, assuming one hole per layer. It is customary that the devices connected to a network cable are trusted and this can be done by having a network administrator supervising all connections. Such is not the case with a wireless network. The wireless signal travels through the air and any device equipped with the required RF equipment can receive the signal, take it somewhere, and analyze it for misuse.

This is equivalent to having an indefinite number of security holes, and a secure device on a wireless network must have a way of closing all these doors to intruders. Securing a wireless network device will, therefore, require either containing the signal (which will bring it in the vicinity of a wired network device, in terms of security, even though such is not quite the case, due to the possibility of hidden devices use) or rendering the mere signal reception useless.

The above discussion brings attention to the need of a protocol plane for security along with the user data, control and management planes of protocol architectures.[4] Two examples of such planes are shown in Figures 9-6 and 9-7, for IEEE 802.11 WLAN-like architectures and for Internet-like architectures.

Since IEEE 802.11 standards can be used for dual purpose, as standalone networks and as access networks, Figure 9-6 can effectively be placed right under Figure 9-7 to replace the lower layers (and redrawing it appropriately).

9.3. SECURE OPERATING SYSTEM (SOS)

In the previous section, we considered the security requirement from a network architecture point of view. Let us now look at a possible requirement of securing an operating system. An operating system consists of processes that are owned by other processes (users being among the type of owner processes). Various user levels (e.g., from guest to root) of an OS are distin-

[4] Welcome to the world of reality. One of the main reasons why we don't have open secure networks is that security can be employed to nefarious communications just as conveniently as to any useful communications.

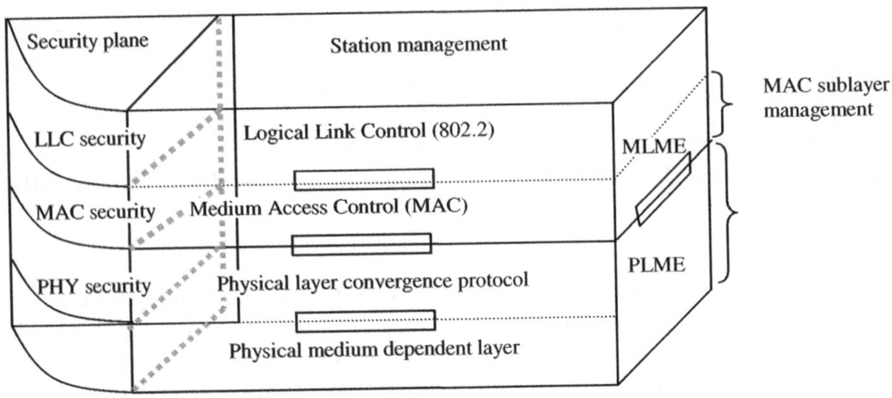

Figure 9-6. Conceptual protocol architecture for *secure* IEEE 802.11WLAN-like standard.

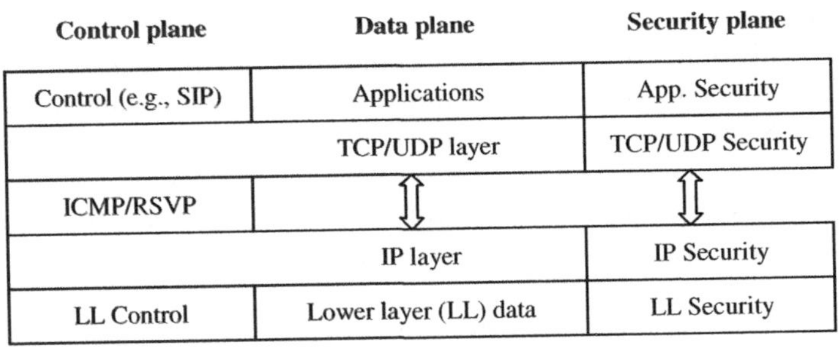

Figure 9-7. Conceptual protocol architecture for *secure* Internet.

guished from one another in terms of which processes they can access and to what extent: that is, use, modify, or remove. Usually, at the highest level of authority, system, root, or administrator level, all processes can be owned to all extent. However, a problem with OSs is that the ownership may be only *skin-deep*, meaning, by changing an ownership variable that the process may check only once anyone can use the process. Is it possible to have the ownership concept built into the fiber of the process? We don't know the answer to that, but the following scenario goes deeper than 'one-stop-shop' risk from an attacker who gets hold of an important password.

We do this by classifying processes according to an attribute that is called *damage*. Let's consider the couplet (n,d), where n is called *influence*, and defined *as the number of processes directly influenced by a reference process*, and d is called *damage*. Of the n processes that this process could influence, let the ith process have the influence value of n_i. Then d, the damage, can be arbitrarily defined as the number of total processes that will be ultimately influenced by this reference process. Remember, if we can list all the processes with their d value, we do not need n to characterize the process in terms of security risk. If we label all the processes of an operating system with the d value, we can conceptually organize them with ascending values of d. Then we can classify these processes into two or more classes, depending on the ranges of d, the damage. As a last step, we set up a password requirement for each next class. In this way, if someone breaks into an administrator's first password, they don't have automatic control of the whole system, but only until the first d-class. To go to the next, another password must be obtained, and so on.

In the above paragraph, we have chosen the definition of *damage* as the number of potential affected processes. The exact objectivity of this can't be ascertained, as we do not know of a scale for damage measurement. However, provided that such a classification can be effected, the resulting operating system will be quite secure as compared with the one that requires one or more passwords used to access everything all the time. This is because the proposed classification can have some potential benefits, such as, if someone breaks part of the hierarchy, but gets stuck at some point, a system log or emergency warning system can be used by the OS to record attacks. Another potential benefit is that the extent of attacker's approach will be known, and a compromised system will not need to be cleaned up entirely. Yet a third benefit is that when an attacker gets hold of a password at a level deeper than the first, it can't be used without having the password for the levels before, and instead can warn of an attacker *a priori*.

9.4. COMPONENTS OF SECURITY SYSTEMS

A network security system is designed for particular functions, or to secure the network or its component against specific threats. In addition to combating the vulnerabilities of the network device, the security system also has to stop intrusion to itself (mainly measured in the form of performance metrics). A secure system consists of protocols, servers, algorithms, keys, certificates, and performance attributes. On a system level, we expect the security systems to provide authentication, authorization, or registration, confidentiality, integrity and resilience measures—the last capability generally not understood as well as others in currently available systems. Another concept, not fully understood yet, is that of self-non-repudiation of a transaction. Non-repudiation is the incapability of an authenticated user/device to back-out from a transaction. This interesting requirement is made controversial by the fact that all that the authentication process proves is the processing of identifying information, and not the presence of a person, such as in a signed document. All these requirements are fulfilled by various components of a network security system. We will discuss a few of these components in this section. We will look at these components of a security architecture, without a particular order.

9.4.1. Protocols

For secure communications between a device on a network and any other device, there has to be a way of exchanging information relating to security. This information, by itself, may not be secure, in general, because if it was secure, there would be no need to add a security mechanism to it. Therefore, the best security protocols have to be shareable and known and yet must provide defense against attacks. A number of protocols have been available in open and proprietary form and they can be implemented in any device on the network. A protocol may include several protocols as part of it, such as authentication, registration, and reconditioning of data. Here's a list of some commonly mentioned protocols and types of protocols. We will come back to some of these at a later stage.

9.4.1.1. Authentication. Authentication protocols are usually part of other protocols. Their purpose is to make sure that a device or user trying to use the network is a legitimate user or device. In wireless networks it is also necessary to authenticate the authenticator, with what is called *mutual* authentication.

9.4.1.2. Association/Registration. Association is the term used in infrastructure type of networks, in which all user devices must communicate through some facilitating device, such as a switch, base station, or an access point. If a network under a common administration consists of more than one access points, a user may be required to go through a particular access point. In such networks, a user device, when switched on, looks for an access point from which it can receive the clearest signal. A registration or association protocol is used for this purpose. Association protocols result in a series of checking and registering information.

9.4.1.3. Re-association/Visitor Registration. If a user or device associated with one central point moves closer to another central point, a re-association process can save the exchange of all the association information again and again. A re-association can consist of exchanging the established association information and confirmation of this association. In other words, re-associations are like registering as a visitor. In a wireless LAN there is usually no home access point, therefore the associations and re-associations are relative to first time the network device connects to the network. A protocol for de-association can help access points de-register those devices that request so, or are no longer served by any access point.

9.4.1.4. Wireline Equivalence Privacy (WEP). WEP is perhaps the most commonly mentioned protocol in relation to wireless LANs. It was designed to make devices (IEEE 802.11 and its extensions) at least as secure as a wired network.[5]

[5] In our view, wireline equivalence privacy is nothing more than a catch phrase. We have not seen an analysis that proves that the probability of an insecure wireline device being compromised is

9.4.1.5. IPsec. IPsec is an IETF protocol to authenticate and encrypt IP packets. The authentication part of the protocol generates a protocol header called authentication header (AH). The data security is provided by encapsulation security payload (ESP). IPsec has a separate key management part. It is a mandatory part of IPv6 and optional for IPv4. IPsec can work by securing the payload of the IP packet (*transport mode*) or securing the whole datagram (*tunneling mode*).

9.4.1.6. SSL. SSL (secure socket layer) is a protocol for TCP/IP socket protection.[6] It uses encryption for socket-to-socket security. Typical use is in authenticating only the server socket. HTTP is the biggest user of SSL, even though it can work with many other application protocols. Most implementations are based on SSLeay. Used with SGC (server gated cryptography), SSL provides strong server security on a per server basis. SGC requires a server certificate from a certificate authority, Verisign. SSL3.1, more popularly called TLS (transport layer security), is an IETF standard. SSL is one of the many application/session level security protocols. The reader is referred to the references section to browse through many others, such as MSP (message security protocol) and S/MIME (secure MIME), PGP (pretty good protocol), and so forth.

9.4.1.7. EAP. EAP (extensible authentication protocol) is an open standard for authentication [1]. The original purpose of EAP was to be used with PPP for IETF remote access application authentication. In the wireless arena it is better known for its application in IEEE 802.1X. Lightweight EAP (LEAP) is a proprietary protocol used in Cisco WLAN equipment. Protected EAP (PEAP) protects the EAP data by using encryption. EAP-TSL uses TSL for two-way digital certificate exchange between client and server (a disincentive, overcome by tunneled TSL called EAP-TTSL).

9.4.2. Algorithms

At the heart of security protocols are algorithms that are used by the network devices to achieve the protocol objectives. Protocols are between two or more entities, but algorithms run on a device using its processing power. Algorithms are required for encryption, integrity protection (hash and checksum), random number generation, digital signatures, certificate generation, and algorithms to generate parts of algorithmic data (e.g., keys). The following algorithm types are frequently encountered in security literature.

the same as that of a wireless device using WEP. In fact, we consider such an assertion more of a way of communicating a comfort level than scientific information.

[6] A socket was originally defined in the UNIX OS as a binding of IP address and transport layer port.

9.4.2.1. Encryption.[7] Encryption of data is possible owing to the properties of numbers (mathematics), and humans (imagination, trust). The reverse process (decryption) is made possible by properties of numbers (mathematics), language (occurrence frequencies of letters), humans (imagination, extent of apathy to secret, infidelity), laws, and events. The art of encryption is to design algorithms that are secret and hard (ideally impossible) to break. The science of encryption is to design algorithms that can be publicized and are still hard to break (ideally to prove that they are impossible, or give a measurement of their resistance to breakage).

An encryption algorithm typically takes as input a string of characters or numbers (data!) called *plaintext* and produces an output called *ciphertext*. The reasons for encryptions have been on the increase ever since the invention of communications; with the ecommerce explosion these reasons exploded too, and with wireless networking encryption algorithms could be the difference between two wireless network standards. Encryption algorithms are of numerous types, depending on the way they are designed and the way they take the input. There is no 'mechanism' of designing encryption algorithms; therefore, we do not know for sure how we can design algorithms with a reliable known hardness to compromise. In general, the harder to break an algorithm, the more expensive it is to *run* it. Commonly used encryption algorithms use a string of characters called a *key* and a set of iterative operations on data and keys. Due to the publicized nature of algorithms, the burden of secrecy is shifted from data to keys. There are two major categories of cipher algorithms with respect to the secrecy of key, the secret-key and the public-key algorithms.

9.4.2.2. Secret-Key Algorithms. In the secret key algorithms, the key is known to the sender/s and the recipient/s only. The key is common among communicators (*shared*, *symmetric*), so the probability of this key getting in the wrong hands is accordingly high. The concept of a secret key algorithm is shown in Figure 9-8, in which the same algorithm can be used by any number of sender/recipient pairs just by having their own secret key.

In actual algorithms, a key is a string of numbers that is used in such a way that the decryption should not be possible without the key. Sometimes, the key is generated from another key or a phrase that the authorized user can easily remember but is complex enough not to be broken by dictionary attacks.

9.4.2.3. Public-Key Algorithms. A public key is one that is known to every one. It is used together with a private key that is not known to everyone, but only to the authorized user or device. The public key algorithms rely on various properties of numbers, for example, various number systems have multiplicative inverses (see below) of numbers that are not obvious. For example, for very large prime numbers in such a system, it could be exceedingly difficult to

[7] We will stick with the term encryption here, even though it must be pointed out that encryption algorithms are actually cipher algorithms, providing both encryption and decryption.

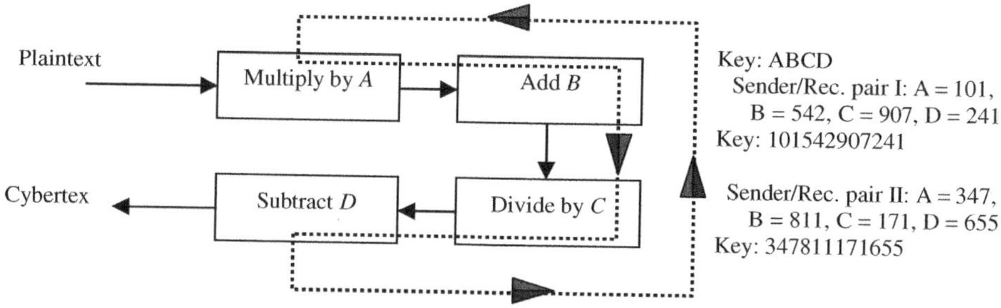

Figure 9-8. Shared key concept. Both sender/recipient must have the same key. However, it does not have to be the same as other senders/receivers.

find the inverse. This is a simplified statement, but the essence of a public key is that there are two keys, one known publicly (*public key*) and the other kept private (*private key*). If encryption is done by using public key, the decryption could be done only with the private key and vice versa. The following example will illustrate how is it possible to have encryption and decryption by different keys.

9.4.2.3.1. How Is Two-Key Cipher Possible?[8] In order to demonstrate that two different code words (keys) can be used to encrypt and decrypt data, we take a simple digression to linear algebra. In this part of the linear algebra, one needs to understand the meanings of number systems to the base-q and *modulo-q* multiplication operation. We will try explaining these concepts without much complexity.

A number system with the base-q can be viewed as a system in which all quantities are represented in terms of q digits. For example, for $q = 16$ (hexadecimal system), all quantities are represented by 16 digits (1, 2, 3, 4, 5, 6, 7, 8, 9, A, B, C, D, E, F, 0). Also, a number system with $q = 8$ (Octal system) has eight fundamental digits (1, 2, 3, 4, 5, 6, 7, 0). We can define a very large q, and represent the fundamental digits in binary numbers instead of decimal numbers. For example, for the base-16 system, imagining that $16 = 2^4$, we can have the following fundamental digits (0001, 0010, 0011, 0100, 0101, 0110, 0111, 1000, 1001, 1010, 1011, 1100, 1101, 1110, 1111, 0000). One of the interesting quantities in all numbers is the *multiplicative inverse* of a number. In simple terms, the multiplicative inverse of a number k is another number, say k', such that the *modulo-q product* of the two numbers is 1. In other words, $k.\ k' = 1$. Lastly, in modulo-q arithmetic operations, the result is always between 0 and $q - 1$. Let's take the example of a number system with $q = 7$. We know that

[8] The author does not regard the discussion in this section mathematically sound. Its purpose is only to convey a concept. For example, there is no mention that (why) we want to have q as prime. Also, there is no mention of the fact that modulo-8 is really an extension of modulo-2 {for coding theory people GF(2^3)}. The author considers such concepts as beyond the scope of the book. Interested readers can find many textbooks on this topic, that will be authentic.

the squares of numbers in base-7 system would be given as $1^2 = 1, 2^2 = 4, 3^2 = 12, 4^2 = 22, 5^2 = 34$, and so on (base 7). Let's call these products as base-7 arithmetic products.

Modulo-q product = Base-q arithmetic product – largest multiple of q.

The largest multiple of q is chosen such that the difference (Base-q arithmetic product – largest multiple of q) is positive or 0.

In terms of Euclidean division, if y is the base-q arithmetic product of two numbers m and n.[9] Then, the modulo-q product of m and n is equal to the remainder from the Euclidean division of y by q.

Example: Let's find the modulo-7 product of 6 and 5. In this case, $m = 6$ and $n = 5$. The base-7 arithmetic product $m.n = y = 42$ (decimal $30 = 4.7 + 2$). Therefore, the modulo-7 product of 6 and 5 is 2.

Therefore, in the above example, the modulo-7 (arithmetic) products are given as: $1.1 = 1, 2.2 = 4, 3.3 = 2, 4.4 = 2, 5.5 = 4, 6.6 = 1$. In fact, we see that 6 is also the *modulo-7 multiplicative inverse* of itself. As another example, the modulo-7 multiplicative inverse of 4 is the number 2 because $4.2 = 11$ (decimal 8) $= 1.7 + 1$, implying that the modulo-7 product of 4 and 2 is 1.

Therefore, one can design a two-key encryption system in which encryption is performed by multiplying data with a large number from base-q system and decryption is performed by multiplying the encrypted data with the inverse. One of the numbers can be made public and the other kept private. This is because in this case:

Modulo-q (PublicKey × PrivateKey) = 1.

The fact that a number can be its own inverse (as 6 in base-7 system) brings us to an important caution, that is, the numbers that should be used (i.e., the *keys*), have to be picked up carefully; just any number will not do. In fact, extensive knowledge is available to determine suitable numbers with inverses that are not easily guessed or determined.[10] What we like to conclude here is that if a message (plaintext) is encrypted by simply multiplying it with a very large number (*key*) in a base-2^s system, where $q = 2^s$, then the message can be decrypted only by multiplying it with the modulo-2^s multiplicative inverse of the key, and not the key itself. Hence, two different keys complete the *cipher* (cipher = encryption + decryption). See Figure 9-9 for a demonstration of this concept.

[9] Euclidean division of y by q is given by $y = c.q + R$, where c is the quotient and R is the remainder.

[10] In author's view, Chapter 2 of [2] presents a fairly low level treatment of the topic of modulo-q algebra.

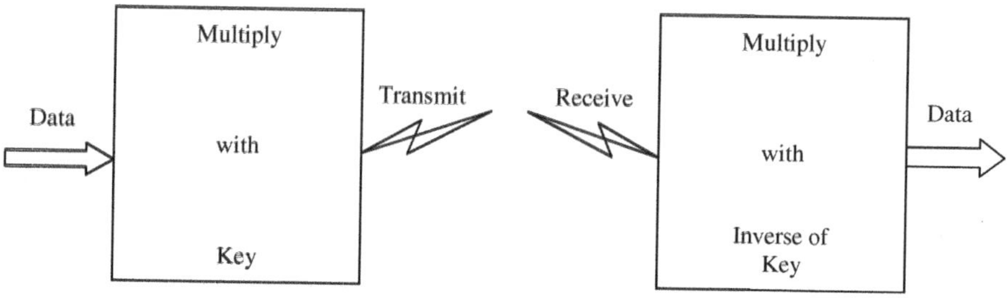

Figure 9-9. In modulo-q systems, a number and its inverse of a number can be used for encryption and decryption.

Actual encryption algorithms are not that simple, but the above example demonstrates that one can recover the plaintext by multiplying it with two different numbers, and the knowledge of one number may not be sufficient for decryption. This is the idea behind the two-key algorithms.

9.4.2.4. Block and Stream Ciphers. In block cipher, the data are fed to the algorithm in blocks of bits, while in stream types, the data are given serially, that is, one bit at a time.

9.4.2.5. Rounds, Key-Size and Data Block. An encryption algorithm is essentially a transformation of data using a key of a given size. The transformation is typically performed by using a function more than once—in *rounds*. Looking at the example of Figure 9-8, the rounds are depicted by a dotted line looping through all the boxes of the algorithms. In actual algorithm, every round adds more complexity to the algorithm. Thus, the higher the number of rounds, the harder it is to break the algorithm. Also, the longer the key, the harder it is to break the algorithm. This is because the attacker has to try all key combinations. For a key using n bits, there are 2^n combinations. A 128-bit key has $2^{128} = 3.4 \times 10^{38}$ combinations, and the algorithm has to be run a maximum of as many times to break it in a brute force attack. In block types cipher, the size of data block is also an important characteristic of the algorithm and information about its size can be used in *cryptanalysis* for breaking the cipher.

9.4.3. Examples of Encryption Algorithms

9.4.3.1. Advanced Encryption System (AES). AES is a U.S. Government standard for unclassified sensitive data (wireless networks qualify for this), which is intended to have universal appeal and application. It employs the *Rijndael* (*pronounced as* Reign-Dahl) algorithm. The algorithm uses block encryption with block sizes of 128, 192 or 256 bits. The key type is symmetric with allowed sizes of 128, 192 and 256 bits. This makes it a highly secure algorithm for the currently available and foreseeable code-breaking technology. It is expected to be used in all public applications and systems of unclassified sensitive data.

The Rijndael algorithm for AES was chosen due to its flexibility and efficiency from among a number of proposals. It is specified as FIPS-197. AES has been already employed for North American 3G wireless standard cdma2000 and IEEE 802.11i, for wireless LAN (IEEE 802.11 suite).

9.4.3.2. Data Encryption System (DES) and Triple DES. DES and triple DES are also U.S. Government standards for unclassified sensitive data. AES has now replaced these algorithms. The original DES, a block encryption algorithm with 64-bit block-size and 56-bit key size (8 character password represented as 7-bit ASCII code), is known since the 1990s to be breakable. It uses 16 rounds for strengthening the encryption. The triple-DES uses triple encryption, twice with the same key and once with a different key, thus strengthening it to a better level. DES has been regarded unsafe and triple-DES is required in its place by many businesses and governments. Both DES and triple-DES are part of Federal information processing standard-43 (FIPS-43).

9.4.3.3. f8 Algorithm. The f8 algorithm is encryption part of Kasumi algorithm. Kasumi algorithm has two parts, f8 for enciphering and another, f9, for integrity check. It is a stream base algorithm and is recommended to be used with UMTS. In the UMTS implementation, a cipher key (CK) is fed to the algorithm along with the user identity and a system time-dependent string to generate a keystream block. This keystream block is logically combined with the plaintext to generate the ciphertext. It uses a symmetric key.

9.4.3.4. RC4. RC4 (name drawn from inventor Ron Rivest of RSA Security—Rivest Cipher = RC) is a stream cipher that was once a trade secret, but appeared on the Internet mysteriously. The algorithm is extremely fast and is basically a random number generator. Just like f8, RC4 generates a keystream that is logically combined with the plaintext to generate ciphertext. It uses a secret key of 256 bit length, which should make it a pretty strong algorithm. However, the first few bits generated by the pseudo-random generator have been known to be predictable. This has affected its marketability and WEP's vulnerability has increased due to this. In WEP, an initialization vector (IV) is transmitted with each MAC packet encrypted using RC4. By hashing the secret key with IV, a new key is generated that was used to encrypt the packet. This is because in stream ciphers a key can be extracted if it is employed more than once.

The same company (and person) also designed RC5, which is a block cipher and uses multiple blocks, and multiple key lengths. RC5, too, is quite simple. The original version suggested a block size of 64-bits, key of 128 bits, and 12 rounds. However, a key anywhere from 0 to 2040 bits can be used with block sizes of 32, 64 or 128 bits. The RSA Security company also submitted an algorithm to NIST for AES. It was called RC6 and was based on RC5. It qualified as one of the five finalists.

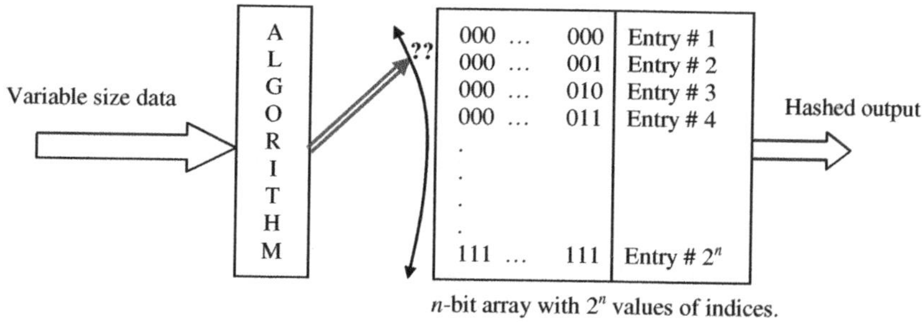

Figure 9-10. A Hash algorithm generates a fixed size, irreversible output for a variable input.

9.4.4. Hash Algorithms

A Hash algorithm assigns an index to a quantity or expression. As an application of a Hash algorithm, suppose an employer uses a Hash algorithm to generate the nine digits confidential identification number (CIN) from information about each employee. At the entrance security booth, the employees give their information (name, DoB, etc.) and the CIN. This information is entered in another algorithm in a terminal. This algorithm generates the CIN by using the hash algorithm and compares it with the offered CIN. Depending on the result of comparison, the output is written on the screen as 'Authorized' or 'Not Authorized'.

Hash algorithms have been used in storing passwords by operating systems. They are useful in securing a key if a key needs to be exchanged for a secure session. The *one-way hash algorithm* would not allow the original data to be retrieved from the hashed value. A hash algorithm must generate a different hash value for different sets of input. A *collision* is the event of having the same hash value for more than one data input. Figure 9-10 shows the concept of a hash algorithm.

In this figure, the Hash algorithm output is always an n-bit string. There are 2^n combinations of n-bit strings. By making n very large, it can be made virtually impossible for any computer to break the algorithm.

9.4.4.1. Message Digest (MD). Message digest (MD) is the term used for the output of the Hash algorithm. Therefore, a 128-bit Hash algorithm takes a variable size data and generates a 128-bit MD.

9.4.4.2. Message Authentication Code (MAC). Sometime, there is a need to transmit the data and the corresponding hash so that it can be figured from the hash if the data have integrity. This works well against random channel errors as well as data injection/tempering attacks. If an attacker gets hold of data, a new hash can be generated. A solution to this problem is to generate hash from a protected key or password. A hash with the secret key or password is called MAC.

9.4.4.3. Digital Signature (DS). Digital signature (DS) is closely related to the discussion on Hash algorithms. The DS identifies a sender of a document. For example, a user, using a one-way Hash with a private key, will generate a bit string and attach it to a document to make a signed document. The recipient, by using a public key, can confirm the identity of the signatures in a secure way.

9.4.4.4. Digital Certificate (DC). A digital certificate is closely related to digital signature. A DC is an identity authentication agreed by communicators or issued by a third party called certification authority (CA). It has many information fields, in addition to the digital signature. In business transactions, when a recipient receives a signed document, this document could be a digital certificate issued to the user by a CA. Typically, the CA has confidence about the identity of the person and has verified it through one or more mechanisms. DC is a descendent of OSI standard electronics directory (X.500) and is specified in X.509 format as well as others. A digital certificate implements nonrepudiation by making an electronic document a signed and authenticated document with CA as the witness. Thus the difference between a signed document and a document with certificate is that the latter has the third-party (CA) verification.

9.4.5. Examples of Hash Algorithms

9.4.5.1. SHA-1. The SHA-1 (secure Hash algorithm 1) [3] produces a 160-bit message digest for any data block of length up to 2^{64} bits. It is a U.S. Government standard and is specified as FIPS 180-1. The collision probability for all computational purposes is close to zero, implying that no change in the data can go undetected. So, if a document and its message digest are transmitted, the recipient only has to generate the MD locally and compare with the received MD for integrity check. The algorithm always does padding of the input message with (1000 . . . 64-bit original message length) so that the total number of bits in the padded message are multiples of 512.

9.4.5.2. MD5. The MD5 (message digest 5) [4] algorithm generates a 128-bit message digest or fingerprint for a variable size input data block. The algorithm is known for its strength. However, [5] reported that it was vulnerable to collision attacks. The algorithm employs the same padding as SHA-1, that is, to make the input a multiple of 512 bits by adding (1000 . . . 64-bit length of original input). It was extended from a predecessor MD4, which is known for its speed and easy of implementation. MD5, however, does not add much in terms of complexity (an extra round and some function modification), but provides a very low theoretical probability of collisions.

9.4.5.3. H-MAC. As is clear from its name, H-MAC [6] is a message authentication code (Hash plus key) and not a Hash algorithm. It can be used with any Hash algorithm, such as SHA-1 and MD5. That also means that H-MAC is as good as the underlying Hash algorithm. The recommended minimum key size is the same as the message digest of the Hash algorithm. A strongly cryptic pseudo-random number generator (PRNG) is recommended to be used to generate the key. The key is recommended to be refreshed periodically. Even though H-MAC is independent of the Hash function, some modifications in the Hash function can strengthen H-MAC, such as making the initialization vector (IV) variable. The suggested method for this is to store the intermediate compression results of the key generation process and later using these results for initializing the IVs. The added complexity relates to the storage of the intermediate results and the way they are used to generate IVs. If this approach is utilized, the stored intermediate values become security-sensitive and have to be protected, just like the key itself. H-MAC, when combined with a specific hash algorithm, adds the algorithm's name as its extension, for example, HMAC-SH1, HMAC-MD5, and so forth. If the output is truncated, there are some advantages to that. However, the truncated output should not be less than half the size of the key length. When truncated output is used, the resulting algorithm is referred with an extension –*t* that shows the output size, such as HMAC-SH1-*t*.

9.4.5. Key

As mentioned in the beginning, the cipher (encryption/decryption) and hashing algorithms must be publicized for open security architecture. That means that anyone with the key could decrypt or decompress the information. This brings key to the spotlight as the most critical part of an implemented algorithm. A permanently allocated key is dangerous and vulnerable to many types of attacks. First of all, such a key provides the intruder to record all data past and present and, once the key is broken, all data are compromised.[11] Secondly, the same key used in all packets or blocks of encrypted data is open to statistical analysis of data. Not surprisingly, a key generation and distribution mechanism is at the heart of all modern encryption and hashing mechanisms. Paradoxically, if key is to be generated and transmitted over the insecure network, then the key needs to be encrypted first. This is especially dangerous for public/private (asymmetric) keys. We will look at some key management (generation, distribution, and protection of keys) mechanisms. There are a large number of key types, each one having different requirements [7]. There are 20 key types listed in [7] for various applications and usages. Table 9.1 summarizes some of these keys. Some of these are long term or *static* and some are short term or *ephemeral*.

9.4.5.1. Key-Generation Algorithms. The key-generating algorithms use the properties of *Modulo-q* algebra and *prime* numbers. Especially, the two-

[11] This does not apply to authentication and authorization, as the past cannot be compromised (one can't get authentication from the past if a key is broken).

TABLE 9.1. Some Important Key Types

Key Name	Purpose	Protection Level
Public authentication key	Used for authentication of user and data. Distributed generally to permitted users. To be used with private key.	Must be validated for association with the private key. Does not require confidentiality measures, but requires integrity measures must be associated with application. It is a long term key.
Long-term data encryption key	Data encryption	Does not need validation. Must have confidentiality and integrity provisions. Must be associated with specific application.
Short-term data encryption	Data encryption	Does not need validation. Must be checked for confidentiality and integrity.
RNG key	To generate pseudo-random numbers.	Does not need validation. Must have confidentiality and integrity provisions. Must be associated with specific application.
Master key	Generate other keys	Does not need validation. Must be completely protected for the lifetime.
Public authorization key	Resource authorization. To be used with pairing private authorization key.	Does not need validation. Specific to application and owner. Integrity must be checked.
Initialization vectors (IV)	To help generate key from the received data.	Does not need validation. Associated only with the received data block, must be checked for integrity and the procedure protected on a long-term basis.
Key transport public key	To protect a key during key exchange. To be used with key transport private key.	Must be validated. Associated with the user. Long-term availability and should be checked for integrity.

key users must generate a pair of keys such that given one (public) key and given the encrypted data should not reveal any clue to the secret (private) key. The keys could be generated by one user and shared using the key-transport mechanisms, both users by sharing some other key-sharing information, or by a server that generates, distributes, and manages keys.

9.4.5.1.1. Diffie-Hellman (DH) Algorithm. The DH algorithm is perhaps the one most cited for generating ephemeral keys. It is used in Windows XP, IPSec, TSL, and scores of other systems and protocols. The account given in the following is according to [8]. Both the parties (say party A and B) agree *a priori* on a large prime number p and the base q of the number. Each party starts with picking a secret number, say a and b, respectively. The two perform the *modulo-p* exponentiation operations with the random numbers ($\alpha = q^a$ *Modulo-p* and $\beta = q^b$ *Modulo-p*). The numbers α and β are exchanged over an insecure channel, such as a wireless network. Using their own random numbers (a and b), the two parties generate the common key k (α^b *Modulo-p* $= \beta^a$ *Modulo-p* $= k$). The strength of the algorithm comes from the fact that a and b cannot be generated using α and β. See Figure 9-11 for a depiction of calculations.

The exchanged number (and, hence, the prime number p) can have 768, 1024, or 2048 bits. The algorithms are grouped into three, based on which number size is used. The groups are named as *group 1* (768-bit group), *group 2* (1024-bit group), and *group-2048* (2048-bit group). The algorithm was originally proposed in 1976 by Whitfield Diffie and Martin Hellman, and has not been reported to be broken so far, in the knowledge of the author.

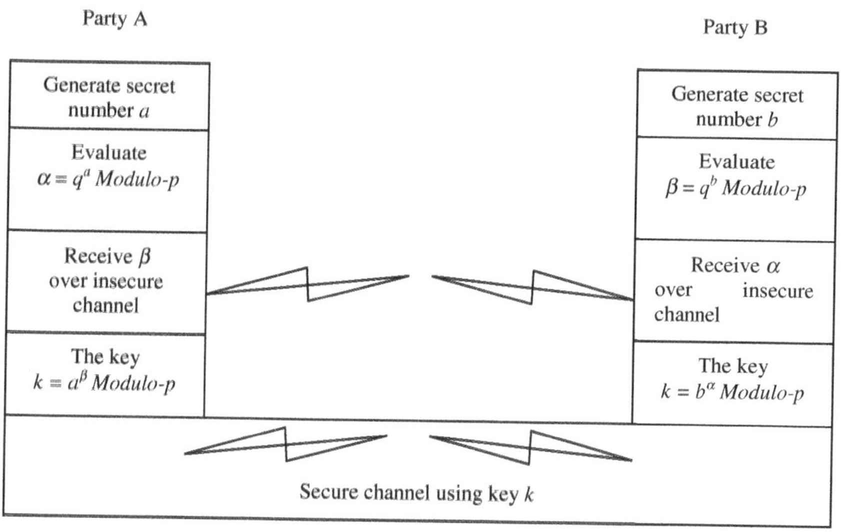

Figure 9-11. Using DH for key sharing.

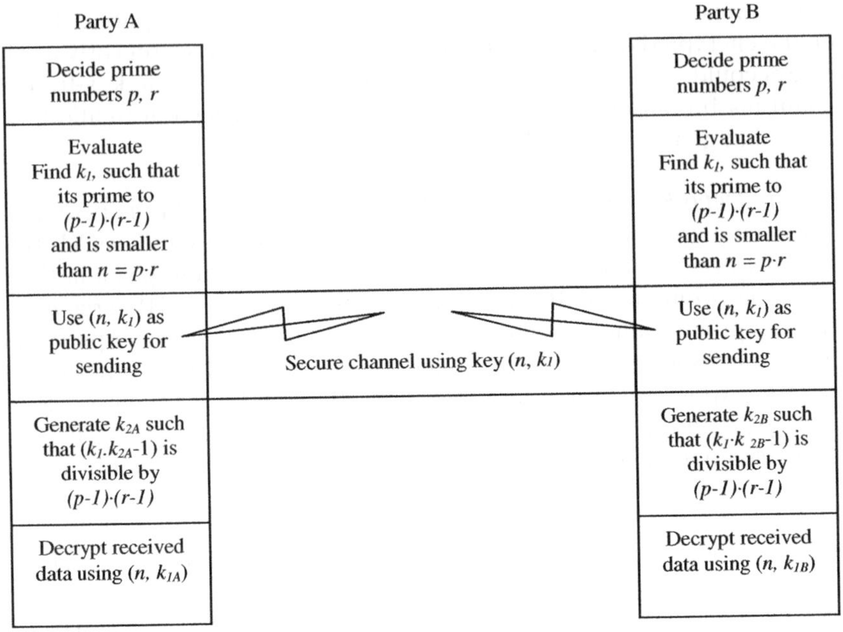

Figure 9-12. Using RSA for public and private key.

9.4.5.1.2. RSA Algorithm. This algorithm can be used to generate a key pair and use it for public key encryption as well as digital signature verification. The following description is based on the one given on the RSA Security (the company who originated it) website.[12]

Suppose parties A and B use RSA to exchange data using public/private key pairs. The two parties have agreed to use two large prime numbers p and r. Let k_1 be a number such that it is:

(i) less than $n = p \cdot r$;
(ii) prime with respect to $(p - 1) \cdot (r - 1)$.

Let k_2 be another number that such that $(k_1 \cdot k_2 - 1)$ is divisible by $(p - 1) \cdot (r - 1)$. Then the public key is given by (n, k_1) and the private key is given by (n, k_2). Figure 9-12 is a depiction of the algorithm.

The algorithm was proposes in 1977 by Rivest, Shamir, and Adleman (R.S.A.). The strength of the algorithm resides mainly in the factorization of the product $p \cdot r = n$.

9.4.5.2. Server-Based Key Management. The key-generation and sharing algorithms run on every user machine. Therefore, part of the ultimate security

[12] www.rsasecurity.com/salabs/node.asp?id=2214

of these algorithms also lies with how secure the individual systems are, in terms of authentication, authorization, and operating system security. Key management using a server dedicated for this purpose could provide better guarantees with respect to the individual system security.

Another issue that can be addressed by a server is non-repudiation. If a recipient receives a message purported to be from sender X, how does the recipient know that it is *really* from X and not from an attacker (*man-in-the-middle*), who stole the private or public key and is impersonating the sender X. A solution to this is that a sending user digitally signs the document and a server certifies the authenticity of the signature. A certifying authority (CA), a certificate protocol (e.g., X.509), and a server infrastructure is used for this purpose. We will briefly discuss the services and components of public key infrastructure (PKI) in this section. It has evolved over time for digital signature verification and certification authorization. Also, refer to 'Managing Certificates' in [9] and [40] for further discussion on the following discussion.

9.4.5.3. Public-Key Infrastructure (PKI). PKI is a complex system of issuing and managing authentication and encryption certification by a certificate authority system managed by the user company or a third party. Some of the companies and administrations that provide PKI services are VeriSign, Thwate Consulting, Internet Publishing Services (IPS), CertiSign, and BelSign.

1. *Issuing certificate.* A certificate authority (CA) uses its own criteria to issue certificates. A business or organization can have its certificate server, thus acting as a CA. In that case, the certificate can be issued at the time of hiring an employee. If a CA is a subscription service, then the issuance may depend on the type of contract between the client organization and the CA. For individual users, identifying information may be requested by CA, which could vary anywhere from e-mail address to personal identification done by visiting an office and showing social security card and other identifying documentation. Figure 9-13 shows the information in the X.509 v3 certificate. Every CA does not

Data Section	Signature Section
• Version • Serial number • Information • Public key and algorithm • Distinguished name of the CA • Validity period • Subject name • Extensions (type, etc.)	• Encryption algorithm • CA signature

Figure 9-13. X.509 certificate.

have to request all the information fields. In fact, even the same CA does not have to have all the same fields of information from all users.

2. *LDAP (Lightweight Directory Assistant Protocol).* LDAP is a protocol designed to access directory services for the X.500 protocol. It has a simple interface and provides commands to manage certificates. LDAP servers can store critical information about the certificates, retrievable using a simple set of commands, such as *bind/unbind* to connect/disconnect, *read/search/compare* for locating and reading information, and *add/modify/delete* for updating the information.
3. *Key Management.* Key management involves issuing the key, stamping the user identity on the key, taking care of a compromised key, and storing the key. Since the entire business of certification depends on confidentiality of the key, this is the most crucial component of PKI. In some cases, more than one certificate can be issued for the same user, with keys varying in strength so that the user can use a different certificate for each security level required. Keys can be generated locally at user terminals or centrally and distributed using LDAP. The former is stronger in terms of non-repudiation, but as safe as the operating system password in some cases. Key storage has to be done securely, using a Hash algorithm, for example. In some cases, backups of keys may be necessary so that they can be retrieved in case of loss of keys. We will address the issue of compromised keys (certificates) separately.
4. *Compromised Certificate.* The whole business of CAs runs on trust. This trust could be either based on a web of trusts or following a chain of authorization. If a key managed by a CA is compromised, it is imperative to revoke the key and broadcast the information about such a key. In practice, instead of broadcasting the key revocation information, a certificate revocation list (CRL) is maintained in some centrally accessible location. When a CA server receives a certificate, it should check the CRL before issuing authentication for the certificate.
5. *Registration Authority (RA).* The RAs are trusted authorities that offload the information verification process from the CAs. The relation between RA, CA, and a user directory (e.g. X.500) is shown in Figure 9-14 [10]. There are many problems (pointed out by [10]) about using X.509 PKI. These relate to the complexity of the infrastructure, lack of

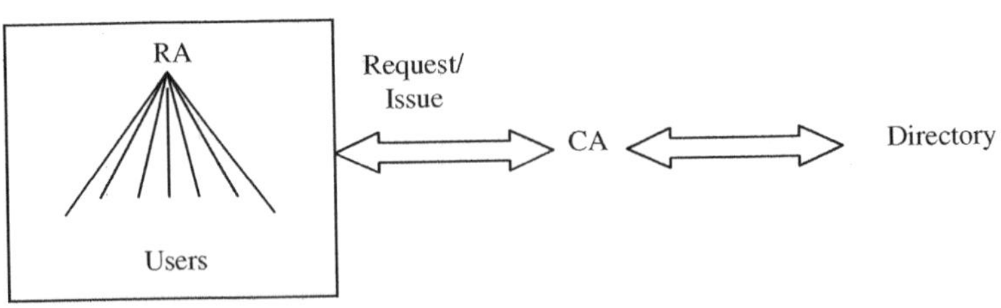

Figure 9-14. RAs offload CAs.

experience of the networking community, and the weak or no certification revocation.

9.4.5.4. Other Key Infrastructure. The X.509 is based on an ITU standard that is expected to have international acceptance. However, there are many other certificates as well as key management infrastructures. Among these, simple public key infrastructure (SPKI) certificates bind a key to an authorization by using the *m* out of *n* authorities principle (e.g., *m* out of *n* administrators signing the authorization). Certificates by PGP (pretty good protocol) are key-based instead of identity-based. PKIX is the Internet PKI profile, FPKI, the Federal PKI (U.S. Government), MISSI, U.S. DoD profile, ISO15782, the ISO profile for financial institutions, SEIS (secure electronic information in society), TeleTrust/MailTrusT (German), ISIS (Industrial signature interoperability specifications, German), PKAF, SIRCA (U.S. Securities industry association).

In the remainder of this chapter we will discuss wireless data networks security mechanisms. We start with WEP and discuss three WLAN security architectures. Following these, we will have a brief discussion on security architectures for cellular 3G systems.

9.5. WIRELINE EQUIVALENT PRIVACY (WEP)

WEP is a WLAN security architecture that was part of the original standard IEEE 802.11. It was not mandatory, leading to implementation issues that resulted in interoperability problems.

9.5.1. WEP Architecture

The IEEE 802.11 standard provides open and shared-key authentication. These are briefly discussed in Chapter 5 on WLAN medium access control. The core function provided by WEP is data encryption. The encryption mechanism is shown in Figures 9-15 and 9-16.

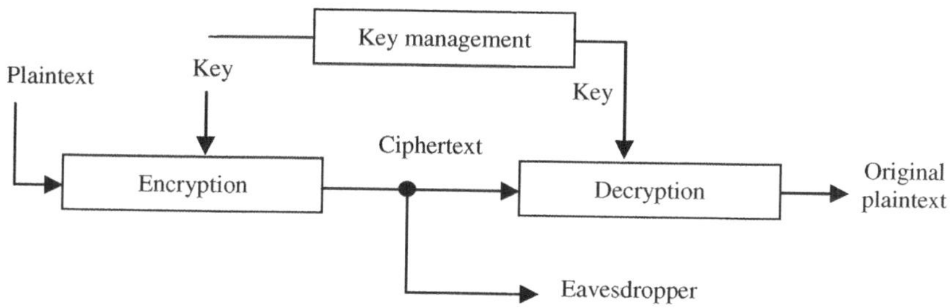

Figure 9-15. An encryption channel [11].

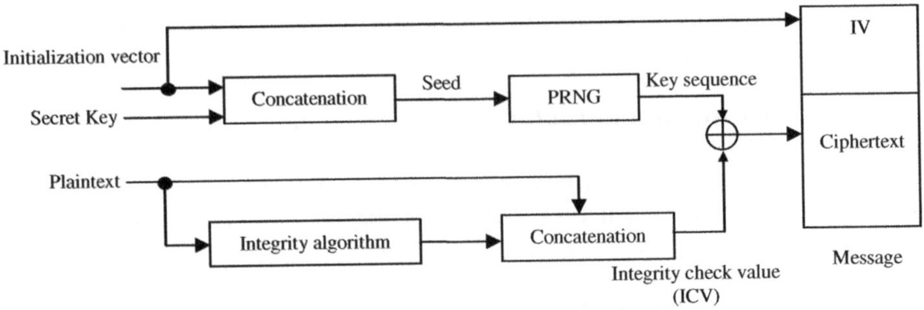

Figure 9-16. WEP encryption block diagram as per the IEEE802.11 standard.

Figure 9-15 is a generic block diagram for the WEP-like encryption and Figure 9-16 is WEP-specific. As seen from Figure 9-16, an initialization vector and a pseudo-random number generator change a fixed shared key into a key-sequence that can be changed dynamically. An integrity algorithm (CRC-32) generates the integrity check vector (ICV), which protects data against bit-flipping (both malicious and channel errors). The IV, ICV, and ciphertext are transmitted as a composite message. The encryption algorithm used is RC4, which effectively generates a highly random pseudorandom number. At the receiving end, ICV is used to determine the accuracy of the ciphertext. The IV and the shared key are used to decrypt the ciphertext.

9.5.2. WEP Vulnerabilities

The study on the vulnerabilities of the original WEP are so well documented [12] that it is now regarded largely as an ineffective protocol. Almost everything is a source of compromise; so much so that an otherwise strong encryption protocol RC4 by RSA Security is not foolproof in WEP. Here is a list of WEP vulnerabilities.

1. The lack of the standard for requiring implementing WEP for compliance resulted in the shipped equipment being without security features turned on. Initially an overwhelming majority of the installed devices were without any protection whatsoever. The SSID, which might provide some secrecy, is broadcasted unless turned off.
2. The shared key in the original WEP specifications is easily compromised if a large number of users are told about it. In a hotspot environment, it is impractical to have a shared key protected.
3. The initialization vector (IV) is sent as plaintext. The initialization vector size (24-bits) is small enough that more than one packet with the same IV can be easily collected by an attacker. Logical analysis of these packets can reveal their plaintext. This attack can also be formed by sending known plaintext from adversary's station, collecting the corre-

sponding ciphertext along with the IV, and thus discovering the plaintext of all subsequent packets using this IV. By having enough number of plaintext and ciphertext pairs, the encryption can be broken [13]. The IV is employed due to the static nature of the key and it forms part of key for each packet. IV was a way to change the key dynamically. It could be more effective if it were never repeated, and if it were encrypted during transmission.

4. Linearity of integrity check vector (ICV) results in the attacker having the capability of changing the ciphertext *and* CRC-32 bit, so that the CRC reflects no changes in the ciphertext. The ICV is included to protect the message against changing the message bits. A number of tools are available to audit WEP, such as AirSnort, BSD-Airtools, WEPCrack, WAP Attach, WEPWedgie, and so forth [14].

9.6. WI-FI PROTECTED ACCESS (WPA)

The Wi-Fi Alliance forum [www.Wi-Fi.org] reacted to the vulnerabilities of WEP by offering a software enhancement called WPA [15]. The agenda of WPA specification was quite simple, that is, to remove WEP vulnerabilities. An added goal was to make it compatible with the future standard from the IEEE802.11*i* Working Group. WPA strengthened security of WLANs (based on IEEE 802.11 and its enhancements) by adding to both authentication and encryption. Effectively, WPA wraps RC4 in four new components [15]

1. Extended IV and IV sequencing rules;
2. A message integrity code (MIC), *Michael*, and countermeasures for forgeries;
3. Key derivation to defeat man-in-the-middle attacks; and
4. Temporal Key Integrity Protocol (TKIP) to generate per-packet keys.

9.6.1 Temporal Key Integrity Protocol (TKIP)

A natural weakness of a shared secret is its vulnerability to the passage of time. The 'time' does not have to be in years, months, or even days. If the same key is used in a large number of packets, it provides a 'common domain' for statistical analysis. '*Temporal*' integrity takes care of this problem by retaining the key validity to a short lifetime. This lifetime can be as long as the whole session duration, or as short as a single packet duration.

The IEEE 802.11*i* (see later) has defined a hierarchy of keys. From among these, three keys are of importance for this section, (1) the master key, (2) key encryption key, and (3) the temporal key. The last of these is used by TKIP. A pair of temporal keys is used in each direction, one for encryption (128-bit key) and one for data integrity (64-bit key). A 2-bit *WEP-key ID* is used by

TKIP to identify these four temporal keys. Under TKIP, WPA provides four main additions to WEP [16]. These include (1) *Michael*, a message integrity code, (2) a new IV sequence discipline, (3) a *key-remixing* function, to decorrelate IVs from weak keys and, finally, (4) a *rekeying* algorithm to regenerate keys.[13]

9.6.1.1. Michael. Michael is a message integrity code similar in paradigm to digital signature verification using Hash algorithms. It is apparently much stronger than WEP ICV. In WEP ICV, the culpability of CRC, which any changes in message bits, are reflected at known locations in CRC, weakens the message integrity. Michael [17] uses a 64-bit key and a tagging function with simple logical iterations to generate a tag. The tag is sent just like a digital signature along with the message. The recipient calculates the tag locally from the message part. If the received and calculated tags match, the message is considered to be authentic.

Michael, though much stronger than CRC-32, provides a weak point for the attacker. Therefore, a counter measure is included in TKIP, in case of attempts to break the key of Michael. A station in an 802.11 network, on detecting two failed forgery attacks, de-associates, deletes keys, and stays out of the network for some time before associating again. This greatly limits the forgery attacks, but can't defend against replays. A new IV sequence discipline is introduced as a measure against such attacks.

9.6.1.2. IV Sequence Enforcement. The TKIP reuses WEP IV as sequence numbers by initializing them at the sender and receiver end with every fresh key. For the reason that made IV weak, the same sequence numbers are not used once a sequence space has been exhausted for a given key use. The receiver simply monitors the sequence numbers (IV fields). As long as the IVs are in ascending order, they are acceptable. If an IV is received in duplication or is lower than an earlier one, a replay attack is assumed and the packet is discarded.

9.6.1.3. Key Mixing. The key mixing function provides a means to generate a unique *intermediate key* for each station from the stations MAC address and the temporal key. The intermediate key has a lifetime equal to the temporal key. The intermediate key is used to encrypt the packet sequence number to generate a 128-bit per packet key. WEP-based concatenation of per-packet key and IV is avoided by masking portion of IV, thus saving RC4 weak key problem.

9.6.1.4. Rekeying. Rekeying is included to avoid the problem of rollover of sequence numbers in Michael.[14] New temporal keys must be generated before

[13] Same as a message authentication code (MAC), but MAC is used in IEEE 802 Working Groups (and largely in this book) for Medium Access Control.

[14] The exact definition as of [16] is not completed.

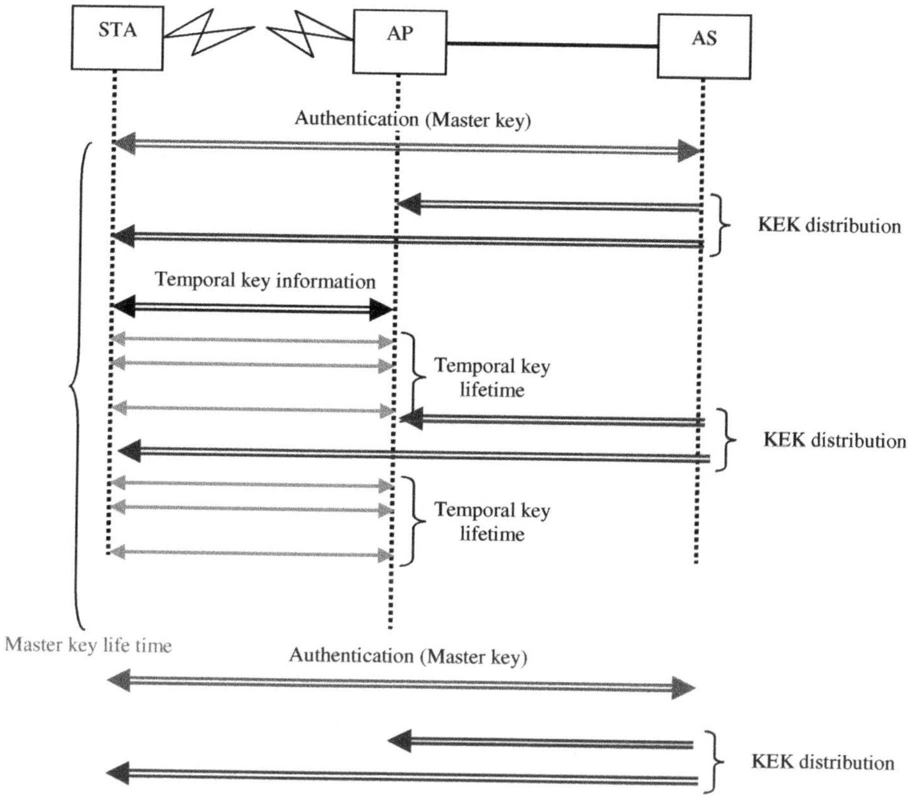

Figure 9-17. The relative use of temporal key, key encryption keys, and master key in TKIP.

the sequence rollover occurs. This is accomplished by allowing the access point and station to exchange key generation information. The key generation information is protected by key encryption keys (KEKs). The KEKs are pushed to the access point (AP) and wireless station (STA) by the IEEE 802.11X authentication server (AS). The AS and AP, being on wired network already, have a trust relation. A key is needed to complement a similar level of trust relation between the AS and station (STA). A master key (MK) is used for this purpose. Figure 9-17 shows the relation between the three key types.

9.6.2. TKIP Encapsulation Process

The above four functions and algorithms result in the secure encapsulation of TKIP-secured data, as shown in Figure 9-18.

Notably, the MIC is added to the MSDU (instead of a fragment), following which regular IEEE 802.11 fragmentation function is invoked. Each fragment thus gets its own sequence number for a true per-packet key in phase II mixing. The result of phase II mixing is in the per-packet key, represented as a WEP

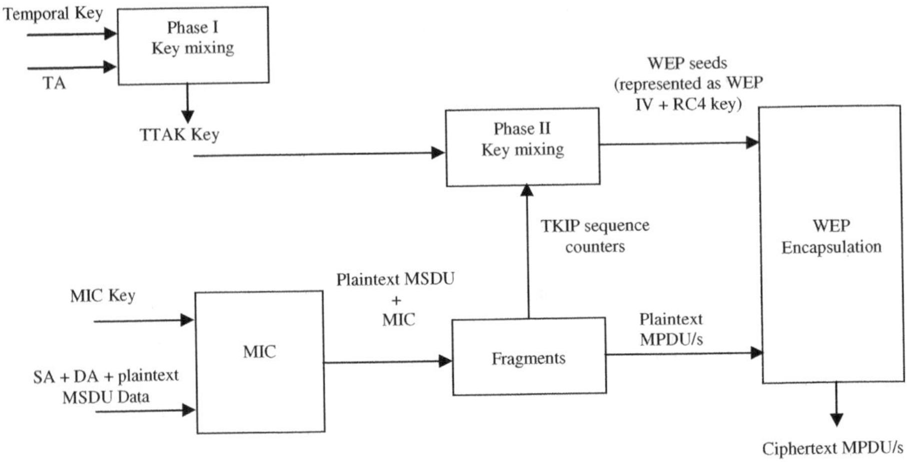

Figure 9-18. TKIP reinforcement to WEP.

IV and base key. The rest is encrypted following the original WEP, thus encrypting data as well as integrity vectors (now ICV and MIC).

9.6.3. WPA Authentication

The second component of WPA is IEEE 802.1X-based authentication. This takes care of the weak authentication in the original WEP. At best, the WEP included an option for a one-way shared key authentication. WPA provides two choices of authentication, a server-based (e.g., using RADIUS), for an enterprise network and a password-based, for home networks. The latter is also called pre-shared key (PSK) authentication.

9.6.3.1. RADIUS-Based Authentication. In this infrastructure, a trusted relationship exists between the access points (AP)s and a RADIUS server. The server and station (STA) use the master key to have an authentication relation. Once the STA has been authenticated, TKIP distributes key encryption keys to be used to generate temporal keys by the STA and AP. This mechanism is applicable to enterprise networks, which can afford to have a RADIUS or similar server (e.g., DIAMETER) for key distribution.

9.6.3.2. Pre-Shared Key (PSK) Authentication. This authentication mechanism is for home and small business WLANs. It does not require the use of an AS. The AP and STA are configured with a secure password that is used to generate the PSK. Mutual authentication is performed using PSK. The key is derived from the user credentials that have been entered in the AP.

9.7. IEEE 802.11*i*

IEEE 802.11*i* defines a robust secure network (RSN) architecture to provide security capabilities for IEEE802.11-based WLANs and their extensions to include authentication, data encryption, key management, fast roaming (pre-authentication), and many other features (e.g., key-caching for quick reconnect, compatibility with IEEE 802.11*e*). For infrastructure netwoks, it requires the use of an authentication server (AS), such as RADIUS, using authentication protocol such as IEEE 802.1X using EAP over LANs (EAPoL), master key (MK) and pairwise master keys (PMK)s. Figure 9-19 shows the relation among various components of an IEEE 802.11*i* architecture [18].

Following is a description of the functions of various key types.

9.7.1. Master Key (MK)

Defined only per session between an AS (not AP) and a STA, the MK is required to establish authentication between the AS and the wireless station. Authentication using MK guarantees access to an AS.

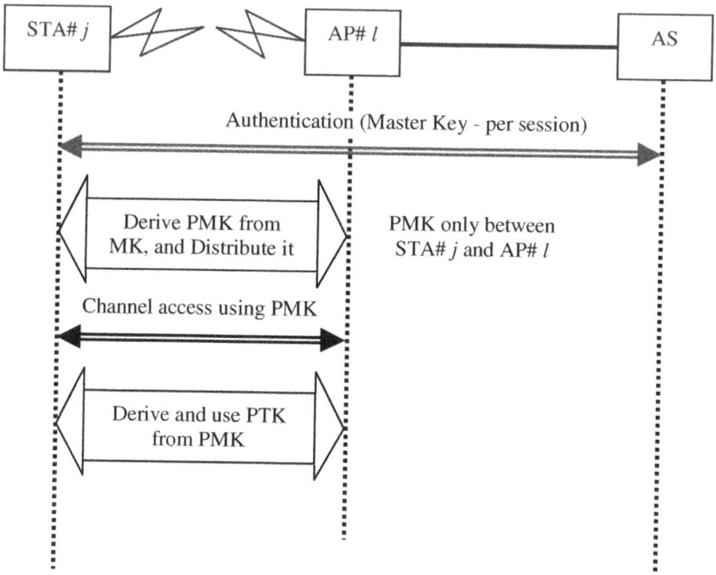

Figure 9-19. Relative position of master key (MK), pairwise master key (PMK), and pairwise transient key (PTK) in IEEE 802.11*i*.

9.7.2. Pairwise Master Key (PMK)

This is generated after a trust between AS and STA has been established (assuming a trust between AS and AP already existed). It is generated from the information sent by RADIUS (or non-RADIUS) AS to AP and STA. PMK is used to access the wireless medium.

9.7.3. Pairwise Transient Key (PTK)

PTK is a collection of keys known as key confirmation key (KCK), key encryption key (KEK), and temporal key (TK).

KCK is used to prove the possession of PMK, thus binding PMK to STA/AP. KEK is used to distribute group transient key (GTK) to be used in multicasting and broadcasting. TK is used for data encryption. Figure 9-20 shows protocol architecture for security for authentication.

IEEE 802.1X is the transport for EAP over LAN (EAPoL). IEEE 802.1X remains within the wireless LAN. Authentication transport between STA and AS is provided by higher-level (end-to-end) EAP-TLS (Extensive authentication protocol—transport level security). What IEEE 802.1X does for LANs is similar to what RADIUS does for IP network, that is, to provide a transport for EAP.

Figure 9-21 shows various operational phases of IEEE802.11*i*. The required data encryption algorithm is a version of advanced encryption system (AES).

In addition to providing a fairly comprehensive hierarchy of key management, IEEE 802.11*i* also has some features to enhance the networking capabilities of the stations. Two of these that need special mention are, pre-authentication for fast roaming and key-caching for fast reconnect. In a wireless LAN infrastructure served by a number of access points, a STA using IEEE 802.11*i* could spend a significant amount of overhead in authentication while moving from close to one AP to another. The handoff between the APs can be rendered quicker by allowing a STA to pre-authenticate with a prospective next AP before actually establishing an authorization relation with it. The STA does this through its current AP. Since APs are connected through a wired

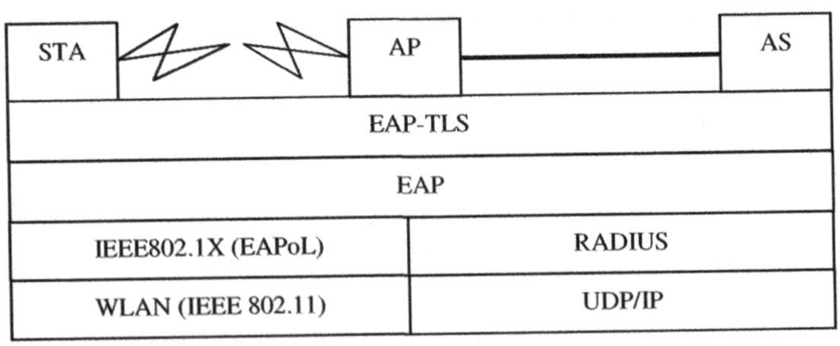

Figure 9-20. Protocol architecture for IEEE 802.11*i* authentication.

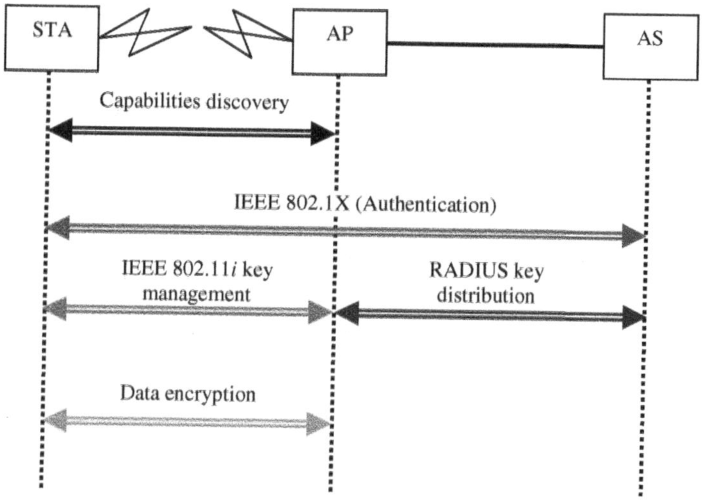

Figure 9-21. Operational phases of IEEE 802.11*i*.

infrastructure, pre-authentication adds an element of security in addition to quickly allowing authentication once the handoff process begins. The *key-caching* feature of IEEE 802.11*i* allows an AP to keep the key information of the STAs that recently de-associated. If a station tries to reconnect while cache entry about its previous connection is still valid, it can be quickly authenticated and associated.

9.7.4. IEEE 802.11*i* and WPA

The WPA is an interim standard and it was projected to be compatible with the then future IEEE 802.11*i*. However, IEEE 802.11*i* has specified a different encryption algorithm (AES) from WPA (RC4). This has created an interoperability problem between the two. There are some commonalities, such as in authentication (IEEE 802.11*i* provides pre-shared key as well for home networks). Table 9.2 (available at various locations, amended from [18] [15]) lists a comparison between WEP, WPA and IEEE 802.11*i* encryption and integrity protection.

The IEEE 802.11*i* is also dubbed as WPA2. However, it is not considered to be available in the existing devices, as it needs hardware upgrade.

9.8. SECURITY IN CELLULAR NETWORKS

In terms of data security, cellular networks pose a scenario quite different from WLANs. Due to wide area roaming capability and international market for operators, the network interoperability requirement and export constraints for security algorithms make a unique paradox. With the all-IP based networks,

TABLE 9.2. Comparison of Encryption and Integrity Attributes of Three Leading WLAN Security Architecture

Attribute	WEP	WPA	IEEE 802.11*i*
Cipher algorithm	RC4	RC4	AES
Key sizes	40-bit or 104-bit	128-bit encryption and 64-bit MIC	128-bit
Key lifetime	24-bit IV	48-bitIV	48-bit IV
Packet key integrity	Concatenation of IV	Mixing function	Not needed
Header integrity	None	Michael	CCM
Data integrity	CRC-32	Michael	CCM
Replay protection	None	Use IV	Use IV
Key management	None	EAP-based	EAP-based

the cellular security architecture will gain tremendously from a rather strong and maturing Internet security architecture. However, export restrictions may again pose a roadblock on the way to a uniform end-to-end security. The Wassenaar Arrangement among industrialized countries has put the key size limit for export to 56 bits [19] [20]. However, since export rules are a political and trade issue, they are subject to change, placing an extra requirement of flexibility and scalability on all security systems. In the WCDMA [21]-based networks (i.e., 3GPP) criteria for cryptographic algorithm, this fact has been kept in view. The North American cdma2000 (3GPP2) follows the same principles.

9.8.1. WCDMA Security Architecture

Figure 9-22 shows the security architecture for WCDMA networks. The security architecture is deigned to provide mutual authentication (user and network), and a safe air-interface with data encryption and immunity to modifications. User confidentiality is a part of earlier generations, and has been considered important for the 3G networks as well. The 3GPP is supposed to improve on the GSM security [24] of weak authentication (due to a weakness in the authentication algorithm COMP128 and A5 being breakable to reveal cipher key), a key-length of 32-bit for data integrity, no mutual authentication (attacker can fake as network) and cipher key transmission as plaintext. As seen from Figure 9-22, the security architecture divides the overall system security into four domains, namely:

1. **Network access domain**: Network access domain extends from user equipment to the serving node (SN). The message exchanges labeled as 'I' in Figure 9-22 are for network access domain security. These messages are exchanged among user equipment (ME, Mobile Equipment), Smart-Card (USIM for Universal Subscriber Identity Module) and Home network Environment (HE).

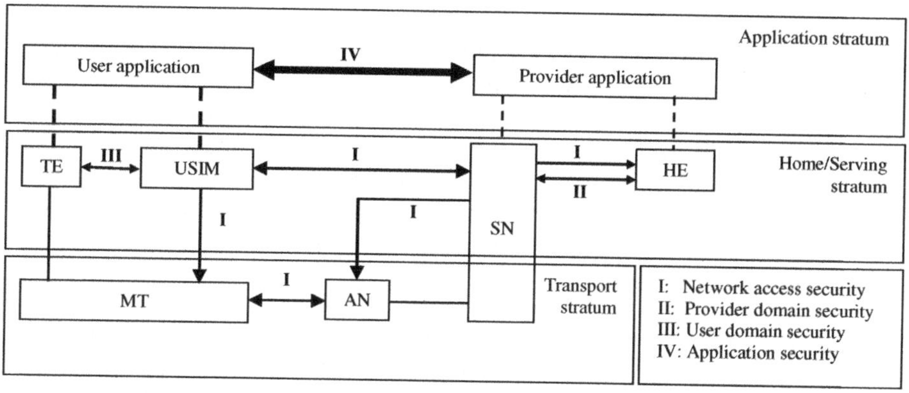

Figure 9-22. 3GPP security architecture. (Source [22] and [23].)

2. **Provider (WISP or Operator) Domain**: Labeled as 'II', the provider domain security message flow between the home environment and supporting node.
3. **User Domain (Home/Business)**: The user security domain is between the SmartCard and the terminal equipment. SmartCards are a portable mechanism of carrying the user's security attributes including keys and other information. SmartCards can be easily protected by a personal identification number (PIN). They have to provide a mechanism of user identification besides their own protection against stolen card.
4. **Application Domain (e.g., HTTP)**. Application domain security is required for ascertaining the authenticity of the applications, both receiving and sending. For a breakdown of security needs in individual domains, reader is referred to [25]. In the following we will highlight the security components of the 3GPP system summarized from [26].

9.8.1.1. User Confidentiality. User confidentiality measures stop potential eavesdropper from gaining access to user identity (International mobile subscriber identification number—IMSI), location, and subscription profile. The confidentiality is achieved by visited network allocating temporary MSI (TMSI).

9.8.1.2. Mutual Authentication. An authentication and key agreement (AKA) phase allows both the user and the network authenticate each other. Also, two keys are generated during this phase, one for encryption algorithm, called the cipher key (CK) and the other for data integrity algorithm, called the integrity key (IK). The message exchange for AKA uses a secret shared key between the authentication center (AuC) and the universal SIM (USIM) [27]. When roaming, the visiting network derives users secret key from the encrypted message received from users home network.

9.8.1.3. Data Integrity and Encryption. The data integrity and encryption algorithms are part of a type of Kasumi algorithm.[15] Using IK, the sending elements employs the f9 algorithm from Kasumi to generate the message authentication code. The recipient uses the same key (symmetric key) to cross-check data integrity with the help of f9. Data confidentiality (encryption) employs the f8 algorithm of Kasumi, which generates ciphertext with the help of CK. The function f8 generates a keystream that is logically mixed with plaintext at the sending end and with ciphertext at the receiving end, to provide encryption and decryption.

9.8.1.4. Flexibility. The user is allowed to decide a range of capabilities to configure security services. This includes USIM authentication, accepting/rejecting non-cipher incoming/outgoing data, or a choice of encryption algorithm.

9.8.2. Security in cdma2000

The North American standard for 3G, that is, cdma2000 (also, now being developed under 3GPP2) has the unique advantage that it is evolved from original CDMA (IS-95). The developers of CDMA technology always claim that the technology provides a privacy element inherent due to the nature of technology (spreading using a long code mask derived from the equipment serial number plus scrambling). For details, the reader is referred to [28] and [29] and many other articles, books, research papers, and specification documents available on this topic.

At the center of CDMA security apparatus is the 128-bit A-key, which is used for data integrity (to generate a 64-bit key) as well as data confidentiality (to generate another 64-bit key). After Release C, the Authentication Key Agreement (AKA) is to be adopted by 3GPP2, to provide the same level of key generation and protection infrastructure as 3GPP.

Temporary Mobile Subscriber Identification (TMSI) provides protection against eavesdropping, just like in 3GPP. Also, a countermeasure for cloning requires that the network and the user equipment maintain and exchange a 6-bit call history. The clone will not have the history, and thus can be caught unless the counters happen to have the same value.[16] Once a cloning attack is reported, the user can change the shared keys. For this purpose, OTASP (over the air service provisioning) could be employed to send a new authentication key (A-key). OTASP uses a 128-bit Diffie-Hellman signature for A-key transmission.

9.8.2.1. Using the A-Key. The A-key is used by an algorithm called CAVE (cellular authentication and voice encryption) along with a random number

[15] Derived from 'Misty' algorithm, (Kasumi, in Japanese, means 'Mist'). The Misty algorithm was designed by Mitsubishi.

[16] The probability of this happening is $1/2^6 \approx 1.5\%$.

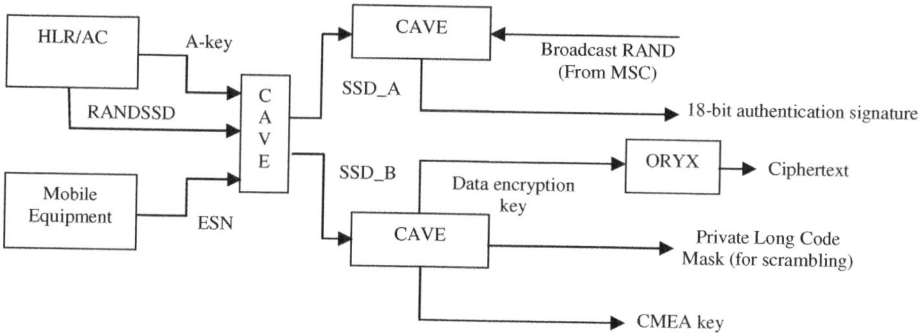

Figure 9-23. Various keys generated using A-key, ESN, and random numbers in CDMA.

generated by HLR/AC and the ESN to generate 128-bit 'sub-key' called *shared-secret data* (SSD).[17]

SSD is split into two parts, SSD_A and SSD_B, each of size 64 bits. SSD_A is used for authentication signature and SSD_B is used for generating keys for (signaling) data and voice encryption. Figure 9-23 shows the use of the A-key.

The data key is a 32-bit key used by an encryption algorithm ORYX. CMEA (The cellular message encryption algorithm (CMEA) key is a 64-bit key used to encrypt cellular signaling messages. The PLCM (private long code mask) is used in data scrambling. It is not known to anyone except the user and the network, and thus adds another layer of protection.

The A-key is programmable and can be changed in the event of its secrecy being compromised. However, in the event of roaming, it is the SSD that is shared. When a roaming customer returns, a new SSD is generated using a different RANDSS (the HLR/AC generated random number).

9.8.2.2. Amendments from Earlier Generations. In addition to adopting AKA, the cdma2000 is projected to use stronger keys for data integrity as well as confidentiality. A 128-bit SHA-1 is used for message authentication and 128-bit AES for data protection.

9.9. FINAL WORD

Placing Figures 9-5, 9-6, and 9-7 together and looking back at this chapter, one can get an appreciation of the fact that security planes are available for the whole network architecture, for wireless and wired networks. Following are some of the reasons that may still prevent future standards from being integrated with security.

[17] AC in cdma2000 = AuC in WCDMA.

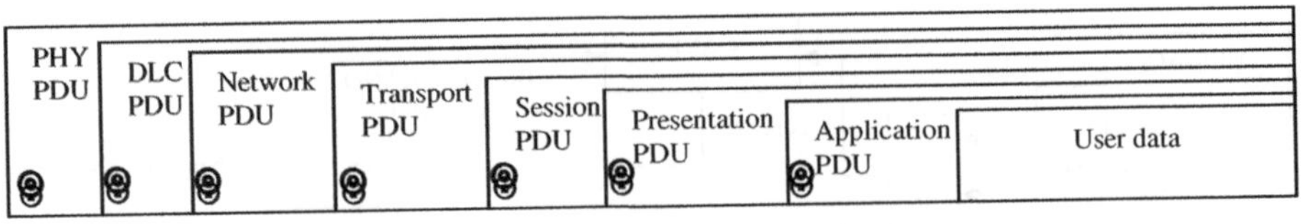

Figure 9-24. Secure physical layer PDU encapsulating six secure PDUs.

1. *Export laws* governing security algorithms limit an integrated approach of security architecture with network data and control architecture. These laws are subject to change, depending on trade relations, and tend to be different among different nations. This favors a patched approach, as has been taken so far. The only issue with the current approach is that it is not in league with network deployment, especially in WLANs. IEEE 802.11*i* seems more of a reaction to WEP vulnerabilities than the result of a thought process perceived with the original standard.
2. The *need of security* varies from device to device, user to user, and application to application. This closes many doors for an integrated approach. Even if a wireless data network has been officially regarded as a sensitive device among unclassified data devices, that does not easily convince everyone to spend more on security. Enforcing security requirements is like forcing everyone to live in a secure, walled house.
3. Looking back at Figure 9-5, the PHY PDU from a secure network will look like that shown in Figure 9-24. All PDUs have a security lock that must be opened before that PDU can be interpreted. For a network like OSI, this results in seven security locks. They all must have different form and keys. In symmetric key algorithms, where keys are generated after authentication, the key management generates *extra traffic* on the network. Processing a number of strong encryption algorithms per session in a WLAN AP will have processing overload. According to some estimates, current processing powers of APs are such that the utilization is above 90%. Obviously, security is going to slow down these devices. Perhaps, networks with minimum protocol layers could prove to be more efficient for this reason.

9.9.1. Alternative View

One of the main reasons for networks being compromised, attacked, hacked, and 'infected' so easily is that we still do not understand network security enforcement. Until an enforcement mechanism is invented, security solutions have to make slow progression. However, we understand from circuit switching that the user data network can be operated completely independent of a supervising (signaling) network. The common channel signaling network, such as SS#7, defines its architecture, resources, components, and protocols separately from the telephone or cellular networks that it supervises. A futuristic

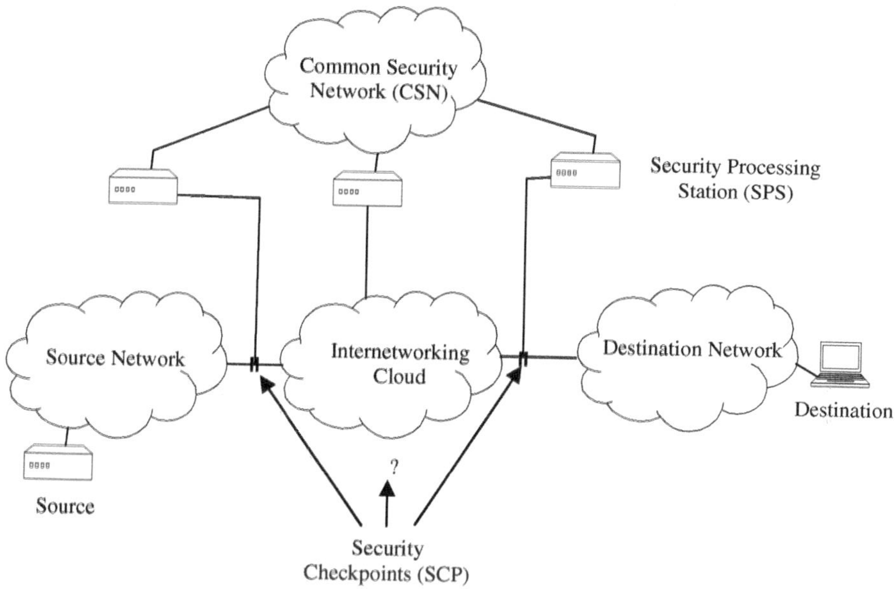

Figure 9-25. A common security network (CSN) may be a future solution.

view of network security can be hypothesized from the relation between SS#7 and PSTN. Maybe we can define, design, and enforce security through a separate, public network that the user does not even have to know. Figure 9-25 explains the idea.

By appropriately defining the architecture of a *common security network* (CSN), we may be able to distribute the security function among its components, such as *security checkpoints* (SCP) and *security processing stations* (SPS).

REFERENCES

In addition to the reference made inside this chapter, we consulted the documents listed below but not cited. We apologize for any missing references above and below.

[1] B. Aboba, Blunk, L., Vollbrecht, J., Carlson, J., and Levkovets, H., 'Extensible Authentication Protocol', *IETF RFC 3748*, June 2004.

[2] Shu Lin and Costello, Daniel J. Jr., *Error control coding: fundamentals and applications*, Prentice-Hall, Englewood Cliffs, 1983.

[3] D. Eastlake, 3rd and Jones, P., 'US secure hash algorithm 1 (SHA1)', IETF *RFC 3174*, September 2001.

[4] R. Rivest, 'The MD-5 message-digest algorithm', *IETF RFC 1321*, April 1992, Errata June 1992.

[5] H. Dobbertin, 'The Status of MD5 After a Recent Attack', RSA Labs' CryptoBytes, Vol. 2 No. 2, Summer 1996. http://www.rsa.com/rsalabs/pubs/cryptobytes.html

[6] H. Krawczyc, Bellare, M., and Canetti, R., 'HMAC: Keyed hashing for message authentication', *IETF RFC 2104*, February 1997.

[7] NIST, 'Key Management Guideline', Workshop Document, Nov. 2001, Available from http://csrc.nist.gov/cryptoToolkit/kms/key_management_guideline.(workship).pc

[8] David Carts, 'A review of the Diffie-Hellman algorithm', Available from www.sans.org/rr/papers/20/751.pdf

[9] Netscape, 'Introduction to public key cryptography', http://developer.netscape.com/docs/manuals/security/pkin/contents.htm

[10] Peter Gutmann, 'Encryption and Security Tutorial' (9-part slides), *University of Auckland*, www.cs.Auckland.ac.nz/~pgut001

[11] IEEE P802.11/ANSI, 'Part 11: Medium Access Control (MAC) and Physical Layer (PHY) specifications', *ANSI/IEEE Std 802.11*, 1999.

[12] William A. Arbaugh Shankar, Narendar, and Wan, T.C. Justin, 'Your Wireless LAN Has No Clothes', available from www.cs.umd.edu/~waa/wireless.pdf

[13] Nikita Borisov Goldberg, Ian, and Wagner, David, 'Security of the WEP algorithm', www.Isaac.cs.Berkeley.edu/Isaac/wep-faq.html

[14] Tech FAQ, 'What is WEP (Wired Equivalent Privacy)', www.tech-faq.com/wireless-networks/wep-wired-equivalent-privacy.shtml

[15] Michael Disabato, 'Wi-Fi Protected Access™: Looking Down the Link', *Wi-Fi Protected Access™ Web Cast*, June 2003.

[16] Jesse Walker, '802.11 Security Series: Part II: The Temporal Key Integrity Protocol (TKIP)', *Intel Corp.* www.ida.liu.se/~TDDC03/literature/wireless/links.html and also Intel website.

[17] Niels Ferguson, 'Michael: An improved MIC for 802.11 WEP', *IEEE 802.11*, January 2002, http://grouper.ieee.org/groups/802/11

[18] Nancy Cam-Winget More, Tim, Stanley, Dorothy, and Walker, Jesse, 'IEEE 802.11*i* Overview', http://csrc.nist.gov/wireless/S10_802.11i%20Overview-jw1.pdf

[19] 3GPP TR T-12, '3G Security; Criteria for Cryptographic Algorithm Design Process (Release 1999), *ARIB TR-T12-33.901*, Version 3.0.0, July 1999.

[20] 3G TS 33.102, '3rd Generation Partnership Project (3GPP); Technical Specification Group (TSG) SA; 3G Security; Security Architecture.

[21] 3G TS 21.133, '3rd Generation Partnership Project (3GPP); Technical Specification Group (TSG) SA; 3G Security; Security Threats and Requirements'.

[22] Bart Vinck, '3GPP security architecture', *IRR Fraud and security conference*, London, March 2000.

[23] Nam Yul Yu, 'Security features of W-CDMA systems: Part I', www.comsec.uwaterloo.ca/~nyyu/Security%20features%20of%20W-CDMA%20systems-partI.pdf

[24] Roy Campbell McKunas, Dennis, 'Analysis of Third Generation Mobile Security', *Annual Motorola Project Report*, June 2002, choices.cs.uiuc.edu/MobilSec/posted_docs/3G_Security_Annual_Report.ppt

[25] John K. Zao, 'Use of IPSec and IKE in Universal Mobile Telecommunication System', *IPSec 2000*, Paris October 2000. Presentation at: www.csie.nctu.edu.tw/~jkzao/Publication/Talk,IPSec Use in 3G-UMTS (IPSec2K 200110).pdf

[26] Myagmar Gupta, '3G Security Principles', August 2001. Available from choices.cs.uiuc.edu/MobilSec/ as 'Overview of 3G Security'.

[27] Bart Preneel, 'Mobile Network Security', February 2004, www.iaik.tu-graz.ac.at/teaching/00_angewandte kryptografie/slides/MobileSecurity.pdf

[28] Christopher Wingert and Naidu, Mullaguru, (contacts), 'CDMA 1xRTT Security Overview', *QualcommIncorporated White paper* available from www.cdg.org/technology/cdma_technology/white_paper/cdma_1x_security_overview.pdf, August 2002.

[29] Andrei Mihaila, 'Security in CDMA based networks', available in δ*doc form from www.cs.joensuu.fi/~amihaila/work/dswc/Security_in_CDMA_based_networks.doc

[30] B. Schneier, *Applied Cryptography*, 2nd Ed., John Wiley and Sons, New York, 1996.

[31] D. Coppersmith, 'The data encryption standard (DES) and its strength against attacks', *IBM Journal of Research and Development*, May 1994.

[32] Arch Red Oy and Huhtanen, Karri, 'WLAN Security is in the Architecture', *ICEFIN Workshop*, April 2004.

[33] John D. Howard and Longstaff, Thomas, 'A Common Language for Computer Security Incidents', *SANDIA Report SAND98-8667*, October 1998.

[34] Guy Pujolle, 'Security and Mobility Management in the Imbedded Internet', *University of Paris 6 Guy.Pujolle@lip6.fr*

[35] Sami Uskela, 'Security in Wireless Local Area Networks', *Helsinki University of Technology, stu@iki.fi*, December 1997.

[36] Marco Casole, 'WLAN Security—Status, Problems and Perspective', *Ericsson AB, marco.casole@ebc.ericsson.se*

[37] Tom Karygiannis and Owens, Les, 'Wireless Network Security: 802.11, Bluetooth and Handheld Devices', *NIST Special Publication 800-48*, November 2002.

[38] Bernard Aboba, 'IEEE 802.1X Pre-Authentication', *IEEE 802.11-02/389r0*, June 2002.

[39] Xing Haitao, 'Flaws in the Security Architecture of IEEE 802.11', *Research Topics in Networks and Distributed Systems*, Xing_haitao@msn.com

[40] Peter Gutmann, 'Encryption and security tutorial', *University of Auckland*, Available from www.cs.auckland.ac.nz/~pgut001, Part 2: Key management and certificates'. See also [10].

CHAPTER 10

ROUTING IN WIRELESS LANs

There are subtle differences in routing protocol requirements for wired and wireless networks. The requirements vary from one network type to another. There are the mobile, wired networks, infrastructure networks, ad hoc networks and cellular networks[1]. The wired networks that allow mobility (i.e., *portability*) are not wireless in general. However, due to portability the location of a terminal may change from time to time. This makes the use of existing IP protocol implementations inefficient. Wireless infrastructure networks add another dimension by loosening the 'last hop' and allowing roaming. The independent, ad hoc networks add yet another factor of requiring at least some stations to have relay/routing capability. The cellular networks, though centralized like infrastructure networks, accentuate the routing problem due to highly dynamic environment as compared to all other network types.

It is a common belief that connectionless IP is the integrating protocol of the future wireless networks. Since routing in IP networks is matured already, a lot of knowledge and experience is available to achieving this end. We will look at the phenomenon of routing in wireless data networks in this chapter. Our main emphasis is on independent wireless LANs.

[1] Personal Area Networks (PANs), too, are getting noticeable presence lately. However, instances of their cluster sizes requiring routing are rare.

10.1. ROUTING IN INFRASTRUCTURE NETWORKS

Infrastructure WLANs have successfully been deployed all over the world. The term 'hot-spot' is also used to refer to such networks in busy places like airports, hotels, and conferences. Part of the reason of their success is the ease with which routing can be performed in such networks. Since most access points are connected to a broadband fixed network, IP with the help of dynamic host configuration (DHCP) is a natural choice for routing.

The IEEE 802.11 suite of WLAN protocols have left room for the distribution function (DF) in order to define customized sets of infrastructure networks called extended service set (ESS), as shown in the Figure 10-1.

The distribution function (DF) that could help route 802.11 frames from one station to another through two or more access points in an extended service set (ESS). However, two mobile stations do not have to be in the same extended services set (ESS) in order to need to communicate via a routing mechanism. If the access points are connected to an IP network, then the access point could simply forward every non-local packet to a gateway router, which can use an Internet Protocol layer routing protocol to find the destination WLAN access point. In this way, IP can be used to route data among infrastructure WLANs. Similarly, if the WLANs access point can be connected via a virtual circuit network (e.g., ATM, as in HIPERLAN), switched or permanent virtual circuit identifiers (VCIs) can be defined among the access points for a connectionless routing among WLANs. In short, infrastructure networks lend themselves to a manipulation in a number of ways, due to the access point

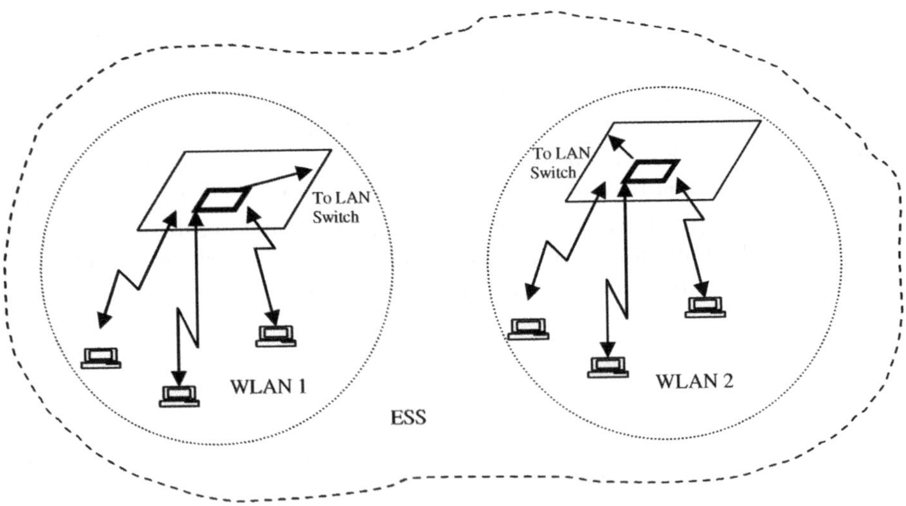

Figure 10-1. Extended service set (ESS) can employ roaming and routing mechanisms to a set of WLANs in IEEE 802.11 networks.

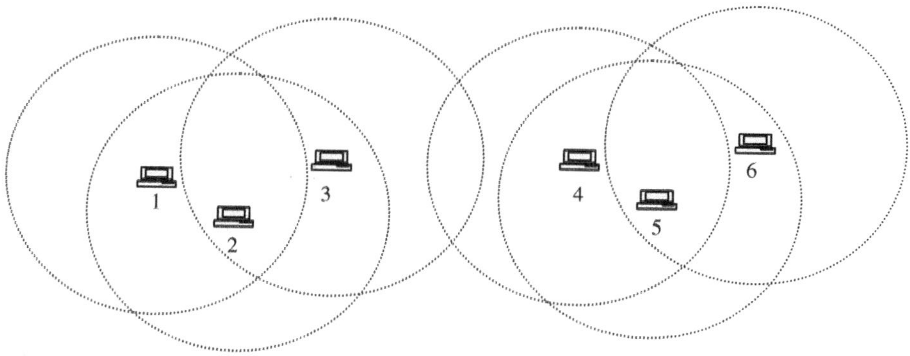

Figure 10-2. Ad Hoc Network with six Stations. Circles show the communications reach of a Station. More than one Stations in a circle means that the stations can exchange packets.

connected to the fixed network. In ad hoc or independent WLANs, there is no access point. If a station has a packet to send to another station outside the directly accessible range of the sender, the sending station has to depend on one or more stations between itself and the destination.

10.2. AD HOC WIRELESS NETWORKS

Figure 10-2 shows an ad hoc network with six terminals at the same locations as in Figure 10-1. The difference between the two figures is that, due to the absence of access points in Figure 10-2, each terminal could send or receive only within the area shown by dotted circles.

Having more than one Station in a circle shows that the corresponding stations can communicate with each other. It is obvious from the figure that:

Station 1 can communicate with Station 2 directly.

Station 1 can communicate with Station 3 only if Station 2 can forward packets from Station 1 and 3.

Station 1 can't exchange packet with Stations 4, 5, and 6. If there was a station in the overlapping areas of Stations 3 and 4, then all stations could exchange packets. From this figure, it is obvious that, in order for ad hoc network stations to communicate with each other, every station should have the capability of packet forwarding. Even that is not sufficient, because (as seen here), a group of stations can be isolated from another group. Thus, routing in Mobile Ad Hoc Networks (MANETs) is a very complex function and could affect the station as well as the whole network.

10.2.1. Characteristics of MANETs

The mobile ad hoc networks have four distinguishing characteristics that separate them from other wired or wireless networks. These are:

1. Dynamic topologies: The *diameter* of a MANET, in number of hops, is the maximum of the minimum distances between all node pairs in the network. Ad hoc networks have a variable diameter.
2. Bandwidth constraints: It has been shown [2] that the capacity of a wireless network reduces with the increasing number of participating nodes.
3. Energy constraints: Generally, the nodes in an ad hoc wireless network are battery operated. This imposes operational constraints on the use of a device as a routing or relay node.
4. Limited security: By its very nature, the ad hoc network stations are vulnerable to intruders.

10.2.2. Goals of the IETF MANET Working Group

The MANET WG has set many goals. These include considering a wide networking context including many environment and network sizes, support traditional, connectionless IP service, provide a standard 'protocol or mode of discovery' algorithm to help new terminals understand the network mode and design effecting routing protocols in wake of topological changes.

10.2.3. Sources of Failure in MANETs

The MANETs are inherently more vulnerable to failures than other network types. There are failure sources common with other network and also ones specific to MANETs. Some of these sources are as follows:

10.2.3.1. Topological Failures. These are the failures due to ad hoc changes in topology. There are three sources of failures in this category.

1. Link failure due to power drainage of one or more relay nodes;
2. A node is powered off; and/or
3. A node moves out of the reachable area.

10.2.3.2. Channel Failures. One or more links could have a long fading spell, or simply an additive of multiplicative noise signal. These failures are shared with other network typs.

10.2.3.3. Protocol Failures. These, too, may be shared by other networks. Protocol failures include:

1. Lack of congestion mechanisms on one or more links.
2. Lack of scalability to the number of users in an ad hoc group.
3. A relay node has a data filter (e.g., a firewall).
4. Security failures owing to unprotected access to any terminal in the network.

10.3. CHARACTERISTICS OF A GOOD ROUTING PROTOCOL

The performance of a MANET routing mechanism may be derived from its characteristics. These include the following [10]: routing overhead, user control over route selection, automatic load balancing capability, robustness to packet and node losses, Loop-free routing, Use of unidirectional links, soft-state routing, recovery from topological changes, multicasting, QoS provisioning, and robustness to battery level variations of nodes. Other characteristic properties that determine the strength or weakness of MANET routing protocols include security and sleep period operation.

10.3.1. Performance Metrics

In view of the above qualities in a good routing protocol, the performance metrics consist of end-to-end data throughput and delay, route acquisition time, efficiency—either in the form of overhead or throughput versus input traffic. Security, of course, remains of paramount interest.

10.3.2. Networking Context

Size, connectivity, topological change rate, link capacity, fraction of unidirectional links, traffic patterns, mobility, fraction, and frequency of sleeping nodes, all determine the actual performance of a routing protocol in operation.

Additionally, the effect on battery drainage plays a critical role in designing routing protocols for networks allowing any kind of mobility. Consequently, the least cost does not translate into the shortest path, as mostly is the case of traditional packet-switched networks. Energy saving routing in active and passive operational modes may have differing implications, as discussed in [1]. Active mode requires power control and path selection based on minimum energy transmission. Passive mode operation requires a power-save mode, in which stations switch-off or transmit low power in case of not being active.

Another factor considered to be crucial is the load balancing. Under high loads, terminals have to transmit at higher power to combat interference. Therefore, routing protocols with load distribution help balance energy utilization across a network. Depending on these and other factors, there are a large number of proposals for ad hoc network routing protocols. The URL http://www.wikipedia.org/wiki/Ad_hoc_protocol_list lists a large number protocols of routing in ad hoc networks under 'Pro-active', 'Reactive', 'Hierarchical' 'Geographical' 'Power aware' 'Multicast', 'Geocast (Geographical Multicast)' and 'Other' categories. Also, http://www.cmpe.boun.edu.tr/~emre/research/msthesis/node1.html lists and explains several protocols along with some figures.

10.4. CLASSIFICATIONS OF ROUTING PROTOCOLS

The routing protocols for MANETs can be classified in many ways, depending on their adjustment to route failures (pro-active, reactive), mechanism of route usage (link state, distance vector), route maintenance mechanism (periodic, on demand).

10.4.1. Pro-active and Reactive Routing

Since the locations of wireless nodes are not fixed, the route in a wireless LAN is subject to topological variations. Even though mobility in a typical WLAN environment is low, whenever an intermediate station moves, it creates the potential for a routing update. The routing update could occur either at regular instants in prediction of route changes, or it could be done on discovery of a route change. The former is called pro-active routing and the latter is called as reactive routing (or on-demand routing). In pro-active routing, a route maintenance and update mechanism is adopted while in reactive mechanisms routing decision is changed only on discovery of a route change impacting a delivery. Pro-active routing is expensive, both in terms of battery life and traffic overhead. However, it is more efficient in terms of end-to-end delay. Reactive routing saves much from the need to maintain route, but could result in excessive delay.

10.4.2. Link State Versus Distance Vector

Routers in the Internet have traditionally employed one or both of these types of protocols. In link state protocols (such as open shortest path first—OSPF), routers broadcast their neighbors' states to the network and receive similar information from other routers. Based on the information thus collected, and using Dijkstra's shortest path algorithm, the routers chalk out the best path to all destinations available. In distance vector algorithms, the routers broadcast the router-to-destination paths calculated using the Bellman-Ford algorithm. Route information protocol (RIP) is an example of distance vector algorithms.

10.5. ROUTING PHASES

A typical MANET route protocol is expected to have a multi-phased routing. Various phases may include a set of the following.

1. Route discovery: Process of finding one or more paths between source and destination node.
2. Route caching: Process of storing routes for quick and easy retrieval.
3. Route maintenance: Process of checking the validity of routes and finding alternative routes in case of a route losing its validity.
4. Cache maintenance: Forming new route caches with changes in topology. Adjusting timers for the cache entries.
5. Route deletion: Clearing entries from routing tables for routes that are not valid any more.

10.6. ROUTING MECHANISMS

In the following, we will briefly describe several routing mechanisms proposed for ad hoc networks in addition to summarizing performance comparison as derived from [3]. A detailed treatment of some very important protocols in beyond the scope here. We have selected one protocol for relative details, that is, the dynamic source routing (DSR) protocol.

10.6.1. Zone Routing Protocol (ZRP)

Proposed in [4][5][6], this routing mechanism breaks up the routing function into two parts, intra-zone routing and inter-zone routing. It is a hybrid of proactive and reactive techniques and adjusts itself according to the changing operational environment. ZRP is based on the (realistic) assumption that a change of topology does not affect the whole network in the same way as it affects a local area. A local area, or a zone, is defined in terms of a number of hops (zone radius). An intra-zone protocol is used within the zone for active updates and route maintenance. Each node in a zone keeps a record of all nodes so that the intra-zone protocol is sufficient for communications within the zone. If the destination node is outside a zone, then the inter-zone protocol is invoked. The inter-zone protocol relies on communications between peripheral nodes of adjacent zones. Each peripheral node searches for the destination within its zone (just by checking its routing table), and forwards the request to the peripheral node of the next zone in case of a failure to find the destination in its own zone. In this way, the final route consists of each hop equal to the zone radius. Figure 10-3 shows this mechanism.

In this figure, 12 stations are shown in a MANET, eleven of which make two zones. Nodes 1 and 2 are peripheral nodes. Suppose Node 8 has a packet

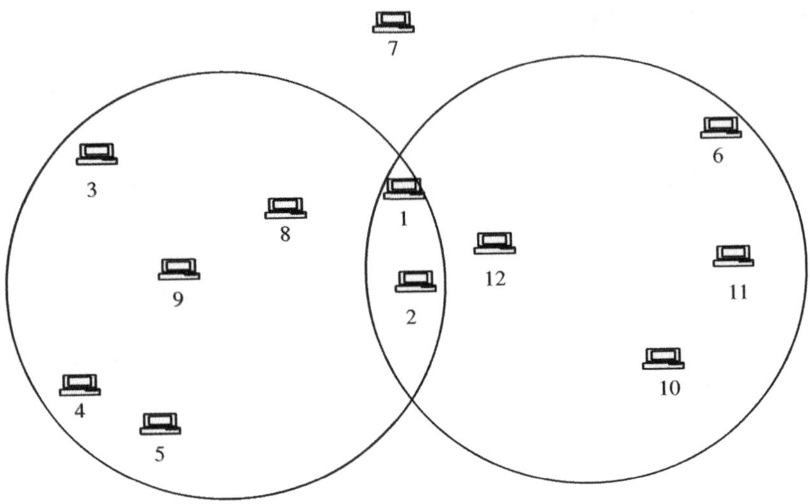

Figure 10-3. Ad Hoc Network with 12 Stations. Circles show the routing zones for intra-zonal routing. For routing between two stations in different zones, an inter-zone routing mechanism is employed.

to send to Node 12. It will first look in its table (that was made using an intra-zone protocol) for an entry for Node 12. If such an entry does not exist (which is the case), Node 8 will inform its peripherals (e.g., Nodes 1 and 3) to look for Node 12. Peripheral nodes will use a second protocol (inter-zone) protocol for peripheral-to-peripheral communication. Each receiving peripheral node will look into its table and respond if destination is found. In this case Node 2 (peripheral for Zone 1) will find Node 12 in its table and will return a response accordingly.

10.6.2. Dynamic Source Routing (DSR)

Dynamic Source Routing (DSR) protocol adds the use of route caching and dynamic updating to source routing. Source routing was proposed in IP networks as an option to prevent IP datagrams from taking unwanted routes. Source routing is done by pre-programming the complete route in every IP packet. In terms of protocol overhead, this puts DSR to an advantage over link-state and distance vector like protocols that require periodic exchange of information.

DSR requires each node to maintain a route-cache of all known self-to-destination pairs. If a node has a packet to send, it attempts to use this cache to deliver the packet. If the destination does not exist in the cache, then a route discovery phase is initiated to discover a route to destination. If a route is available from the route-cache, but is not valid any more, a route maintenance procedure may be initiated.

We have chosen DSR for more detailed description due to its adaptability to topology and load changes, and interoperability with other protocols.

10.6.3. Destination Sequenced Distance Vector (DSDV)

DSDV [7] is a hop-by-hop distance vector routing protocol requiring each node to advertise routing tables periodically. Each route is tagged with a sequence number. A route with a higher sequence number is considered to have higher priority. This protocol uses the Bellman-Ford algorithm adapted to the wireless environment.

10.6.4. Ad Hoc On-demand Distance Vector Routing (AODV)

As explained in [3], AODV uses some features from DSR and some from DSDV to have monotonically increasing sequence numbers of the route entries of DSDV and updating the route entries using a route discovery protocol (similar to the one in DSR). It is more of a reactive protocol in which the path discovery begins when a node has a packet to send. The protocol results in a bidirectional route.

10.6.5. Temporally Ordered Routing Algorithm (TORA)

TORA is based on 'link-reversal' idea. Each node has a 'height' with respect to the destination. If this height is above the destination, the data can flow downstream to the destination, otherwise, it can flow upstream from the destination. A sequence of nodes with the ordered 'heights' makes a route from source to destination. TORA is layered above Internet MANET encapsulation protocol (IMEP). IMEP [8] is a common protocol that could be used by other protocols to provide network address resolution and inter-router secure authentication.[2]

The above mechanisms are for general application. Due to the specialized characteristics of MANETs, there are proposals that emphasize MANET characteristics. Since there is no theoretical framework to consider these factors, we have included a separate section for a theoretical framework later in the chapter. At this point we will briefly mention some of these protocols.

10.6.6. Wireless Routing Protocol (WRP)

WRP emphasizes on loop-free paths and routing table updates to accommodate topology changes. Another hierarchical routing algorithm is presented in reference [5] of [4]. This uses two-layered routing for clustered networks; uses a spine for intra-cluster routing and link-state between clusters. Spine is computed using *minimum connected dominating set* (MCDS) algorithm.

[2] Internet-drafts are works-in-progress and may not be considered complete.

10.6.7. Mobile Multimedia Wireless Network (MMWN)

MMWN assumes that the endpoints make cells connected to cell-heads or switches. These switches form a hierarchy of cluster each of which forms a multihop packet radio network. There could be clusters of clusters. Link state is used as routing algorithms.

10.6.8. Transmission Power Optimization

10.6.8.1. Flow Augmentation Routing (FAR). This routing algorithm assumes static topology and minimizes the sum of link costs, each given by: $[e_{ij}]^{x1}$, $[E_i]^{x2}$ and $[R_i]^{x3}$, whereas;

e_{ij} = Energy cost per unit of transmission over the link (i, j).
E_i = initial energy of transmitting node.
R_i = Residual energy of transmitting node.
x_1, x_2, x_3 = Exponents of the energy parameters.

10.6.8.2. Online Max-Min Routing (OMMR). OMM finds the shortest (minimum energy P_{min}) path by using Dijkstra's algorithm. Then, it defines a set of paths not deviating more than zP_{min} from the shortest path. From, these, it chooses the one that maximizes minimum residual power. 'z' is the tradeoff between the max-min path and min-path.

10.6.8.3. Power-Aware Localized Routing (PLR). This mechanism depends on selection of appropriate neighbors without global route search.

10.6.8.4. Minimum Energy Routing (MER). MER includes the power levels that should be used by each intermediate node. These levels are calculated during initial phase when each receiving intermediate node calculates the required power from the knowledge of transmitted power and received power. The algorithm has eight options, some in firmware and others implemented in software.

10.6.8.5. Retransmission-Energy Aware Routing (RAR). This protocol considers the effect of link retransmissions on power consumption.

10.6.8.6. Smallest Common Power (COMPOW). This protocol requires each node to make routing tables for each of a discrete set of transmission powers ($P_1, P_2, \ldots, P_N$) and select the minimum of these that result in a connected graph.

10.6.9. Load Distribution Protocols

Load distribution protocols do not look for minimum energy routes, but least used intermediate nodes to increase the lifetime (most heavily loaded node is the lifetime) of the network.

10.6.9.1. Localized Energy-Aware Routing (LEAR). LEAR is like DSR, except that intermediate nodes, on getting a route request check the requested energy against their residual threshold. Also, if a route from an intermediate node to destination already exists, it is not automatically used without first getting confirmation from all intermediate nodes using a route-cache message. The threshold value is not fixed. The following protocols are for inactive stations.

10.6.10. SPAN Protocol

SPAN is based on selection of master nodes. These are the nodes that can access two or more mutually inaccessible nodes.

10.6.11. Geographic Adaptive Fidelity (GAF)

GAF allows each node to belong to a grid with one master. The grid association is done via a global positioning system (GPS). A station in a grid is in one of the three states, inactive, discovery, and active. In discovery, it looks for a master. If it does not find one, it becomes the master itself.

10.6.12. Prototype Embedded Network (PEN)

This protocol does not have a master. Every sleeping node wakes up and listens to the beacons.

10.7. PERFORMANCE COMPARISON

Work on performance analysis of the routing protocols for MANETs is still in progress. In Ref. [3], the authors use simulation to compare DSDV, TORA, DSR, and AODV. The following is a summary deduced from their results.

The paper uses a 'random waypoint' movement model with pause times on destinations and random speed. 10, 20, 30, constant bit rate (CBR) sources of rates 1, 4, and 8 packets (64 and 1024 octet) are used. Result summary, starting with the best:

Fraction of the application packets successfully delivered: DSR, AODV, DSDV = TORA

Number of routing packets sent per pause time: DSR, AODV, DSDV, TORA

Packet delivery ratio as a function of pause time: DSR, AODV, TORA, DSDV

Routing overhead as a function of pause time: DSR, AODV, DSDV, TORA

Fraction of application packets successfully delivered: DSR, AODV, DSDV, TORA

Comparison of the number of routing packets sent (speed 1 m/s): DSR, AODV, DSDV, TORA

Contrasting routing overhead in packets and in bytes:

In Packets: DSR, AODV, DSDV, TORA

In Bytes: AODV, DSR, DSDV = TORA

The paper compares these routing strategies by simulation in *ns*. Almost all the results show that DSR is superior to all others.

10.8. MULTICASTING

We do not consider multicasting in this chapter. We will base this section on a paper on this topic. The paper referenced in Ref. [9] claims that multicasting is a neglected area of research in MANET until the writing. *Hyper-flooding* is one of the ideas that can be used for multicasting in MANETs. In hyper-flooding, an intermediate node can rebroadcast a packet if it acquires a new neighbor. State is maintained only for those packets that are within a time to live (TTL). The paper holds that various sizes of MANETs require different multicasting mechanisms. Highly dense and large MANETs can be partitioned into clusters and hierarchical routing can be performed. Intra-cluster multicasting can use hyper-flooding, while inter-cluster multicasting can use unicasting. A seamless multicast routing operation in an integrated (fixed and mobile) environment translates into many requirements, such as mechanism for active on-the-fly switching among routing mechanisms as the host passes through various networks. However, each mobile can carry only one multicast algorithm, and would have to download new ones as it moves among networks.

10.8.1. Mobility Support Using Multicast IP (MSM-IP)

MSM-IP supports host mobility. A mobile is allocated a permanent multicast address and, as it moves to another area, it registers with the multicast server. For networks with fixed infrastructures, mobility is restricted to the last or the first hop only. One suggestion in Ref. [4] of [9] is to use mobile IP tunnels either between the sender and home agent or between the receiver and foreign agent.

10.8.2. Multicast Routing in MANETs

The only reported work is the shared-tree wireless network multicast (ST-WiM), which adopts PIMs (Ref. [6] of [9]) sparse mode algorithm to MANETs. A simulation of this algorithm confirms that the tree maintenance mechanisms' performance degrades due to mobility.

The Wireless Internet Gateways (WINGs) project at University of California, Santa Cruz, developed wireless IP routers to connect MANETs with fixed IP networks. These routers use an extension of RIP and RIPv2 called WIRP. It does not address multicasting.

10.9. DYNAMIC SOURCE ROUTING (DSR) PROTOCOL

In this section, we will describe the DSR protocol, which is a likely candidate for next-generation IETF protocols for MANET routing. It was originally proposed by David Johnson [2][6], and is described in a recent Internet-draft [10][3]. Table 10.1 lists the characteristics of DSR and their implications.

10.9.1. Protocol Operation

The DSR protocol is triggered by a packet generated at the source node (SN) for a destination node (DN) whose IP address is/(can be) known to the SN. The protocol has the following phases: (0) Route caching, (i) Route discovery, (ii) Data exchange, (iii) Route maintenance, (iv) Route deletion.

10.9.1.1. Route Caching. The SN checks for a route to the DN in its route cache. If there are one or more routes available, the phase (ii) starts. Otherwise, protocol signaling is triggered from phase (i).

10.9.1.2. Route Discovery. In this phase, the SN copies a control packet called Route Request to a buffer called Send Buffer (SB). It also transmits Route Request to the network. At the same time, it initiates a timer for the duration of SendBufferTimeout (SBT) within which it expects a response. The Route Request bears the address of SN, a unique identifier given by SN and the address of the DN. The Route Request is broadcasted to all nodes within the transmission range of SN. It is received by all nodes within a radius of one hop.

When a node receives a Route Request, it checks whether this Route Request has already been through it. There are two types of information that can be used for this purpose. First, in the Route Request packet there is a list

[3] Internet drafts are subject to changes. This one is valid until 19 January 2005. Care should be taken in consulting them. They should not be considered as reference material for actual standards.

TABLE 10.1. DRS-Characteristics and Their Implications

DSR Characteristic	Meaning	Implications
Soft state	Node/Route states discovered and re-discovered on as-needed basis.	A dropped relay node can rejoin the route after rebooting.
Route cache	Multiple routes available.	1. A new route discovery is not needed if existing route fails as long as there is a valid route in the route-cache. 2. Load balancing possible by selecting more than one simultaneous routes for a single communications session. 3. Loop-free operation possible by selecting non-overlapping routes.
On-demand	Reactive type protocol, responding to adjustments as the need arises.	Saves routing overhead of periodic route maintenance.
Unidirectional route support	A route is valid even if it can be used only for outgoing or incoming packets.	Nodes and WLANs that have only clear channel in one direction can participate in routing.
Route reversal	Once a route from source to destination is discovered, use the same route from destination back to source.	Quickens the route completion process in networks that require link level ACK or bi-directional communications (such as IEEE 802.11).
Hop limit	Specify TTL field to control maximum number of hops to destinations.	Can be used in delay-sensitive sessions to limit the end-to-end delay to a given number of hops.
Cached Route Reply	Route discover ending at an intermediate station which already has a route to destination.	1. The route discover phase is shortened. 2. If there is a duplicated relay node in both routes (from source and to destination), loops could be created.
Packet salvaging	Save packet from being discarded in case of a broken intermediate link.	1. Alternative route to destination (if available) is used. 2. If destination is down, there is possibility of false salvage attempts.
Automatic route shortening	Eliminate intermediate nodes if possible.	1. Network latency reduces. 2. Route optimization without going through discovery process again.
Routing with foreign network	Exchange packet with other DSR or Non-DSR networks.	1. The scope of routing can be arbitrarily extended through Internet. 2. Interoperability with other ad hoc routing mechanisms.

of all the nodes that have inserted themselves in the route for this Route Request. Second, the recipient node can check for a previous entry of this packet in a Route Request table. This is done by comparing its unique ID with the IDs of the Route Request packets listed in the Route Request table. If it already exists, then the packet is discarded, thus avoiding chances of loops. If it is a new Route Request packet, then the DN address is checked to see if the current recipient is the DN. If it is not the DN, then this recipient is a relay node (RN) and it adds its own IP address to the existing list of route IP addresses starting from SN. It makes an entry in the Route Request table and broadcasts the Route Request to within its own one-hop radius.

When the DN receives a Route Request, it checks for a previous entry in the same way as the RNs do. If this is the first packet from this sending node, it records the route from source to itself and responds with a Route Reply packet that carries the list of IP addresses of all nodes from source to destination. This is, however, assuming that (1) either a route to the source node already exists in the route cache of the destination node, or (2) all links from the source to destination are required to be used only as bidirectional links, such as in IEEE 802.11-compliant stations.

The DSR allows for using unidirectional wireless links, so that a different route in both directions can be used if required. In that case, the DN, on receiving a Route Request will initiate its own Route Request reversing the roles of source and destination nodes. In this Route Request, it will piggyback the route list of SN to DN to avoid an infinite cycle of Route Requests from both sides. There can be more than one route between a source destination pair.

At the end of a successful Route Discovery phase, both the SN and DN will have at least one cache entry for each other.

If a SendBufferTimeout expires before a Route Reply is received, then more Route Request packets can be used. In such a situation, an exponential backoff algorithm is used to spread the latter Route Requests by doubling the time between successive broadcasts. Figure 10-4 shows the Route Discovery phase.

10.9.1.3. Data Transmission Phase. Once a (set of) route(s) has been established, data packets can be exchanged by SN/DN over it. Each packet contains the complete sequence of the IP addresses from SN to DN as source route option. A relay node, by checking the list and comparing it with its own route to DN, simply broadcasts the packet to the next hop. Since transmission is on a link-by-link basis, a mechanism is used for mapping next hop IP into next hop link (MAC) address. The address resolution protocol (ARP) can be used for this purpose. The ARP cache can be updated pro-actively by reading the link address for a given IP address from any packet bearing that IP address instead of initiating an ARP procedure for every outgoing packet. For packets using the same route on both sides, usually a link layer ACK mechanism already exists. There are two more mechanisms for 'obtaining' an acknowledgement. These are *passive ACK* and *soft ACK*.

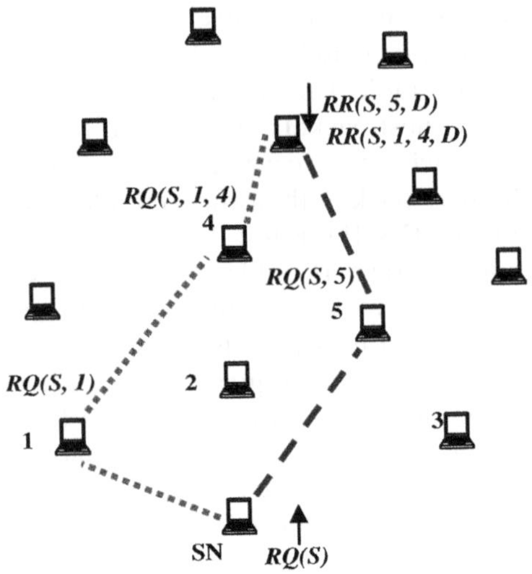

Figure 10-4. Route Request (RQ) and Route Reply (RR) with two possible routes between source (**S**) and destination (**D**) stations.

A passive ACK is possible due to the broadcast nature of the medium. Referring to the ad hoc network in Figure 10-4, suppose SN send a packet to node number 5 (NN5). NN5 receives it successfully, and forwards it to DN. Since SN is also within the transmission range of NN5, it can hear the packet relaying and get a passive ACK. Figure 10-5 illustrates this.

A soft ACK is explicitly requested as part of DSR Route Maintenance Procedure, discussed next.

10.9.1.4. Route Maintenance. The third phase of DSR protocol is route maintenance. Route maintenance is required by all routing protocols, especially the ones for MANETs due to very high probability of routes being lost. If an explicit or passive ACK is received by a node for every transmission, it keeps using the route(s). In case of non-availability of a MAC layer ACK and not overhearing a passive ACK, a node may request for an ACK using DSR ACK request. This is a *soft ACK* and is recommended not to be requested for every transmission. In the event of non-receipt of a soft ACK, a retransmission for ACK request can be done up until a certain number of times.

Route Error. A station, on failing to receive an ACK after certain number of ACK requests, assumes that the corresponding link is broken. It removes the link from all route entries in its cache (wherever it is used) and sends a Route Error packet to all its neighbors. On receiving the Route Error, neighbors follow suite. When the source node receives the Route Error on one of the links, it deletes the corresponding route from the cache. If there is another

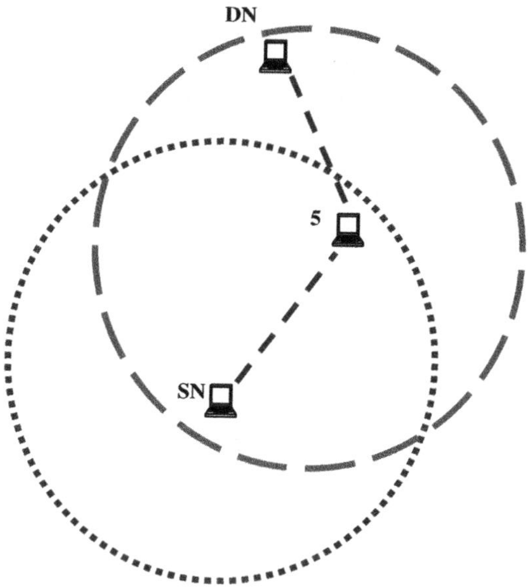

Figure 10-5. SN receives a passive ACK by eavesdropping relaying of packet by NN5 to DN.

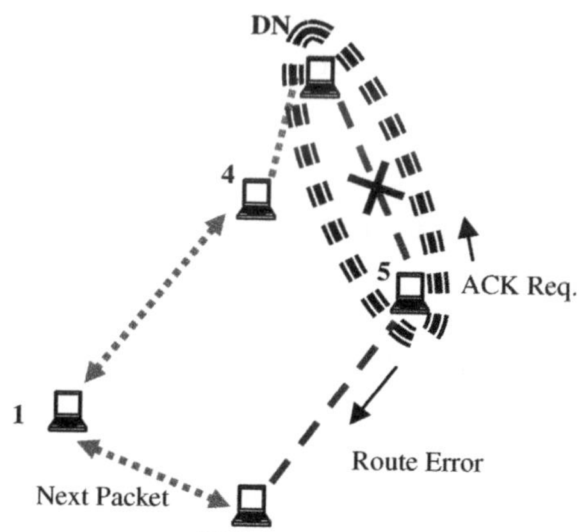

Figure 10-6. NN5 sends Route Error to SN on failing to get a soft ACK after several requests. On receiving Route Error, SN adopts new route through NN1.

route available to the same DN, the new route is assumed automatically and transmission continues un-interrupted. ACKs for packets received from DN by SN can be sent by using another higher-level protocol, such as TCP. Figure 10-6 shows breaking of the route (S,5,D) and adoption of (S, 1, 4, D).

10.9.2. Flow State Option

The normal DSR operation requires a complete list of all nodes en route from source to destination on every packet. In many instances of using an ad hoc network, this may not be necessary. An example is of a multimedia conference in an enterprise where wireless signal arrives in every office and all attendees are at their fixed places for most of the time. The flow state extension allows for a short cut to sending route list in every packet. This is done by establishing a soft state with a flow ID using DSR options header. The first packet that carries this option also carries the source-route. All the intermediate nodes create a soft state for this flow for a duration included in the flow-setting packet. Once a soft state has been established, the subsequent packet can use the flow ID without the source route list of all relay nodes on the route. In the example of multimedia conference, elimination of source routing can expedite the end-to-end packet delivery.

10.9.3. DSR Packet

The DSR is an option in IP. Its existence is recognized from the protocol field in IP packet. Thus, DSR header is encapsulated just like a transport layer protocol. The DSR header is a multiple of 4 octets. Every node detecting a DSR header from the Protocol field initiates the DSR procedures unless it is indicated that the node should not execute the options.

Figure 10-7 shows the DSR options header. The options field is defined separately for each option. The Internet-draft on DSR [10] defines the options listed in Table 10.2.

10.10. SELECTING THE BEST ROUTE

The DSR protocol provides the capability of using multiple routes between a source-destination pair. It, however, does not provide a mechanism (or framework) to prioritize these routes. The prioritization of routes may require some state information about the relay nodes. Therefore, more DSR options may be required to implement it. Three factors that are especially important for a wireless network (especially MANET): random topology changes, effect of

Next Header (8 bits)	F (1 bit)	Reserved (6 bits)	Payload Length (16 bits)
Options (Payload Length)			

Figure 10-7. DSR options header.

TABLE 10.2. DSR Options as Defined in [10]

Number	Name	Purpose
1.	Route Request	Initiate route request.
2.	Route Reply	Response to Route request.
3.	Route Error	Report link outage.
4.	Acknowledgement Request	Request a soft ACK.
5.	Acknowledgement	Response to a soft ACK request.
6.	DSR Source Route	List of route addresses.
7.	Pad1	To make options length multiple of 4 octets (not enforced)
8.	PadN	To make options length multiple of 4 octets (not enforced). For padding multiple consecutive octet.
9.	Timeout for Flow state	TTL for a flow ID.
10.	Destination and Flow ID	Used to indicate to intermediate node that this is a flow type packet and not source route packet.
11.	Unknown Flow	To report an invalid flow ID in Route Error.
12.	Default Flow Unknown	Reporting lack of a flow ID expected from a node to process.
13.	Previous Hop Address	ACK request extension.

The column 'Number' is just a listing

mobility and effect of battery level on a node's availability. In this section, we present a theoretical framework to incorporate these factors. We will define the node or link availability as a function of these quantities, which can be combined to define link availability. We consider the effect of random changes in topology, followed by battery level and discharging rate and then mobility. These can be incorporated in any routing protocol by defining appropriate parameters for collecting the required information during route discovery and maintenance phases. The treatment given to routing is of mathematical nature and requires the knowledge of probability distributions.

10.10.1. Topology of Fixed Ad-Hoc Networks

We start with the assumption that the physical layer capacity of a node j is fixed at Cj regardless of (i) propagation conditions, (ii) number of active nodes in the neighborhood, (iii) mobility, (iv) topological variations and (v) residual battery time.

According to [2], the number of nodes n acts to cut down the network capacity as $1/\sqrt{n}$ even in ideal situation of traffic and node clustering. Thus, if a station j can directly access n_j stations within a disk of radius $1/\sqrt{\pi}$, its capacity is proportional to $1/\sqrt{n_j}$.

Consequently, every node in the neighborhood of a node can have different capacity depending on n_j. For power-controlled nodes, if power range is (S, P_0), the node j has a capacity bounded by $\left[\sqrt{8n/\pi}W/\{(\beta S/P_0)^{1/\alpha}-1\}\right]_j$. In this expression, α is the propagation exponent and β is the signal power to total noise power ratio to successful reception. A subscript of (j) is simply a reminder that these quantities are different for each node in general, and denote the node number j.

In deriving the capacity expression, [2] considered no topological changes, due to mobility or due to the stations changing between ON and OFF states. If ρ_T is the probability that a station is active (ON), then, n, the number of active stations out of a total of K stations, has Bernoulli distribution. In that case, even the average network capacity is not static and has a stochastic behavior. In particular, the probability that the capacity is as given in [2] $\left[\sqrt{8n/\pi}W/\{(\beta S/P_0)^{1/\alpha}-1\}\right]_j$ could be well below 1. The average capacity in such case would be given by:

$$\overline{C}(\rho_T, K) = \sum_{n=0}^{K} \binom{K}{n} \rho_T^n (1-\rho_T)^{K-n} \left[\sqrt{\frac{8n}{\pi}} \frac{W}{\left(\frac{\beta S}{P_0}\right)^{\frac{1}{\alpha}} - 1} \right] \tag{10.1}$$

10.10.1.1. Topology Index. To understand the effect of activity factor, we define topology index $\zeta_T(\rho_T,K)$ as the ratio between the average capacity of an ad hoc network as a fraction of the same network capacity with all the K stations being active. We can write the equation for $\zeta_T(\rho_T,K)$ as follows:

$$\varsigma_T(\rho_T, K) = \frac{\overline{C}(\rho_T, K)}{\overline{C}(1, K)} = \sum_{n=0}^{K} \binom{K}{n} \rho_T^n (1-\rho_T)^{K-n} \left[\sqrt{\frac{n}{K}} \right] \tag{10.2}$$

As seen from Eq. (10.2), the topology index depends only on the distribution of the terminals being active and the total number of stations K. Figure 10-8 shows that the situation is not very good except for very high values of ρ_T.

10.10.2. Effect of Mobility

Work on mobility modeling has been done in terms of Doppler's effect as well as statistical modeling of mobility. Of special interest is the work reported in [11]. The authors define the concept of mutual mobility between two moving MANET stations. This model has been validated by simulation in the same and other works by same authors. It departs from earlier mobility models, such as random way point in that the prior models were done mainly for cellular

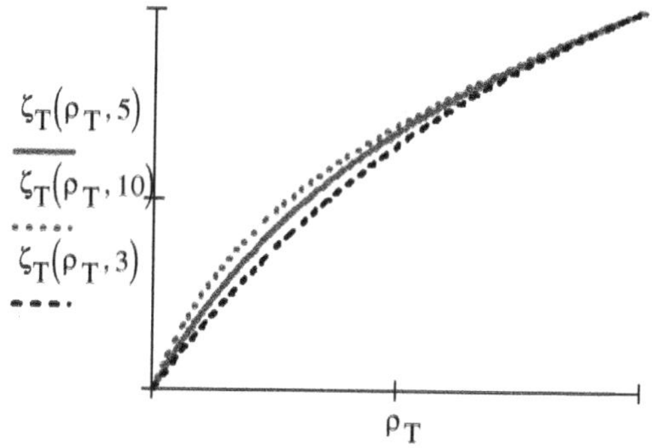

Figure 10-8. Topology index for various values of nodes in a network.

networks with fixed base stations and mobile nodes. The paper defines a relative mobility vector D_{lm} between nodes l and m. In a related work [12], the same authors have shown that D_{lm} has approximately Rayleigh distribution with the direction of motion uniformly distributed $(0,2\pi)$.

The apparent effect of mobility is a reduction in the availability of an established path. However, counter results have also been reported. Even though [2] claims that the throughput degenerates as $1/\sqrt{n}$, the authors of [13] show that by adding mobility, the throughput can be maintained even with increasing n. The reasoning for this behavior is that, when the nodes are mobile, there is a high probability that there will be a node close to destination. Thus, if a packet is relayed to a number of relay nodes, at least one of them or a subsequent relay node will be able to deliver the packet to the destination. This benefit is really about a specific routing mechanism, which may have its own disadvantages, such as the reduced throughput due to packet duplication. We hold that this conclusion does not apply to an existing route between two stations, unless the route is defined only dynamically and perhaps for each packets transmission.

In [14] the authors introduce the concept of topology change rate (TCR) as the rate at which the link changes for a given node and the paper determines the relation between expected duration and length of a transition. All the above mentioned models are too complex to be used in a simple manner in routing mechanisms.

10.10.2.1. Mobility and Displacement. The random way-point mobility model used in [14] and widely accepted, will be amended in this section in a more realistic way for demonstration purpose only. Instead of assuming the mobile station velocity to be uniformly distributed between (v_{min}, v_{max}), we assume that $v_{min}, = -v_{max}$. The amount of displacement from a station's position depends on mobility profile and the station's distance from the preceding relay station. If this distance is d and if a propagation loss due to d has an average of $-10.\alpha.\log(d)$, then the probability $Pr[.]$ of a station remaining part of a route after moving 'away' by $\hat{d}$ units of distance is given by the expression:

$$P[\text{Station retained}] = P\left[-10.\alpha.\log\left(d+\hat{d}\right) \leq \text{some threshold}\right], \qquad (10.3)$$

α being the propagation exponent.

Equation (10.3) is not a very simple equation to evaluate for the following reason. Even if we assume the d is constant and the distribution of loss $-10.\alpha.\log(d)$ is normal, $\hat{d}$ is random and depends on the mobility profile of the station. Suppose that we can get help from a GPS system to track $\hat{d}$ as well. That does not solve the problem either because average loss is not a linear function of distance. We use *excess loss* $L_d(\hat{d})$, defined as the equivalent loss when a distance $\hat{d}$ is added to the existing distance d. The following expression gives the excess loss.

$$L_d(\hat{d}) \cong -\frac{10\alpha}{\ln(10)}\frac{\hat{d}}{d}. \qquad (10.4)$$

10.10.2.2. Mobility and Path Loss Models. There are multiple ways in which node mobility is known to affect a communications adversely, for example, by changing the distance between the communicating nodes and by causing Doppler effect. If a station can measure accurately the amount of Doppler effect, it can find out the *equivalent mobility* that would be causing it. This concept of equivalent mobility could, in fact, offer the best solution of measuring mobility of a MANET node, as the net effect of mobility should be inclusive of the mobility of the surroundings as well. This topic is beyond the scope of this section. Instead, we like to elaborate on the concept of mobility index presented originally in [15]. The mobility index can be used to gauge the extent of damage done to a route due to mobility of users. Since the effect of mobility on propagation loss should also be considered, we will integrate the two in the following analysis.

Suppose that a station receives a strong signal during route discovery procedure, when the transmitted signal is P_0, average loss is L_0 and the signal receiver sensitivity is S. The difference $P_0 + L_0 = \Delta P_0$ is the average power margin. Power control may be used to adjust the signal to the minimum possible, such that the power received is as close to the receiver sensitivity as possible. The difference $(L_d(\hat{d}) + \Delta P_0)$ is the power margin remaining after the station has moved $\hat{d}$ units. Since $\hat{d}$ can be positive or negative, the new margin can have arbitrary range. However, when it is below 0, the signal received is below the receiver sensitivity, and the station is not retained in a route. In other words,

$$Pr[\textit{Station retained}] = Pr\left[\left(L_d\left(\hat{d}\right) + \Delta P_0\right) > 0\right]. \qquad (10.5)$$

Assuming that the signal power has lognormal distribution, power in dB (and hence the power margin) will be normally distributed with an average equal to $(P_0 + L_0 - S)$ dB. In that case $Pr[\text{Station retained} \mid \hat{d}]$ is normally distributed with an average of $[(P_0 + L_0 - S + L_d(\hat{d})]$ dB. In other words:

$$Pr[\text{A station retained}] = \int_{\in \hat{d}}\int_0^{\infty} \frac{1}{\sqrt{2\pi}\sigma_L} e^{\frac{(x-(P_0+L_0-S+L_d(y))^2}{2\sigma_L^2}} dx f_{\hat{d}}(y)dy. \quad (10.6)$$

We regard the above equation as the Mobility Index of a MANET station. It can be calculated for various mobility profiles. Let's use the symbol $\zeta_M(P_0, S)$ to denote it. Thus

$$\zeta_M(P_0,S) = \int_{\in \hat{d}}\int_0^{\infty} \frac{1}{\sqrt{2\pi}\sigma_L} e^{\frac{(x-(P_0+L_0-S+L_d(y))^2}{2\sigma_L^2}} dx f_{\hat{d}}(y)dy. \quad (10.7)$$

The quantity $f_{\hat{d}}(y)$, pdf of the displacement, can be calculated from the knowledge of inter-packet arrival times (or a session of packets) and the distribution of the velocity of the MANET station. For inter-packet distribution of $f_\tau(t)$, another integral will be added to the above equation.

Figure 10-9 shows that a little displacement in one packet-inter arrival time can result in losing the node, as long as this displacement is in the positive direction (away). The mobility index remains 100% for movement toward the transmitting station (negative values of $\hat{d}$). The curves are drawn for three values of receiver sensitivity (–50, –60 and –70 dBm) and without considering a particular mobility profile. What these curves show is that for channels with a propagation exponent of 4, station mobility can result in an extremely fragile network path and the best strategy for routing could be to find a route for

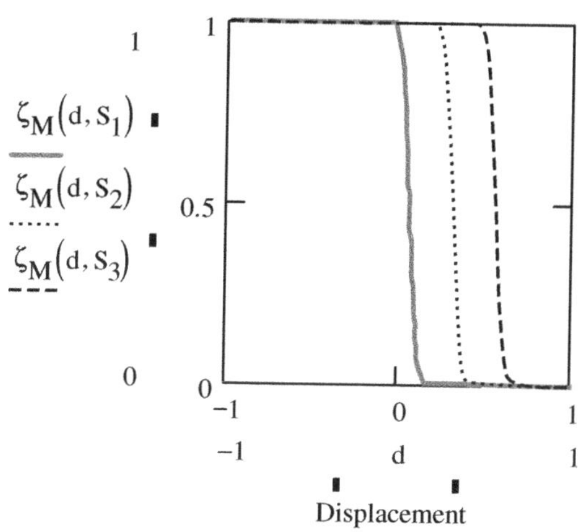

Figure 10-9. Mobility index.

each packet before transmitting it. Otherwise, the network throughput will plunge due to retransmissions and the timer going off too often.

10.10.3. Residual Battery

Two routes with same number of hops could result in different failure probabilities if the residual energies of station batteries are different. Assuming that the ad hoc nodes that are connected to the mains can always replenish the battery fast enough to continue transmitting packets as fast as they are generated, there will be a fraction, ρ_B of nodes that have no connection to the mains. Let's assume that these nodes can participate in a routing protocol as long as their residual energy is above a certain threshold γ_{Th} and continue as relaying nodes as long as it is above another threshold $\gamma_B < \gamma_{Th}$ (both thresholds are measured as fractions of the maximum battery output). The capacity of a route with respect to residual battery could be defined as the minimum number of packets an intermediate station can forward before its battery reaches γ_B. In an actual MANET, this quantity is subject to measurement of actual values of residual power. The battery index [15], defined as the probability that a node (or route) is not discontinued due to battery drainage, depends on several factors, including battery discharging mechanism and the thresholds. This topic is not exactly the same as designing energy-conserving routing protocols for which a number of researches have been reported recently[4]. Of relevance is the work in [16] in which authors use Markov chain analysis to model the battery capacity. Among other things, the results show that the packet generation capacity is a strong function of the packet inter-arrival times. By admitting delay, these batteries can generate more packets than the continuous-drainage-capacity as long as they recharged during idle time.[5] However, a fraction of ad hoc stations operate without any mains; therefore, for them, the number of packets generated cannot exceed the capacity. As a simple illustration, we borrow the concept of time constant from electrical circuit theory (the time when an exponentially decaying function reaches $e^{-1} \cong 0.37$ fraction of its full capacity). More realistic models are too complex for a demonstration. Suppose that the battery, with starting voltage level v, discharges continuously as $e^{-t/T}$, where T is the discharge time constant. Then, the probability, $\zeta_B(.)$, that a MANET node is retained after time τ, is given by:

$$\zeta_B(T, \gamma_{Th}, \gamma_B | \tau) = Pr[ve^{-\tau/T} > \gamma_B | v > \gamma_{Th}, \tau]. \tag{10.8}$$

If the packet inter-arrival times have a distribution of $f_\tau(t)$, the battery index is given by the following equation:

[4] Work related to energy-savings protocols, reported in many papers such as [17], presents and simulates various ways of power saving in IEEE 802.11 and HIPERLAN.
[5] The number of packets that can be transmitted without recharging the battery.

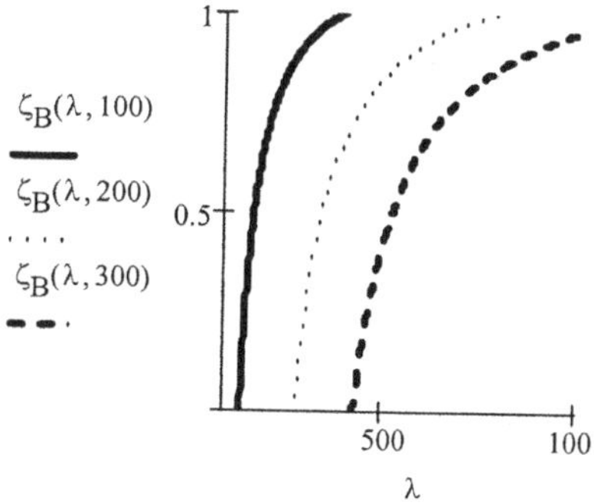

Figure 10-10. Battery index for T = 100, 200 and 300 as a function of arrival rate.

$$\zeta_B(\lambda, T, \gamma_{Th}, \gamma_B) = \int_0^\infty P[ve^{-\tau/T} > \gamma_B \,|\, v > \gamma_{Th}, t] f_\tau(t)dt. \qquad (10.9)$$

Example: Suppose the packet interarrival time is exponentially distributed with a parameter λ, the probability that k packets are generated in a given time has a Poisson distribution, and;

$$\zeta_B(\lambda, T, \gamma_{Th}, \gamma_B) = \int_0^\infty P[ve^{-\tau/T} > \gamma_B \,|\, v > \gamma_{Th}, t] f_\tau(t)dt = \int_0^\infty \lambda e^{-\lambda t}\left(\frac{1-\gamma_B e^{t/T}}{1-\gamma_{Th}}\right)dt;$$

$$\zeta_B(\lambda, T, \gamma_{Th}, \gamma_B) = \frac{1 - \dfrac{\gamma_B \lambda}{\lambda - T}}{1-\gamma_{Th}}. \qquad (10.10)$$

Figure 10-10 shows a plot of Equation (10.10) as a function of packet arrival rate for various vales of the time constant T. The parameters γ_{Th} and γ_B have values of 0.4 and 0.3 respectively.

As seen from Figure 10-10, the probability of losing a node due to battery exhaustion (the *battery index*) goes quickly to 1. There are other discharging residual power relations available in literature, such as the one given in [18]. The paper proposes a weight function $f(v) = \dfrac{1}{1-g(v)}$, where v is the measured voltage and $g(v)$ is the remaining normalized lifetime (equal to $1 - \gamma_B$ in the above example).

10.10.4. Example of Application of Above Results

Let's consider the ad hoc network of Figure 10-11 reproduced from Figure 10-4 for an example application. Each station is labeled with the $\{\Delta P_0 - S, \gamma_B, \rho_M, \rho_T\}$, where ΔP_0 = Power margin, which implicitly includes distance.

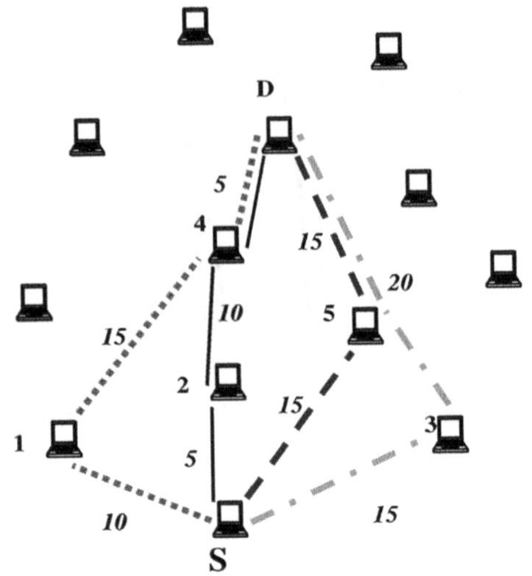

Figure 10-11. Four possible routes between source (**S**) and destination (**D**) stations. *Italics* are distances.

TABLE 10.3. Parameter Exchanges During Route Setup

Node Number	$(\Delta P_o - S)$ dB	γ_B	ρ_T	ρ_M
1	40	0.25	0.9	0
2	60	0.35	0.8	0.5
3	55	0.4	0.9	0.7
4	80	0.0	0.6	1
5	57	0.6	0.7	0.6
D	55	0.8	0.5	0.3

S = Station sensitivity.

γ_B = The battery threshold for disconnection as a relay node. A value of 0 means that the station is connected to the mains.

ρ_M = Probability that the node is mobile, to be defined in one of the many ways, for example, from the history of the station operating as a MANET node.

ρ_T = Probability that a station remains in the ON condition during the specified time.

Table 10.3 shows example values of the parameter for the relay nodes.

There are a total of nine links. In order to calculate the probabilities of link availability, let's assume a time of five packets on the average ($\lambda T = 5$) and the battery threshold of 50% for accepting a node. Let the mobility profile be such

TABLE 10.4. Link Availability

Link	ζ_B	ζ_M	ρ_T	Link Availability
S-1	1	1 (no mobility)	0.9	0.9
S-2	1	$0.5*1+0.5*1=1$	0.8	0.8
S-3	1	$0.3*1+0.7*0.489$	0.9	0.578
S-5	0.668	$0.4*1+0.6*0.892$	0.7	0.437
1–4	1	$0+1*0=0$	0.6	0
4-D	0.334	$0.7*1+0.3*0.272$	0.5	0.086
2–4	1	$0*1+1*1$	0.6	0.6
5-D	0.334	$0.7*1+0.3*0.489$	0.5	0.1413
3-D	0.334	$0.7*1+0.3*7.1*10^{-4}$	0.5	0.1169

TABLE 10.5. Route (Path) Availability for Each of the Four Possible Routes in Figure 10.11

Route	Availability	Remarks
S-1-4-D	$0.9*0*0.086=0$	Even though the route is initially available, after λT time, its availability drops to zero.
S-2-4-D	$0.8*0.6*0.068=4.1\%$	Available only with very low probability
S-5-D	$0.437*0.1413=6.2\%$	Slightly more available.
S-3-D	$0.578*0.1169=6.75\%$	Close to S-5-D

that the normalized displacement be equal to 0.2 (Figure 10-9). Then, Table 10-4 can be realized by the source or destination node before making a final selection of the route.

With the link availability given by Table 10.4, the route availability for all possible routes (4 in this case) is given by Table 10.5.

10.10.5. Discussion

Placing the results of Tables 10.4 and 10.5 together, we make some important observations. First of all, mobility of a single node can alone result in the unavailability of a path that otherwise has some stable links. Second, the number of hops is not always a good measure (one path with two hops is worse than another with three hops). Third, even for some links having very clear and stable links, we cannot say anything about a route between two stations in a MANET. In the above case, even though some link availability probabilities are very high (90%, 80%), the maximum path availability is less than 7% in all cases. Lastly, we make the observation that one way to improve the probability would be route diversity. In the above case, selecting routes #3(S-5-D) and #4(S-3-D) results is the overall path availability of above 10%.

10.11. WLAN ROUTING THROUGH CELLULAR NETWORK INFRASTRUCTURE

Until this point, we have discussed two mechanisms of routing data among stations in a wireless LAN, namely, IP, for example, as a distribution function (DF) in IEEE 802.11 for the infrastructure WLANs, and ad hoc network routing protocols for ad hoc networks. Third-generation wireless systems' shift toward end-to-end IP includes WLANs as among the access methods. Two terminals in two different WLANs can exchange IP packets through the cellular architectures by having relevant subscription services. Some of the initial hitches in doing so relate to the earlier lack of call admission control, security, authentication, authorization, and accounting mechanisms in WLANs. Cellular operators, however, have kept working on providing an interface between WLANs and cellular networks due to the mere fact that an increasing number of notebooks and PDAs come factory-ready with WLAN cards. In this section, we will describe aspects of a network architecture employed by Nokia to connect IEEE 802.11 infrastructure networks with GSM cellular networks. For a fuller description of the architecture, the reader is referred to Ref. [19].

10.11.1. Introduction to OWLAN

An operator's WLAN (OWLAN) provides subscription-based WLAN access by combining it with GSM subscriber management and billing mechanisms. OWLAN enables IP roaming between different operator access networks. The solution is available for any WLAN terminal with a GSM SIM card reader.

The reference system was implemented as part of R&D and was piloted in a real mobile operator network in 2000. The first commercial system was launched in July 2001.

10.11.2. Design Objectives

The following is a list of some of the design objectives:

1. Future mobile operator network to combine GSM/GPRS/3G/WLAN technologies. OWLAN provides a unified access mechanisms for subscribers with user equipment loaded with both technologies, the cellular and WLAN.
2. For smooth roaming and seamless service, a single user identity is crucial to utilize a virtual home-like environment.
3. OWLAN to be optimized for terminal originated IP data as most of the data originated for networking use IP.
4. ETSI TS 101 393 defines a charging mechanism for GPRS that could be used for WLANs as well.

10.11.3. OWLAN System Architecture

The OWLAN consists of a public LAN access and cellular operator site that communicate over the IP backbone. See Figure 10-12.

The main design challenge was to transport standard GSM subscriber authentication signaling from the terminal to the cellular site using IP framework. For this purpose, the network contains four physical entities: (1) authentication server, (2) access controller, (3) access point, and (4) mobile terminal. Each system component has a counterpart in the GPRS system, as shown in Table 10.6.

As opposed to GPRS, only control signaling data are transported to the cellular core for this project. In order to have the capability of using routing via the cellular infrastructure, data packets must be able to use cellular transport.

10.11.4. System Elements

See Figure 10-13 for system elements and interfaces and Figure 10-14 for interface protocols.

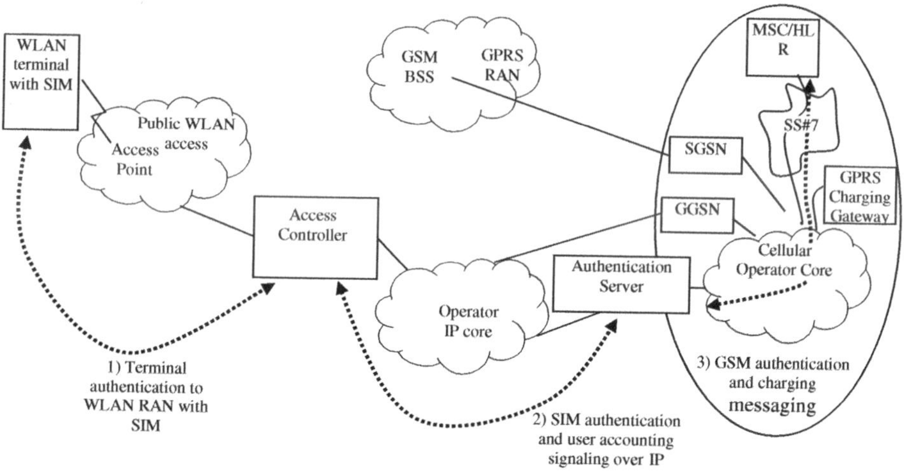

Figure 10-12. Operators' WLAN (OWLAN) network layout.

TABLE 10.6. Equivalence of OWLAN and GPRS Network Components

OWLAN	GPRS
Authentication server (AS)	SGSN
Access Controller (AC)	GGSN
Access Point (AP)	BTS
Mobile terminal (MT)	Mobile phone

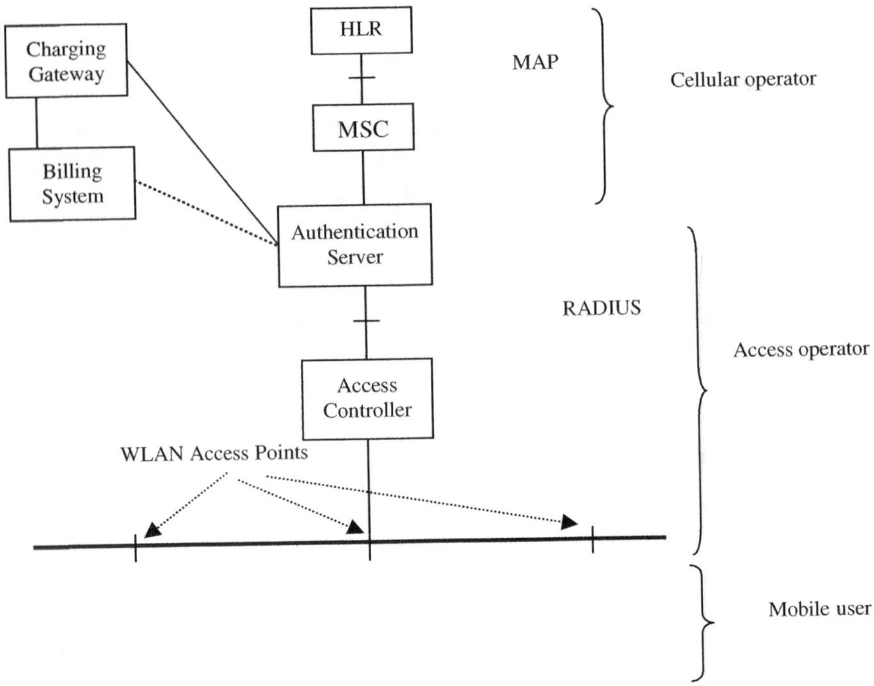

Figure 10-13. OWLAN system elements and interfaces.

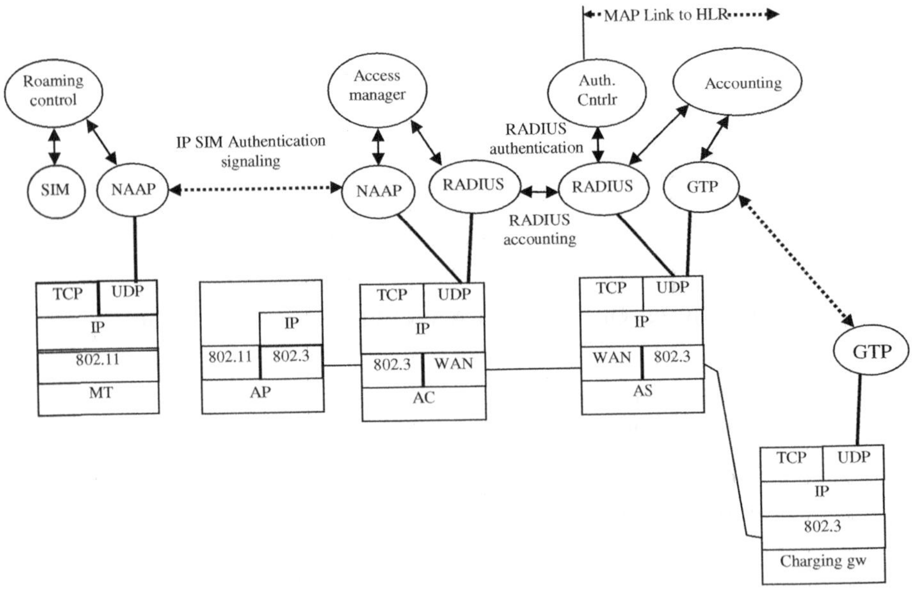

Figure 10-14. OWLAN interfaces and components.

10.11.4.1. Authentication Server (AS). The AS is the main control point of OWLAN subscriber management. A single AS may support several access controllers (ACs) and provide authentication and billing services to thousands of roaming users in different access zones. AS communicates with AC using RADIUS protocol. RADIUS is a *de facto* authentication, authorization, and accounting (AAA) protocol in the IP industry. It collects data from AC and converts it into GPRS format at call termination. The cellular network identifies the user with IMSI. AS hides the cellular infrastructure from the access network. An IP-compliant vendor-specific protocol carries authentication requests from AS to MSC. Alternatively, AS could be implemented with a native MAP interface. In the future the service profile could be used to include QoS of the WLAN user.

10.11.4.2. Access Controller (AC). The access controller acts as a gateway between RAN and the fixed IP network. It allocates and maintains a list of IP addresses so as to allow to pass through it only authorized IP datagrams. The access controller uses a WLAN link-layer specific MAC address to filter unauthorized IP users. It also collects billing information. Nokia's IP router IPSO330 series IP router was used for this purpose.

10.11.4.3. Mobile Terminal (MT). The MT requires a WLAN card, a SIM reader, and SIM authentication software. Some laptops will have SIM integrated with WLAN in the future. There is a list of roaming networks where the terminal can operate. The OWLAN terminal may detect the correct WLAN using this list. WLAN profiles may be distributed by operator via SIM or web.

10.11.5. System Operation

SIM-specific signaling messages are transported using the IP protocol. This makes OWLAN independent of the WLAN standard. IP-packet filtering is computing intensive—it resides on the access network side. This allows distributing processing among a number of controllers.

10.11.5.1. MT (Mobile Terminal). Core operational components of MT include:

Roaming Control: Provides GUI (graphic user interface) to roaming services. It communicates with the SIM card for this purpose.

Network Access Authentication and Accounting Protocol (NAAP): Encapsulates GSM specific signaling messages in IP packets. NAAP uses UDP with retransmissions.

10.11.5.2. AC (Access Controller). Access Manager is the key component of AC. It controls IP routing and collects accounting statistics. Accounting data are collected via standard RADIUS accounting attributes. RADIUS carries SIM-specific authentication parameters inside vendor-specific attributes.

10.11.5.3. AS (Authentication Server). The key component of AS is the authentication controller (AuC). AuC communicates with the core GSM and handles RADIUS authentication messages. The accounting module receives and stores the accounting information from the access network. There is no uniquely defined *GTP/billing protocol*. The AuC provides an open FTP interface to transfer accounting data directly to various billing systems.

10.12. ROUTING IN PERSONAL AREA NETWORKS

Due to short packet size and typically small memory size, the Bluetooth™ devices cannot make effective use of the routing protocols designed for MANETs [20]. To add to the difficulty, the topological restrictions, such as the master-slave paradigm and a limited allowed number of nodes per piconet, work against defining zones in terms of hop sizes. However, the fact that a node can be a slave in more than one piconet within a scatternet could be potentially useful in designing routing algorithms for Bluetooth scatternets. In Ref. [20], a binary tree-like topology is used to define the self-routing Bluetooth scatternet, in which every node is the root of up to seven children. Thus a hierarchical mechanism allows efficient packet flooding from any node to any other through roots. In another method, the Routing Vector Method, the piconets are represented by local identification numbers (LoIDs). This helps route a packet among piconets, until they reach the one with destination node. The reader is referred to Ref. [10] for details on this mechanism. Even though Bluetooth comes built-in with many personal communication devices these days, the real need for scatternet is still not proven. Consequently, work in this direction mainly consists of filling in the gaps that exist if routing protocols for ad hoc networks were used for Bluetooth. The underlying assumption is that the Bluetooth networks are similar to ad hoc networks, as they are not bound through an interconnecting backbone network. However, Bluetooth networks are unique in the sense that they have the characteristics of both the infrastructure as well as ad hoc networks; infrastructure type because they can exist only as a root/leaf of a piconet and because two leaf nodes cannot communicate except through the root (master-slave paradigm). This is shown in Figure 10-15.

10.13. SUMMARY

Routing in WLANs may sound like a misnomer due to the LAN protocols, defined below the network layer. In practice, however, it is an important capa-

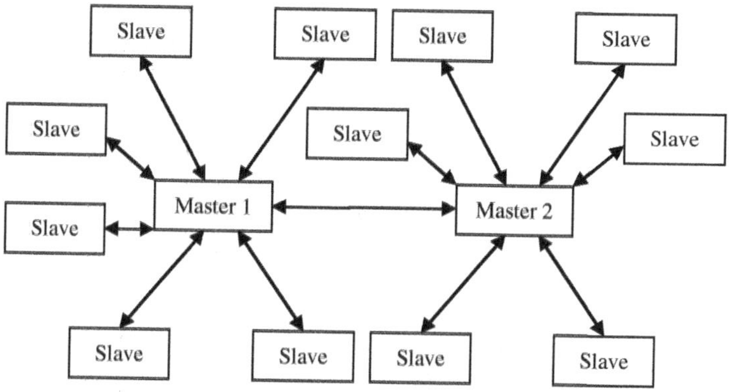

Figure 10-15. Bluetooth networks have characteristics of ad hoc and infrastructure networks.

bility. Relaying and routing can be performed in infrastructure and ad hoc networks. We have discussed various options, mainly paying attention to mobile ad hoc networks (MANETs). This chapter does not consider cellular data routing, enterprise data network routing, and barely mentions routing in wireless PANs defined by Bluetooth. However, we have discussed a routing protocol for MANETs, called DSR, provided a theoretical framework for route prioritization in MANET routing protocols, and included a WLAN architecture (OWLAN) that gets subscription-related services from GSM. The OWLAN may not have originally been designed to use cellular network routing, but once WLAN becomes a standard access network, it will be possible to route WLAN packets through cellular networks.

REFERENCES

[1] Chansu Yu, Lee Ben, and Yoon, Hee Yong, 'Energy Efficient Routing Protocols for Mobile Ad Hoc Networks', *Wireless Communications and Mobile Computing Journal*, Vol. 3, Issue 8, pp. 959–973, December 2003. http://academic.csuohio.edu/yuc/papers/energy_routing_final.pdf

[2] P. Gupta and Kumar, P. R., 'The Capacity of Wireless Networks', *IEEE Transactions on Information Theory*, 46(2):388–404, March 2000.

[3] Josh Broch, Maltz, David, Johnson, David, Hu, Yih-Chun, and Jetcheva, Jorjeta, 'Perfromance Analysis of Multihop Ad Hoc Networks Routing Protocols', *MobiCom 98.*

[4] Z. Haas, 'A New Routing Protocol for the Reconfigurable Wireless Networks', ICUPC'97, San Diego, Oct. 12, 1997.

[5] Z. Haas, Pearlman, M., and Samar, P., 'Intrazone Routing Protocol (IARP)', *IETF Internet Draft draft-ietf-manet-iarp-01.txt*, June 2001.

[6] Z. Haas, Pearlman, M., and Samar, P., 'Interzone Routing Protocol (IERP)', *IETF Internet Draft draft-ietf-manet-ierp-01.txt*, June 2001.

[7] C.E. Perkins and Bhagwat, P., 'Highly dynamic destination-sequenced distance vector routing (DSDV) for mobile computers', *SIGCOM'94*, pp. 234–244, August 1994.

[8] M.S. Corson, Papademetriou, S., Papadopuolos, P., Park, V., and Qayyum, A., 'Internet MANET encapsulation protocol (IMEP) specification', *IETF Internet-draft draft-ietf-manet-imep-spec-01.txt.*

[9] Katia Obraczka and Tsudik, Gene, 'Multicast Routing Issues in Ad Hoc Networks', USC Information Sciences Institute {isi.edu}.

[10] David B. Johnson, Maltz, David, and Hu, Yih-Chun, 'The dynamic source routing protocol', *IETF Internet-draft*, July 2004.

[11] A.B. McDonald and Znati, T., 'A Path Availability Model for Wireless Ad Hoc Networks', *Proceedings of the IEEE WCNC* 1999, pp. 35–40, Vol 1.

[12] A.B. McDonald and Znati, T, 'Link availability models for mobile ad hoc networks', *Technical Report TR99-07, University of Pittsburgh, Department of Computer Science*, May 1998.

[13] M. Grossglauser and Tse, D., 'Mobility Increases the Capacity of Ad-hoc Wireless Networks', *IEEE Infocom 2001*.

[14] X. Perez-Costa, Bettstetter, C., and Hartenstein, H., 'Towards a mobility metric for reproducible and comparable results in ad hoc networks research', Poster abstract, Mobicom 2003, Poster-6, September 2003, San Diego.

[15] A. Ahmad and Kim, M.J., 'Multidimensional Resource Characterization in MANETs', *World Wireless Congress 2004 (WWC-04)*, San Francisco, May 2004.

[16] C.F. Chiasserini and Rao, R.R., 'A model for battery pulsed discharge with recovery effect', Wireless Communications and Networking Conference, 1999. WCNC. 1999 IEEE, 21–24 Sept. 1999, pp. 636–639 vol. 2.

[17] H. Woesner, Ebert, J.P., Schlager, M., and Wolisz, A., 'Power-saving mechanisms in emerging standards for wireless LANs: The MAC level perspective', Personal Communications, IEEE [see also IEEE Wireless Communications],Volume: 5, Issue: 3, June 1998 pp. 40–48.

[18] S. Singh, Woo M., and Raghavendra, C.S., 'Power-aware Routing in Mobile Ad Hoc Networks', *Proc. Mobicom*, 1998.

[19] Juha Ala-Laurila, Mikkonen, Jouni, and Rinnemaa, Jyri, 'Wireless LAN Access Network Architecture for Mobile Operators', *IEEE Communications Magazine*, November 2001.

[20] Min-Te Sun, Chang Chung-Kuo, and Lai, Ten-Hwang, 'A Self-Routing Topology for Bluetooth Scatternet', International Symposium on Personal Area Networks 2002.

[21] Bhagwat, P.; Segall, A. 'A routing vector method (RVM) for routing in Bluetooth scatternets' A. Mobile Multimedia Communications, 1999. (MoMuC '99) 1999 pp. 375–379.

[22] T. W. Chen, Tsai, J.T., and Gerla, M., 'QoS routing performance in Multihop, Multimedia Wireless Networks', ICUPC 1997.

[23] David B. Johnson and Maltz, David A., Dynamic Source Routing in Ad Hoc Wireless Networks. In *Mobile Computing*, ed. by Tomasz Imielinski and Hank Korth, Chapter 5, pp. 153–181, Kluwer Academic Publishers, New York, 1996 (ISPAN 2002).

[24] J. Li, Blake, C., Couto, D. S. J. D., Lee, H. I., and Morris, R, 'Capacity of Ad Hoc Wireless Networks,' *in Mobile Comp. and Networking*, pp. 61–69, 2001.

[25] B. J. Kwak, Song, N.O., and Miller, L.E., 'A mobility measure for mobile ad hoc networks' *Communications Letters, IEEE*, Volume: 7, Issue: 8, Aug. 2003 pp. 379–381.

[26] Abhishek Rai and Kumar, Kapil, 'Performance Comparison of Link State and Distance Vector Algorithms', www. fsl.cs.sunysb.edu/~abba/report2.pdf

[27] C. F. Chiasserini and Rao, R. R, 'Pulsed battery discharge in communication devices', *Proceedings of Mobicom 99*, Seattle, August 1999.

CHAPTER 11

WIRELESS PERSONAL AREA NETWORKS AND ULTRAWIDE BAND COMMUNICATIONS

Wireless networking provides freedom from wires and the ability to roam around while connected. However, much of the communications wire infrastructure that we have to deal with is not about wide range of mobility. The back of today's communications appliances is wrought with a mesh of cables, and quite often, a call to technical support to a communications company ends up in reworking those wires. Wired remote controls have long been replaced with wireless ones. Other wireless short-distance solutions are available that are customized to distance, data rates, and applications. Infra-red data association (IrDA) has been working on the mission of cleaning up the wire bunch for some years. In recent years, the topic of wireless at short distances has gotten more recognition in the form of a network that is different from WANs, LANs, and MANs. It is called *personal area network* (PAN). IEEE committee 802 recognizes a *personal operating space* (POS) as the space around a personal communications cluster where PANs are defined.[1]

In this chapter, we will look at some wireless technologies available for personal area networks (also called wireless PANs or WPANs). We will start the chapter with the most heard-of and perhaps mature PAN technology, Bluetooth.™ The IEEE 802.15 Working Group was set up after Bluetooth had been specified, and it adopted Bluetooth version 1.1 as IEEE 802.15.1 by adding some IEEE interfaces to it. However, the WG didn't stop there, and has

[1] Due to the scalability of many PAN standards, the POS is meeting the same fortune as 'local' in the LAN etc.

Wireless and Mobile Data Networks, by Aftab Ahmad

addressed both sides of the scalability in Bluetooth by announcing a low data rate PAN standard (IEEE 802.15.4) and a high data rate PAN standard (IEEE 802.15.3). Work is in progress to finalize a PHY for very high rate PANs using the ultra wideband (UWB) spectrum. Since Bluetooth, low data rate and non-UWB high data rate PAN standards are all defined for 2.4 GHz ISM, the WG has also specified co-existence specifications [12] in the form of IEEE 802.15.2.

Due to the variety of standards discussed in this chapter, we are brief about each. The reader is referred to the actual standard specifications for exactness and to a number of other publications for a more detailed treatment.

11.1. WIRELESS PERSONAL AREA NETWORKS (WPANs)

The PANs are networks that interconnect a few communications devices within close proximity to each other. They generally do not have an infrastructure and could potentially connect devices 'worn' by the subscriber. An example of PAN is the connection between a computer and its peripherals. Devices in a PAN are within a few meters of each other. Wireless PANs are thus a replacement for the short cables.[2] A PAN between a printer and computer provides the services of the cable at parallel port or network port, a PAN between an MP3 player and headphones provides the services of audio cable, and a PAN between a DVD player and HDTV provides a video cable replacement.

The transceivers for PANs are low-power devices that need to have a short span (<10 m) and line of sight.[3] The channel model thus becomes very simple, power budget easy to calculate, and transceiver design straightforward. The WLAN standards, such as IEEE 802.11 or its extensions, can be used for this purpose. There are, however, WPAN standards specified for this purpose. We discussed the protocol architecture of Bluetooth in Chapter 2. As mentioned in that chapter, IEEE has adopted Bluetooth v1.1 as IEEE 802.15.1. The IEEE 802.15 group has specified other standards as well for PANs. Table 11.1 shows a list of standards of this committee.

In the remainder of this chapter, we will describe these standards for a general understanding. It may be useful to mention that the technologies employing the ISM band will cause interference for one another. One of the task groups (IEEE 802.15.2) has published specifications for co-existence among these various device types in order to help these various technologies co-exist in spite of interference. A discussion on IEEE 802.15.2 is beyond the scope of the book. At the time of writing of the book IEEE 802.15.3a is in limbo on voting for a specification. Two proposals, one for multiband-OFDM PHY led by Texas Instruments and Intel, and the other on DS-SS led by Xtreme and Motorola, are

[2] From this point on, we will use PAN for a Wireless PAN, unless mentioned otherwise.

[3] This is a typical network characteristic. The standards and protocols may provide services for non-line-of-sight situations.

TABLE 11.1. IEEE 802.15 Committee Standards Scopes

Name	Max Data Rate	Typical Use	Max Power (Regulated by Respective Regions)			Span(meters)
IEEE 802.15.1 (2.4 GHz)	723.2 kbps	voice and data	2.4 GHz	902–928 MHz	868 MHz	10
IEEE 802.15.3 (2.4 GHz)	11–55 Mbps	multimedia consumer electronics	USA/Canada: 1 W, EU: 100 mW EIRP or 10 mW/MHz peak, Japan: 10 mW/MHz	USA: 1 W	EU: 25 mW	>70
IEEE 802.15.4 (868/915 MHz and 2.4 GHz)	20, 40, 250 kbps	low power, low data arte devices				10
IEEE 802.153a (3.1–10.6 GHz)	55–480 Mbps	Broadband video cable, IEEE 1394	USA (FCC) −41 dBm/MHz = 79.4 nW/MHz			10, 4, less

in fierce competition. The former is supported by a majority of voters of the task group, but has failed to meet the minimum 75% mark as yet in voting. Also, it is not clear if the proposed multiband-OFDM PHY meets FCC requirements for emissions. The DS-SS proposal has the benefit that it is based on a mature technology and Motorola has technology available to ship products based on it earlier than the competing proposal proponents. We will summarize both, as they may find their niche in the market.

11.2. TERMINOLOGY FOR WPANs

In addition to an array of new networking standards and paradigms, PANs have brought a new set of terminology. A separate section on terminology may not be a good idea because the same term may mean different things in different specifications.

11.3. IEEE 802.15.1 STANDARD

For IEEE 802.15.1, the Task Group adopted the slightly amended Bluetooth v1.1 (see www.Bluetooth.org for copy of the specs). Therefore, the architecture and components of IEEE 802.15.1 specifications [8] are Bluetooth specifications. The IEEE specifications emphasize the equivalent of the OSI PHY and DLC layer. It includes up to the OSI equivalent of PHY and MAC sublayer and adds the interface for IEEE 802.2 (LLC), thus setting the stage for an *access gateway* (AG) between Bluetooth and other IEEE LAN standards. Figure 11-1 shows the protocol architecture of the IEEE 802.15.1 and its equivalence to OSI reference model.

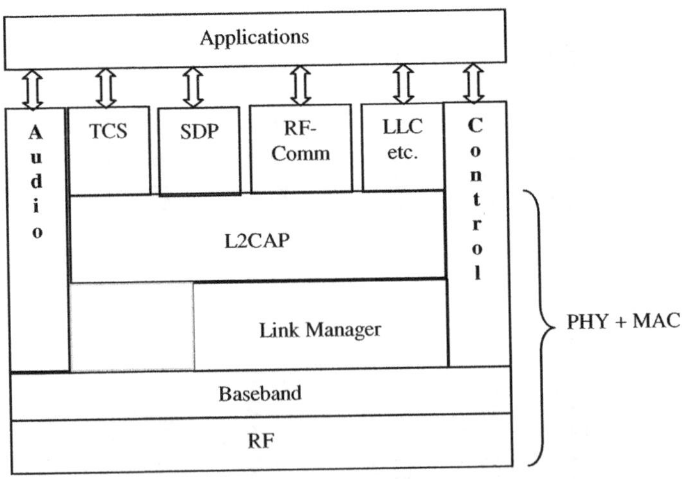

Figure 11-1. IEEE 802.15.1 protocol architecture and OSI equivalence.

11.3.1. Bluetooth Components

A Bluetooth network, though much less complex than typical WLANs, provides a dual mode service architecture for circuit-switched voice with 64 kbps synchronous operation and symmetric and asymmetric connectionless data with time division duplexing (TDD). The Bluetooth special interest group and the IEEE 802.15 group have defined a variety of Bluetooth network components, including station types, service types, connection types, and network configurations. We will look at some of these in this section. The Bluetooth network does not have an inherent infrastructure, and is more like an ad hoc network that operates in a *personal operating space* (POS). The POS is a space surrounding a typical user with one or more personal communications appliances. Typical values assumed for POS are within 10 meters.

11.3.1.1. Bluetooth Stations. There are two fundamental types of stations, Master and Slave. A station can be of either or both types.

A *master station* controls and allocates channel resources. It communicates with the other stations through point-to-point and point-to-multipoint channels. A master device has a frequency hopping spread spectrum (FH-SS) channel allocated to it. A *slave station* is one that is controlled by a master in a temporary network configuration called a *piconet*. A slave station can be *parked* when it is not active but is a part of a piconet. A parked slave is synchronized to the master device.

11.3.1.2. Network Configurations. A piconet consists of a master device synchronized with seven or less active slaves and some parked slaves. Since each piconet is allocated a unique hopping sequence, two or more piconets can co-exist. A master of a piconet can also be a slave in other piconet(s). A slave in a piconet can also be a slave in other piconet(s). Figure 11-2 demonstrates these configurations. A piconet can exist by itself or be a part of *scatternet*. Scatternet is a cooperation of piconets made possible by having common active stations. The piconets in a scatternet are not synchronized; the

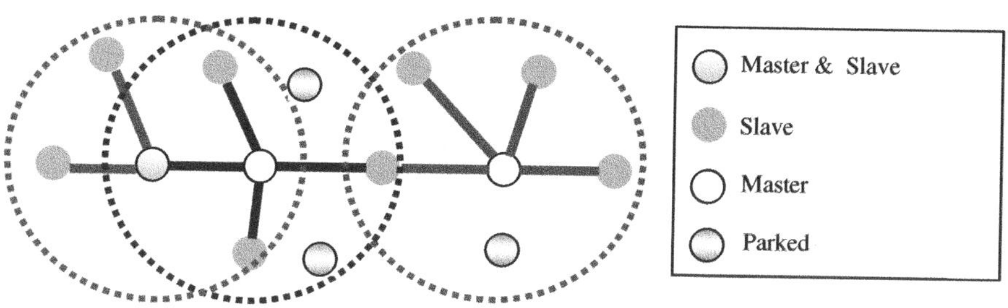

Figure 11-2. 3 piconets, 3 Masters, 1 Master/Slave and 8 active slaves and 3 parked slaves. Dotted circles shows transmission ranges of respective piconets.

inter-piconet communication is possible due to a slave or a master/slave being part of multiple piconets and due to time division multiplexed transmission. In Figure 11-2, the master/slave device shown by double shading could communicate with in its own piconet during one time slot and with the master of the other piconet in another slot.

11.3.1.3. Channel Media. The Bluetooth standard is defined over the 2.4 GHz ISM spectrum. It employs a fast frequency hop spread spectrum (FH-SS) mechanism to combat interference from other ISM devices, including other piconets. A piconet is characterized by a unique FH-SS hopping sequence. Every packet uses a different frequency channel. The channel hopping rate is 1600 hps. North America, Japan, and most of Europe uses 79 channels. France uses a slightly different part of the (2.4 GHz) spectrum with 23 channels. The modulation used is Gaussian frequency shift keying (GFSK) with a symbol rate of 1 Msps. Time division multiplexing is used to divide the channel into TDM slots of 625 μs, corresponding to a sampling interval of Nyquist rate sampled 4 kHz analog signal, such as used in 64 kbps PCM. The slots can be used by master or slave in a time division duplexing (TDD) mode. The standard specifies that the even-number slots are to be used by master device and odd numbers by slave devices. All devices are synchronized to the beginning to slots. The synchronization applies only to a single piconet. Thus there is no need of a system-wide master clock in a system of multiple piconets. All devices have their own free-running clocks with tick rate at least twice the slot rate and they synchronize with the master clock for each instance of a piconet.[4] Due to the synchronous operation of slots, connection-oriented transmission is possible. Certain slots at regular intervals are reserved exclusively for synchronous transmission and the rest can be used in asynchronous connectionless mode. Slot allocation to synchronous or asynchronous transmission is made possible by allocating 27-bit numbers to slots. The bit rate for the synchronous operation is fixed at 64 kbps, whereas it may vary from below 64 kbps to above 720 kbps for asynchronous transmission. The two different types of links are called synchronous connection oriented (SCO) and asynchronous connectionless (ACL), for obvious reasons. Figure 11-3 shows a relation between bandwidth, FH-SS, channel, slots, and physical links.

One or more (up to 5) slots carry a packet. A packet has minimum of one field called *access code*. Access code carries piconet identification (*channel access code*), device identification (*device access code*) and inquiry information (*general inquiry~* and *dedicated inquiry access code*). Packet header (if included) defines various packet types and carries control information. The payload is of variable length, from 0 to 2745 bits. Figure 11-4 shows the three fields and their lengths.

[4] Piconets are temporary. Therefore, every time a device becomes part of a piconet it has to synchronize clocks locally.

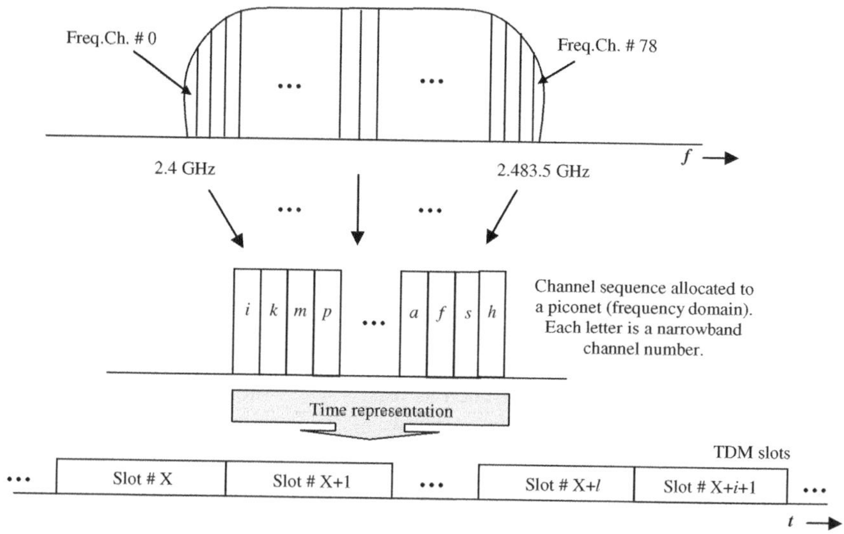

Figure 11-3. Relation between bandwidth, channel sequence, and slots.

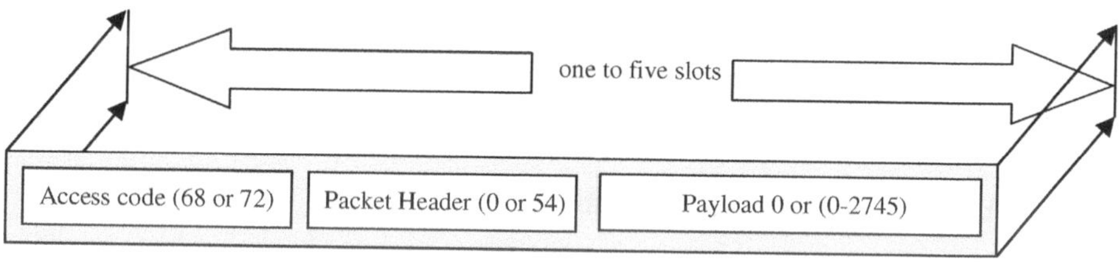

Figure 11-4. Bluetooth general packet format.

Bluetooth defines a number of packet types for link control (four), asynchronous connectionless mode (seven), and synchronous connection-oriented mode (four).

11.3.1.4. Logical Channels. The logical channels are distinguished by the functions they perform. The slots are a type of physical channels (so are narrowband frequency channels). SCO and ACL are types of physical links, the former for connection-oriented traffic and the latter for connectionless traffic. Connection-oriented-ness requires a prior understanding between the communicating devices (master and slave, in this case), whereas connectionlessness does not require a prior understanding. Logical channels are needed when more than one type of information flows among communicating devices, so that a channel can be used for a certain type of information. Logical channels are defined within physical channels, so a slot can contain one or more logical channels. A logical channel can use multiple slots as well. Table 11.2 lists the logical channels defined in Bluetooth. There are two categories of

TABLE 11.2. Bluetooth Logical Channel

Channel Name	Purpose	Realization
UA user channel	Carry user asynchronous (UA) data, not visible to L2CAP.	Baseband layer packets.
UI user channel	Carry user isochronous (UI) data.	Same as UA, except that the higher layer packets are timed properly to reflect the isochronous nature of traffic.
US channel	Carry user synchronous (US) data.	Caries over SCO link. No ARQ for SCO packets.
Link control (LC) channel	Carry low level link control information, such as for error and flow control.	In every packet header except the ID packet.
Link management (LM) channel	Carries link management protocol information between a master and slave.	In data packets

channels; control channels and user channels. *Control channels* carry control information to set up piconet, discover capabilities, and set up links. *User* channels carry user data.

11.3.2. Bluetooth Network Operation

The baseband layer is responsible for Bluetooth network creation, maintenance, and supervision. The Bluetooth network consists of two or more devices on an ad hoc basis and ceases to exist when one of the two last nodes leaves the piconet. Due to the fact that (i) a piconet is started from scratch or is to be joined, (ii) there are a number of transmission options, and (iii) a number of functions (e.g., encryption) are provided, there are discovery and capability exchange phases in a network operation. In this section, we will look at network operation for synchronous and asynchronous transmissions. What is common to all operations is:

- Master is the initiator of the operation.
- All slave clocks must be synchronized with the clock of the master. This synchronization is only in terms of occurrence of clock edges and not with reference to some external timing source. The master clock defines the phase of the frequency channel hopping and timing of the slots.
- Every Bluetooth device is in either one of the two major states (CONNECTION and STANDBY), or in one of the seven interim states (page, page scan, inquiry scan, inquiry, master response, slave response, inquiry response).

11.3.2.1. Access. We assume for the following scenarios that a master will page a slave to set up a connection. The slave is assumed to be in STANDY mode. In this mode, a device occasionally scans the channel for a page. Two types of channels can be used for scanning, namely, predicted hop frequencies with a period of 1.28 s (including just before and after) and a number of wake-up frequencies. If the slave successfully receives a page (step 1), it responds exactly 625 μs after the beginning of the page message and at the same hopping frequency on which it received page (step 2). This aligns the initial hoping sequence and provides a basis for aligning clock to the master's clock. On receiving the slave's response, the master responds with an FHS packet (step 3), which includes the clock information from the master to help slave synchronize to master's clock. The slave sends an ACK for the FHS packet (step 4). The final packet before the slave goes to a CONNECT state is a poll packet (step 5), sent by the master to the slave. After this, the slave uses master's device access code (DAC) and synchronizes to master's clock. If a device does not know which other devices are present in the neighborhood, or wants to know if a certain device is present, it uses inquiry instead of paging. Figure 11-5 shows the successful procedure.

11.3.2.2. Link Establishment. Once in CONNECT mode, a Bluetooth device can be in one of the many modes, such as active mode, sniff mode, hold mode, or park mode. The link manager protocol (LMP) then can establish a link. When a connection is established, an ACL link is available by default.

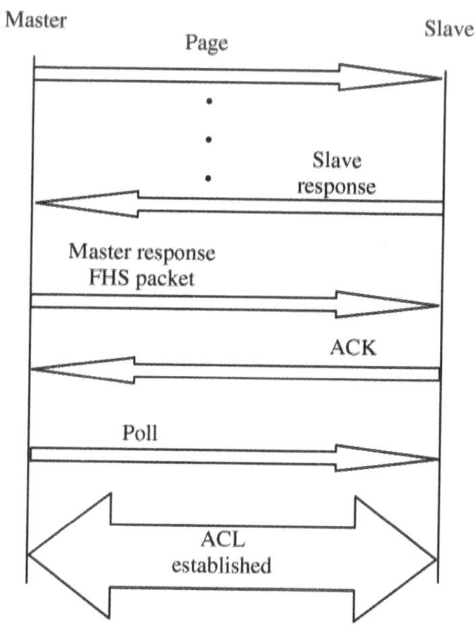

Figure 11-5. Successful change of state to CONNECT through paging by master.

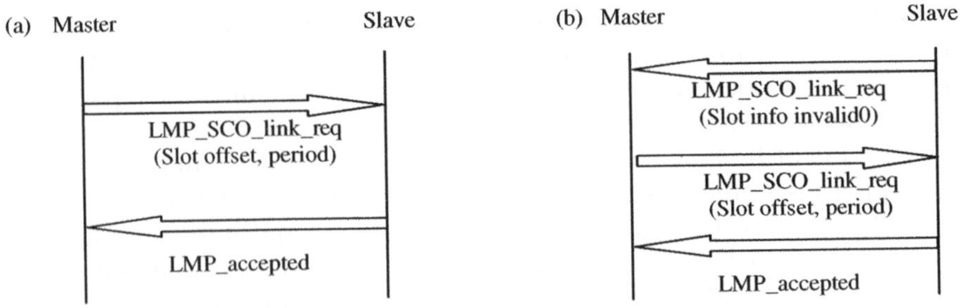

Figure 11-6. SCO link set up. (a) Request initiator: Master; (b) Request initiator: Slave.

ACL is a broadcast link. The master transmits in slots that are used by the ACL link and all slaves hear the transmission to check if it belongs to them. If a transmission does not belong to a slave device, it is discarded. Otherwise, the slave receives the ACL packet.

11.3.2.3. Synchronous Transmission Scenario. In this operation, the master or the slave can request an SCO link. If the master device requests the SCO link, it will send the LMP_SCO_link_req message to the slave. This message contains the time offset for the first SCO channel slot as well as the repetition interval. The slave, on receiving the request, replies with an LMP_accepted message if the parameters and the request are acceptable to it. Otherwise, it replies with an LMP_not_accepted message. If a slave requests an SCO link, then it initiates the request by sending an LMP_SCO_link_req message to the master. If the master can comply with the request, it replies with an LMP_SCO_link_req message containing the required link parameters. On receiving this, the slave replies with LMP_accepted or LMP_not_accepted. Figure 11-6 shows the timing diagram for the two scenarios. Synchronous operation provides the use of three types of voice coding, two PCM-based (μ-Law and A-Law) and the continuously varying slope delta modulation (CVSD).

11.3.2.4. Asynchronous Connectionless (ACL) Mode. The ACL mode of transmission is the default mode at the baseband layer. It is used by the link management layer to set up and supervise SCO. The ACL provides a best effort delivery capability in the Bluetooth network. L2CAP adds the various functions to the ACL link. These functions include multiplexing, SAR (segmentation and re-assembly), and group management for upper-layer protocols. With the help of SAR capability of L2CAP, the higher-layer protocols can send packets of sizes up to 64 kbytes.

The L2CAP is not defined for the SCO in IEEE 802.15.1. L2CAP does, however, provide both connectionless and connection-oriented services to the upper layers using the ACL link. For upper-layer protocols requiring QoS, it translates the upper-layer QoS requirements into baseband piconet parameters. L2CAP depends on the baseband layer for reliability and does not provide its own reliability guarantee, such as ARQ.

11.3.3. Bluetooth Summary

Bluetooth is a set of two parallel architectures, one for providing synchronous connection-oriented service via dedicated slot allocation and the other for providing symmetric and asymmetric asynchronous connectionless service to best effort and connection-oriented, higher-layer protocols.

The PHY defines a set of TDM time slots driven from the 2.4 GHz ISM band by employing FH-SS with 1600 hops per second hopping rate and 1 Msps GFSK modulation. The baseband layer above PHY defines the piconet by allocating a unique hop sequence to a master device. With the help of this hop sequence, other sequences dedicated for discovering devices on the piconet, and its clock, the master can have up to seven active slaves. There are two link types defined between master and slaves, by the link management protocol (LMP). Using these links, the baseband piconet can provide connection-oriented synchronous service at 64 kbps allowing three voice codings (μ-, A-Law and Delta Modulation), and best effort connectionless service for up to a raw bit rate of 723 kbps. The ACL is used by L2CAP to provide packet-oriented services for higher layer protocols with segmentation and re-assembly, QoS transfer, and group management. Both, connection-oriented and connectionless upper layers, can use the services of L2CAP. For detailed discussion on earlier Bluetooth, the reader is referred to [9] and [10].

11.4. HIGHER DATA RATE PANS (IEEE 802.15.3)

The Task Group IEEE 802.15.3 has specified a high-data-rate WPAN standard [11] with a standard data rate of 22 Mbps and a range from 11 to 55 Mbps. The standard specifies the PHY and MAC for the links in the 802.15.3 piconet. The piconet is the network configuration defined by MAC sublayer functions. We divide the discussion into three parts: (1) the network configuration, in which we discuss piconet, (2) PHY, and (3) MAC. To keep the account brief, we will use tables to highlight important attributes. Let's call the 802.15.3 piconet a high-data-rate piconet (HDR-PN), for reference.

11.4.1. High-Data-Rate Piconet (HDR-PN)

The HDR-PN is defined for low-power devices that require high data rates, in access of 20 Mbps, in close proximity from each other. The devices could be of any kind in terms of QoS requirements, as the standard has provision for isochronous and asynchronous transmissions. The HDR-PN is establish on ad hoc basis, just like the Bluetooth piconet. However, it can consist of a single device, a feature not present in IEEE 802.15.1 piconet. The piconet has a hierarchical configuration, with vertical as well as lateral hierarchies. In the vertical hierarchy, two adjacent piconets have a parent/child relation. In lateral hierarchy, the relation is essentially that of neighbors. With the help of the neighbor piconet concept, the HDR-PN can establish a relation with a piconet using a different standard or protocol. There are two generic device types defined, the piconet controller (PNC) and piconet device (DEV). A piconet can have only one PNC and up to 243 DEVs.

11.4.1.1. Piconet Controller (PNC). The PNC provides access control with QoS by generating the basic timing for the participating devices. It has a piconet ID (PNID) that uniquely defines the piconet controlled by the PNC. It allocates device IDs (DEVID) to the devices participating in its piconet. A PNC propagates information about its piconet through a beacon. Alternatively, any device that generates this beacon is a PNC.

11.4.1.2. Piconet Device (DEV). The DEV is a device that is a member of a piconet and is controlled by a PNC. A DEV in one piconet can be a PNC of a child piconet. A DEV can either associate with a piconet or start its own piconet.

11.4.1.3. Piconet Hierarchy. There are four types of piconets defined by 802.15.3 specifications. These are (1) independent piconet, (2) parent piconet, (3) child piconet, and (4) neighbor piconet. A piconet that does not have any relation with another piconet is called an independent piconet. If a piconet has a member DEV with its own piconet, then it is called a parent piconet. The piconet that the member DEV controls as PNC is the child piconet. A neighbor piconet is autonomous from the parent piconet and has its own association and security procedures. It depends on the parent piconet only for a private channel time allocation (see MAC for this). The PNC of a neighbor piconet is not a member of parent piconet. Figure 11-7 shows the hierarchy.

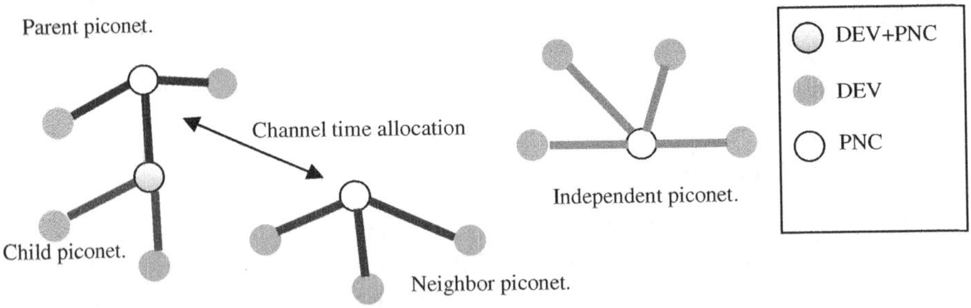

Figure 11-7. Piconet hierarchy for IEEE 802.15.3. A neighbor piconet does not have to be 802.15.3 based.

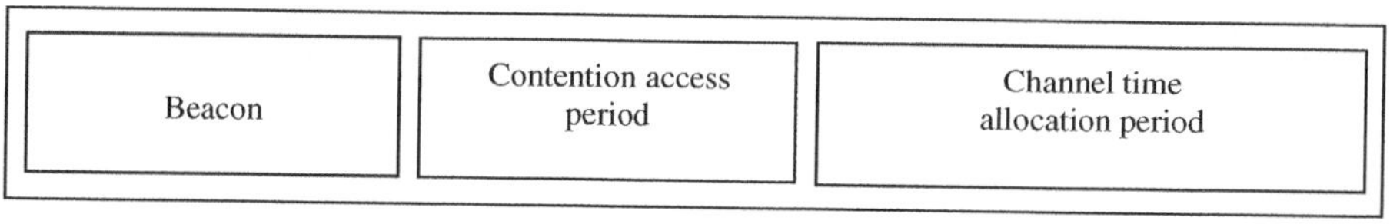

Figure 11-8. IEEE 802.15.3 superframe.

11.4.2. Medium Access Control (MAC) Layer

In this section, we will summarize the functions provided by the IEEE 802.15.3 MAC. The objectives of the MAC sublayer include multimedia capability, fast configuration of piconets, efficient use of resources, and security. Multimedia capability is provided by allowing reservation for delay-sensitive data, security is provided by having encryption option for data, efficient use of resources is provided by having contention-based access for best-effort data, and fast configuration of piconets is provided by a number of procedures for adding, removing, and changing the status of a device in a piconet. Due to the 'unbalanced' nature of piconet, a set of commands and responses is used to invoke, provide, and report various MAC functions.

11.4.2.1. MAC Superframe. All MAC communications between PNC and DEVs is timed through the MAC superframe, shown in Figure 11-8. A superframe is of variable length and can be up to 65.535 ms in resolution of 1 μs.

11.4.2.2. Beacon. Beacon is the timing and command link between PNC and DEVs. It is sent by the PNC and all DEVs are its recipients. It uses a variable number of information elements (IE) for commands, in addition to carrying the piconet synchronization information. This information is organized in a frame called *beacon frame*.

11.4.2.3. Contention Access Period (CAP). CAP is used for carrying commands and asynchronous data, if present. A DEV trying to transmit a packet in CAP follows a carrier sense multiple access with collision avoidance (CSMA/CA) mechanism. This method consists of a backoff timer and some interframe space (IFS). The actual value of IFS depends on the type of packet, and could be one of the four, in increasing order: (1) minimum-IFS (MIFS), (2) short-IFS (SIFS), (3) backoff-IFS (BIFS), and (4) retransmission-IFS (RIFS). The BIFS is required for every data packet transmission. A random value is generated every time a BIFS timer is to be set. Its limit is increased in case an ACK is not received for a transmitted packet.

11.4.2.4. Channel Time Allocation Period (CTAP). CTAP is used for commands, isochronous, and asynchronous data. It includes channel time allocations (CTAs) and management CTAs (MCTAs). MCTAs are for commands and requests. A PNC may use an MCTA for sending commands to DEVs. A

TABLE 11.3. IEEE 802.15.3 MAC Sublayer Functions

Function Name	Purpose	How It Is Provided
Piconets	Start and manage piconets.	A range of piconet services.
Channel access	Allow DEVs to get channel resources.	CTA reservation requests (slotted ALOHA) and contention (CSMA/CA)
Channel timing and synchronization	Synchronize PNC with DEVs. Transmit/receive isochronous traffic.	Superframe.
Fragmentation	Send large higher layer PDUs.	Any DEV can fragment MSDUs.
Acknowledgement	Reliability	Three mechanisms; no-ACK, Imm-ACK (for immediate ACK) and Dly-ACK (delayed ACK).
Interference mitigation	Reduce interference from other 802.15.3 piconets.	Either be part of the interfering piconet, change channel, or reduce power.
Multi-rate support	Support PHy with multiple rate.	Commands between DEV and PNC to discover available rates.
Power management	Control interference and prolong battery life.	By defining four power modes.

DEV may use an open MCTA to send a request to PNC using slotted ALOHA. The CTAs are allocated to individual DEVs and are guaranteed in terms of start time. This allows a DEV to transmit a time-sensitive data packet during the reserved CTAs.

11.4.2.5. Private CTA. A private CTA is not used for communication of data from DEVs to PNC. It is used for signaling information, such as to create a dependent piconet by a DEV. The private CTAs have the same device ID (DEVID) as source, as well as destination. Table 11.3 lists some of the functions provided by the MAC sublayer.

11.4.3. IEEE 802.15.3 Physical Layer (PHY)

This standard is specified for the 2.4 GHz ISM band. Another PHY specification for the ultra wideband (UWB) is currently being considered by the task group IEEE 802.15.3a. We will have a separate section on it. In this section, we describe the salient features of PHY for 802.15.3 in the form of Table 11.4.

TABLE 11.4. PHY Attributes for IEEE 802.15.3

Attribute	Options					Comments
Number of channels	5					Two sets: The first one using 4 and the other one 3 of the five channels.
PHY Frame length	Variable 64 to 2048 octets.					Excluding header.
Header check sequence	CRC-16					Combined MAC + PHY header.
Modulation	DQPSK	QPSK-TCM	16-QAM-TCM	32-QAM-TCM	32-QAM-TCM	Single carrier PHY, DQPSK is the basic modulation. TCM (Trellis coded modulation adds error control capability).
Transmission rates	22 Mbps	11	33	44	55	Corresponding to each modulation. Header always at 22 Mbps. 22 Mbps the basic rate.
Receiver sensitivity	–75 dB	–82	–74	–71	–68	For respective modulations.

11.5. ULTRA WIDEBAND (UWB) SPECTRUM

The ultra wideband spectrum ranges from 3.1 GHz to 10.6 GHz for license-free operation in the United States. Any spectrum with a bandwidth of greater than 1.5 GHz can be regarded as UWB [7]. The 3.1–10.6 GHz spectrum has been used in military communications and by law enforcement devices. Recently, the U.S. FCC allowed the use of this spectrum for commercial off-the-shelf products with three potential applications: (1) high-resolution image communications at short distance, (2) high-speed personal communications systems, and (3) vehicular radar and other measuring devices. The short distance is assured due to the low power regulation (–41.3 dBm per MHz). Figure 11-9 [7] shows the relative power consumption of UWB with some other wireless technologies used in WLAN and WPAN standards.

The real power of UWB is in the possibility of thin baseband pulses used without carrier modulation. Figure 11-10 shows the relation between the bandwidth and 'thin-ness' for a pulse function; the thinner a pulse, the higher is the bandwidth it carries. Alternatively, the broader the available bandwidth, the thinner is the possible pulse.

With UWB, communication systems using pulses of durations from 0.2 ns to 1.5 ns are possible [6]. This has two important implications. First, the pulses reaching a receiver from reflections can be easily distinguished from

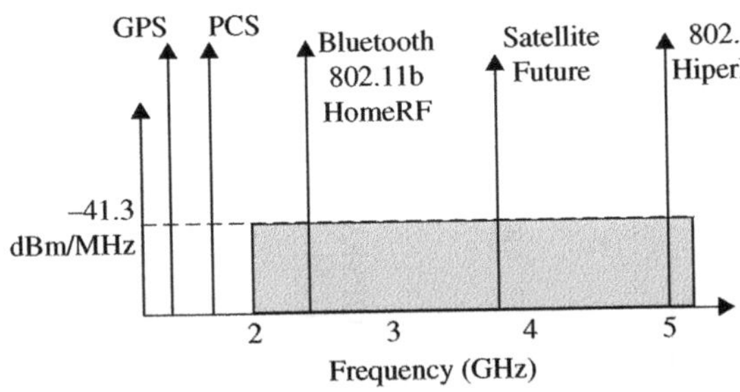

Figure 11-9. Relative power density of UWB [7].

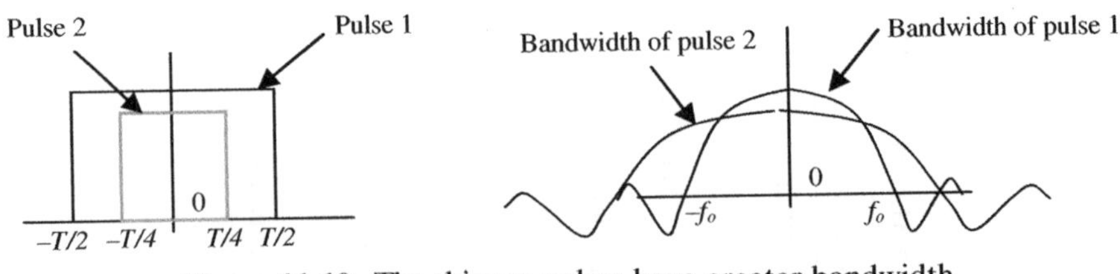

Figure 11-10. The thinner pulses have greater bandwidth.

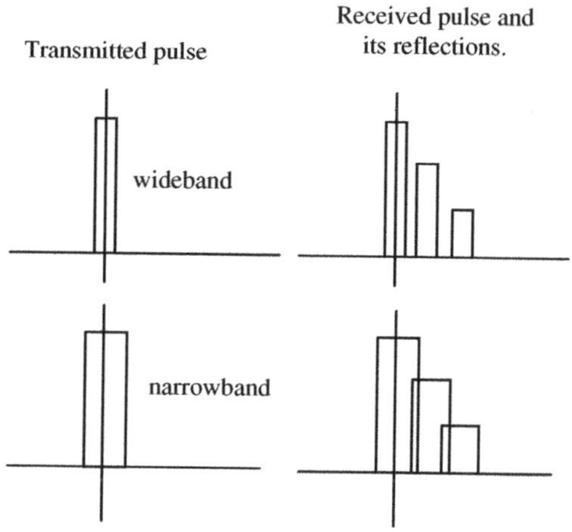

Figure 11-11. Broader pulses (narrowband) overlap when received through multipath.

the one arriving through line of sight (see Figure 11-11).[5] In other words, a RAKE receiver with appropriate number of fingers can solve the multipath problem in a UWB wireless system. Second, spread spectrum technology can be used at the baseband pulse level as a multiple access scheme with extremely low level pulse, virtually undetectable without the spread spectrum code.

Since the wavelength at this frequency range is very small, the UWB signal is not reflected from many surfaces and penetrates through them. This also means that if UWB is used for location services, the signal can give correct location of hidden objects within centimeters of the object.

One very interesting application of UWB is that it can be the 'killer application' for high-speed cellular networks. Companies can install UWB base stations in large buildings. Phones inside a building use a UWB connection. The UWB base station communicates with the cellular network base station (node B for W-CDMA) through a high-speed channel with all the building traffic multiplexed in it. Figure 11-12 shows this concept.

11.5.1. UWB PHY for IEEE 802.15.3a

The IEEE TG 802.15.3a is working [13] to come up with PHY specs for the WPANs with very high data rates (in excess of IEEE 802.15.3) using UWB. The objectives of the group include [6]:

- Data rates: 110 Mbps to 480 Mbps;
- Range: 10 m or less (4 m, 2 m);

[5] The major source of multipath is eliminated due to the fact that UWB is not reflected from the walls.

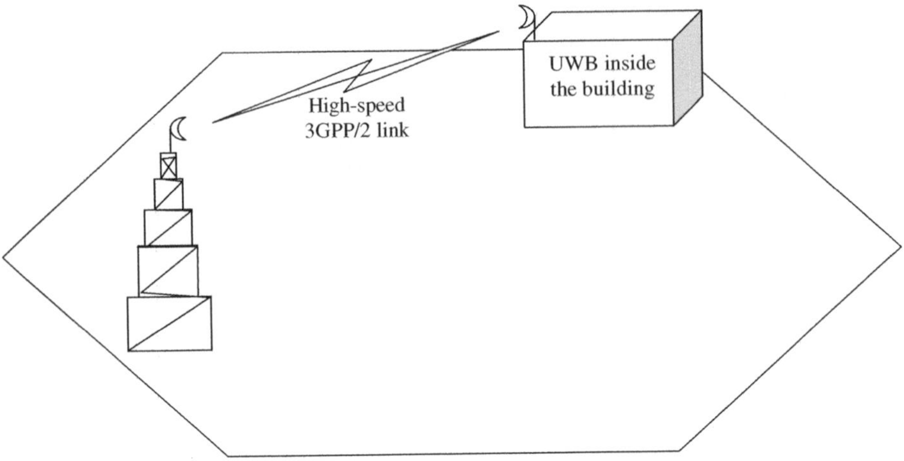

Figure 11-12. UWB can be used to multiplex a number of cellular devices signals from a large building.

- Co-located piconets: 4;
- Maximum transmitted power ratings: 250 mW, 100 mW, less;
- Cost: Comparable to Bluetooth.

The Task Group received a number of proposals. Successive voting brought the number down to 2, one based on OFDM and the other based on DS-SS. In the following, we will summarize each.

11.5.2. DS-UWB (Direct Sequence—Ultra Wideband)

The DS-UWB uses the two parts of the spectrum for two bands of operation, with either one implemented in a UWB device. The bands are defined in the ranges 3.1–4.85 GHz and 6.2–9.7 GHz, effectively eliminating the interference from the license-free U-NII. Figure 11-13 shows these bands. Each band can support up to six (6) piconets, assuming each piconet will use one channel. There are a total of 12 channels, numbered 1 through 12. Support for channels 1 through 4 is mandatory and support for channels 5 through 12 is optional. DS-UWB is based on the familiar direct-sequence spread spectrum technology, tested in two generations of cdma cellular systems. A variable chip rate (1 through 24) provides the capability of a range of data rates.

11.5.2.1. Modulation. Binary phase shift keying (BPSK), with each BPSK-symbol consisting of a number of chips, is employed. The chips are baseband UWB pulses. The pulse shape is root-raised cosine, as shown in Figure 11-14. Table 11.5 summarizes the PHY proposal.

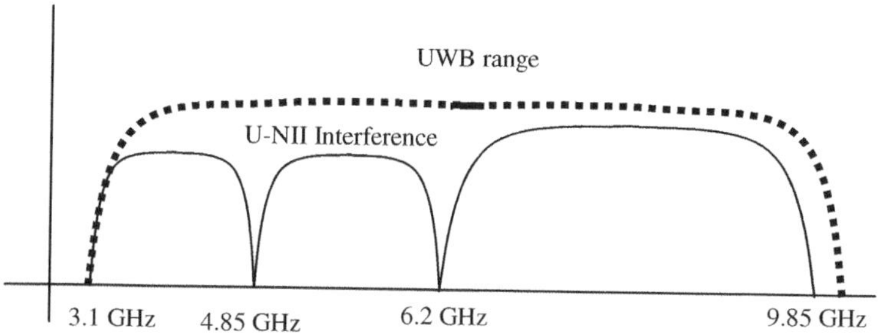

Figure 11-13. Relation between UWB band and the proposed bands.

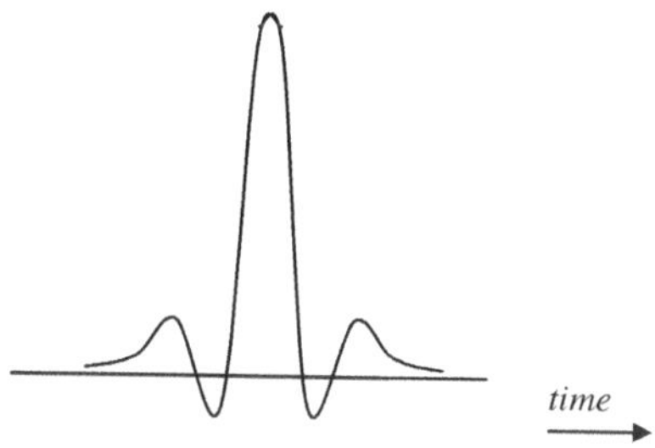

Figure 11-14. Root raised cosine pulse shape.

11.5.3. Multi-Band OFDM PHY Proposal

Supported by a number of big companies [4] (Intel, Microsoft, Texas Instruments, etc.), this proposal originally divided the spectrum into four band groups, three of which could be used for 802.15.3a PHY. Each band had a bandwidth of about 528 MHz, giving a total of 1.584 GHz, thus meeting the requirement for UWB definition. The division of bands is shown in Figure 11-15.

Band group #1 is the lowest band and #4 is the highest (rightmost) band. OFDM is used in each band. In a later update, more bandwidth was added on the right side (higher band). The resulting plan has five groups, each band group having three bands except the group #5, which has 2 bands, as shown in Figure 11-16. This results in a total of 14 multibands across the UWB spectrum, divided into five band groups. For band groups 1 through 4, four logical channels per group are defined, based on time-frequency code (TFC). Of these, band group number 1 is mandatory, requiring all piconet controllers (PNCs) and devices (DEVs) to be able to send or receive beacons in this band. Four time frequency codes (TFCs) can be employed in order to have four co-existent piconets, thus meeting the 802.15.3a minimum requirement. In order to leave the flexibility of world-wide spectrum allocations, bands can be turned on/off dynamically.

TABLE 11.5. Summary of DS-SS PHY Proposal for DS-UWB. [5]

PHY Attribute	Proposed				Comments
Frame format	Includes HCS over PHY and MAC header. Also could include bit stuffing.				The bit stuffing is done to have integral multiples of symbols.
Data scrambling	All except PHY preamble and PHY header.				Called 'randomization', uses $1 + X1^4 + X^{15}$ as randomization polynomial.
FER (forward error correction)	Convolutional with code rates 1/2 or 3/4 mandatory for all DEVs.				With code puncturing and convolutional interleaving allowed.
Modulation	BPSK		4-BOK (quaternary biorthogonal keying)		BPSK mandatory, transmission capability of BOK necessary, reception capability not necessary.
Data rates (Mbps)	Lower band: 28, 55, 110, 220, 500, 660, 1000, 1320	Higher band: 55, 110, 220, 500, 660, 1000, 1320	Lower band: 110, 220, 500, 660, 1000, 1320	Higher band: 220, 500, 660, 1000, 1320	The rates achieved with FEC.
Chip rates	1313 to 2730 Mcps				For 12 channels centered at (3939 to 8190) MHz.

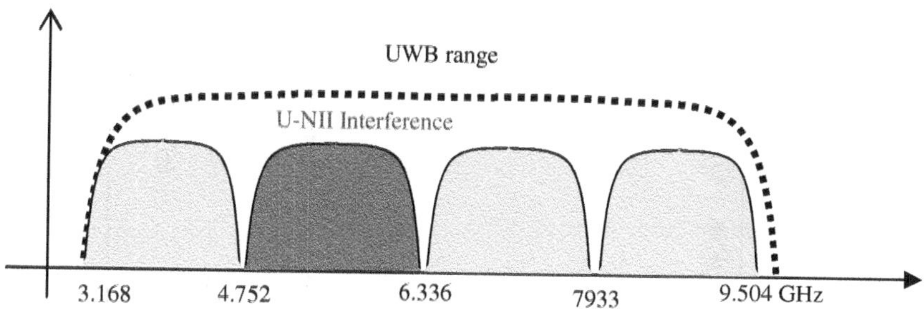

Figure 11-15. OFDM PHY multibands.

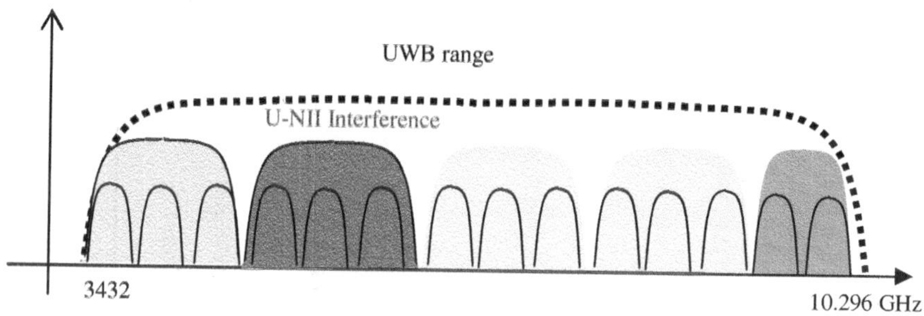

Figure 11-16. OFDM PHY multiband with five band groups. Frequencies shown are central frequencies at the first and the last band.

Table 11.6 summarizes the OFDM process as given in the actual presentation by the proposing group.

Receiver sensitivity is –80.5 dBm for 110 Mbps, –77 dBm for 200 Mbps and –72.7 dBm for 480 Mbps rates. There are some concerns about the FCC interference limitations that are under study at the time of this writing.

For a channel model for UWB, the reader is referred to [3] and the references in it. For UWB application in data communications, refer to [2] and the references contained in it. For a detailed treatment of multi-carrier CDMA and its application to UWB, refer to the slides [1].

11.6. LOW DATA RATE WPANs (LR-WPANs) AND IEEE 802.15.4

IEEE 802.15.4 standard [14] specifies the low-rate version of the WPANs. The target appliances range from very small level (e.g., for sensors), ultra-low power personal devices to toys in the personal operating space (POS). The standard specifies the physical layer (PHY) and medium access control (MAC) sublayer. One of the main differences between LR-WPAN and other

TABLE 11.6. OFDM System Parameters for Various Data Fates [4]

Into. Data Rate	55 Mbps*	80 Mbps**	110 Mbps*	160 Mbps**	200 Mbps*	320 Mbps**	480 Mbps**
Modulation/ Constellation	OFDM/QPSK	OFDM/QPSK	OFDM/QPSK	OFDM/QPSK	OFDM/QPSK	OFDM/QPSK	OFDM/QPSK
FFT Size	128	128	128	128	128	128	128
Coding Rate (K = 7)	R = 11/32	R = 1/2	R = 11/32	R = 1/2	R = 5/8	R = 1/2	R = 3/4
Spreading Rate	4	4	2	2	2	1	1
Data Tones	100	100	100	100	100	100	100
Into. Length	242.4 ns	242.4 ns	242.4 ns	242.4 ns	242.4 ns	242.4 ns	242.4 ns
Cyclic Prefix	60.6 ns	60.6 ns	60.6 ns	60.6 ns	60.6 ns	60.6 ns	60.6 ns
Guard Interval	9.5 ns	9.5 ns	9.5 ns	9.5 ns	9.5 ns	9.5 ns	9.5 ns
Symbol Length	312.5 ns	312.5 ns	312.5 ns	312.5 ns	312.5 ns	312.5 ns	312.5 ns
Channel Bit Rate	640 Mbps	640 Mbps	640 Mbps	640 Mbps	640 Mbps	640 Mbps	640 Mbps
Multi-path Tolerance	60.6 ns	60.6 ns	60.6 ns	60.6 ns	60.6 ns	60.6 ns	60.6 ns

TABLE 11.7. LR-WPAN Features

Feature	Comment
Range	10m (POS)
Data rates	20, 40, 250kbps
Applications	Toys, sensors, low-power personal devices.
Network configuration	Star and peer to peer
CSMA/CA based channel access	For best-effort data.
Guaranteed time slots (GTS)s	For delay-bound data.
ISM bandwidth	2.4GHz (16 channels), 915MHz (10 channels) and 868MHz (1 channel)

802.15 committee standards is that in 802.15.4, peer-to-peer communication is allowed. Table 11.7 lists distinguishing features of a LR-WPAN.

11.6.1. Network Configuration

The LR-WPAN consists of two types of devices with respect to the extent of functionality and two topologies. The two device types are full-function device (FFD) and reduced-function device (RFD). There are three types of devices in terms of their role in the network; (1) *PAN coordinator* (must be a FFD), (2) *coordinator* (must be a FFD) and (3) *device* (could be either FFD or RFD). A LR-WPAN is a PAN that requires at least one FFD and zero or more RFDs. The RFD can't act as a PAN coordinator or coordinator. The two topologies are *star* and *peer-to-peer*. In star topology, a PAN coordinator controls all communications among the devices and controls the entry and exit of other devices to and from the PAN. The devices communicate through the PAN coordinator. In a peer-to-peer network, every device is free to communicate with any other device directly. A peer-to-peer PAN also has a coordinator, but the traffic is not required to be sent through the coordinator in this configuration. The devices can extend beyond the transmission range of each and a routing protocol, such a dynamic source routing (DSR), can be used to route packets from one end of PAN to another. Figure 11-17 shows the concept of two topology types.

11.6.1.1. Star Topology. Each PAN in a star topology is identified by a unique PAN ID. The PAN coordinator uses this ID to communicate with devices (FFDs and RFDs) associated with the PAN. A PAN is independent of other PANs existing concurrently. The process of forming a star PAN starts with a FFD scanning a list of specified channels. A device can use any one of the three scan procedures, each carried in specified logical channels. A FFD can use *active scan* to set up a new PAN and become PAN coordinator. In this case, it will obtain a PAN ID through the active scan procedure. Active scan

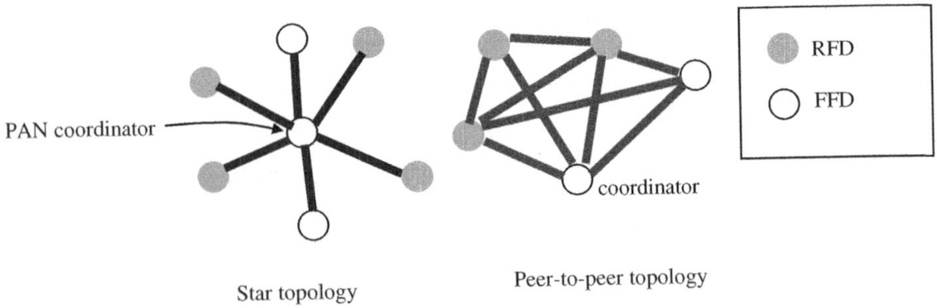

Figure 11-17. Two network topologies in LR-WPAN.

can also be used by a FFD to search for an existing PAN in its area of POS. A *passive scan* can also be used for this latter purpose. An *orphan scan* is used by a device to relocate its coordinator after it has been disconnected with it. Once a coordinator receives an orphan scan, it can *realign* the orphan device with its list of PAN devices by making appropriate checking. An example scenario for star topology is the home appliances.

11.6.1.2. Peer-to-Peer Topology. The peer-to-peer networks are general topology networks that have a mesh of links, as shown in Figure 11-17. The standard provides a mechanism for expanding the network coverage by making *cluster trees* of peer-to-peer clusters. A cluster tree is a hierarchical configuration at the heart of which is a piconet with a PAN coordinator. If a PAN coordinator decides to make a cluster tree, it becomes the cluster head (CLH) and it selects one of its FFDs to start another cluster. It chooses a cluster ID of zero and sends beacons to its neighbors announcing the formation of a cluster. Another device can request the PAN coordinator if it could join as a CLH. If the request is accepted, the PAN coordinator adds the requesting device (e.g., the FFD) as a child and the requesting FFD adds the PAN coordinator as a parent. Since the new device is also a CLH, it can form its own cluster by sending beacons. In this way, the PAN started from a single FFD (now with CID = 0) can be extended indefinitely. Figure 11-18 shows the process. The figure shows a cluster in which a cluster head is a child of a parent cluster head. In terms of attachment, a new cluster head could be attached to a device other than the CLH of the parent cluster. But, since this is peer-to-peer networking, the actual communication occurs in a mesh topology, and not a hierarchical mechanism.

11.6.2. LR-PAN Physical Layer (PHY)

In this section, we will briefly describe the PHY of the IEEE 802.15.4. There are two PHYs specified, at 2.4 GHz ISM band and at the 900 ISM band (915/868 MHz). The fundamental job of the PHY is the transmission and reception of signals. In OSI-RM terminology, it provides services to the MAC

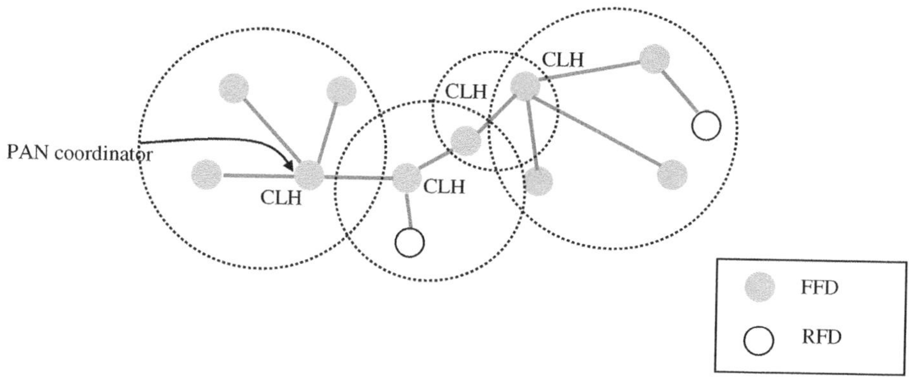

Figure 11-18. Peer to peer networks can be used to extend the coverage through cluster trees.

by exchanging MAC PDUs with peer PHY. Like any modern network architecture, the LR-PAN PHY provides an array of services with the help of many functions specified. These include clear channel assessment (CCA) to MAC sublayer, link quality indicator (LQI) to determine clarity of the channel, energy detection, channel frequency selection and activation/deactivation of transceivers. Table 11.8 lists characteristics of the PHY.

11.6.3. LR-PAN Medium Access Control (MAC)

The LR-PAN architecture is shown in Figure 11-19. The MAC has interfaces with the service specific convergence sublayer (SSCS) for conforming the MAC PDU to logical link control (LLC) sublayer, such as IEEE 802.2. Also, it interfaces with the upper layer protocols to provide services to routing and application layers.

11.6.3.1. MAC Features. The MAC sublayer provides a number of services including:

- Beacon management for creating and controlling PANs;
- Channel access for best effort and time-bound data;
- GTS (guaranteed time slot) management for delay bound data;
- Reliability services for frame validation and acknowledgement;
- Association and disassociation for managing PANs; and
- It also provides security hooks to application layer protocols.

11.6.3.2. Synchronization and Data Transfer. The MAC sublayer provides both contention-free and contention-based access. It does so by optionally defining a superframe to allow a PAN coordinator synchronize itself with the PAN members. The superframe has several parts, discussed below.

TABLE 11.8. LR-PAN PHY Spec for IEEE 802.15.4

PHY Attribute	Option(s)		Comment	
Spectrum	2.4 GHz	900 MHz	902–928 MHz	868–868.6 MHz
Modulation	OQPSK	BPSK	16-ary quasi-orthogonal modulation for OQPSK.	
Chip rate	2 Mcps	600 and 300 kcps	600	300
Bit rate	250 kbps	40 and 20 kbps	40	20
Channels	16 $f_c = 2405 + 5(k - 11)$ MHz; $k = 11 \ldots 26$	10 $f_c = 906 + 2(k - 1)$; $k = 1 \ldots 10$	1 f_c = 868.3 MHz	
PHY PDU (Octets)	Preamble (5) + Header (1) + PSDU ($0 - 2^7 - 1$)			
PN-sequence size	32 chips	15 chips		
Transmitted power density	–30 dBm/MHz	–20 dBm/MHz		
Receiver sensitivity	(–85)+dBm	(–92)+dBm		(x)+ = x or better.
Clear channel assessment	Energy level, carrier detection or both			

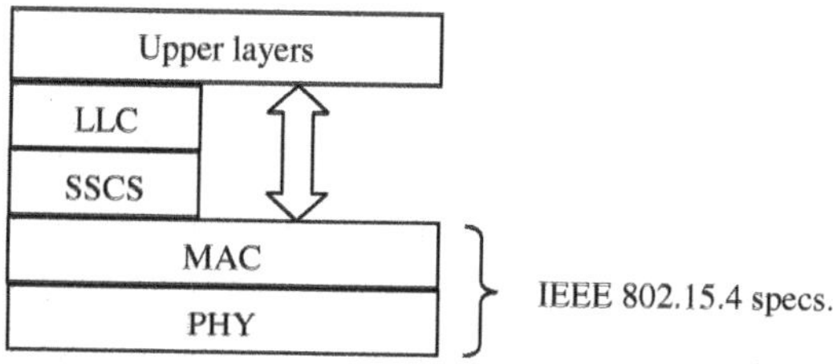

Figure 11-19. Protocol architecture of the IEEE 802.15.4 device.

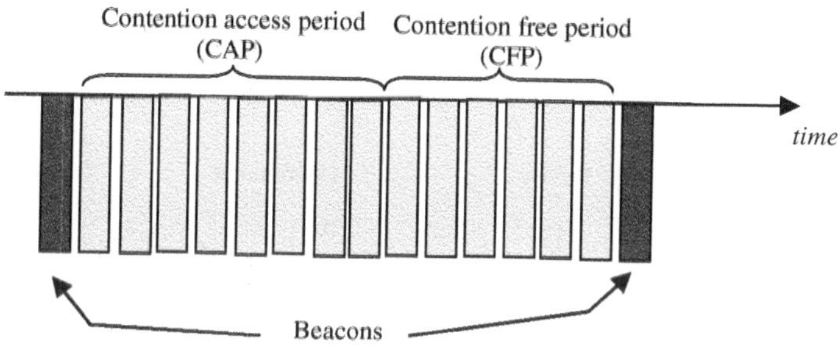

Figure 11-20. Superframe with CAP and CFP.

11.6.3.3. Beacons. The superframe is bounded by beacons. During the beacons the listening devices can synchronize to the coordinator of their PAN. A superframe is identified by a beginning beacon. Therefore, a coordinator who decides not to use superframe option turns off the beacon. Beacon uses first of the 16 slots of equal duration, as seen in Figure 11-20.

11.6.3.4. Active and Inactive Portions. The superframe may have active and inactive portions. The active portions are used for transmission while the coordinator goes to a low power mode during the inactive portion.

11.6.3.5. Contention Access Period (CAP) and Contention-Free Period (CFP). The contention access and contention-free periods are provided by a coordinator by allocating parts of the active superframe portion. During CAP, a device can attempt transmission of data packet using slotted CSMA/CA. However, during CFP, only a device that is allocated the guaranteed time slots (GTS) can transmit only in those slots.

11.6.4. Data Transfer Modes

In order to accommodate star and peer-to-peer topologies, three data modes are defined in MAC sublayer; from coordinator to device, from device to coordinator, and peer-to-peer. The star topology employs the first two, while a peer-to-peer network can use all the three modes since there is a coordinator in a peer-to-peer network as well.

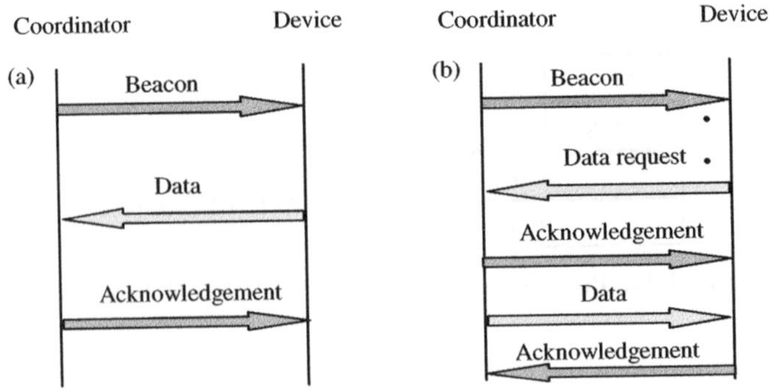

Figure 11-21. (a). Sending data to the coordinator with beacon. (b). Receiving data from the coordinator with beacon.

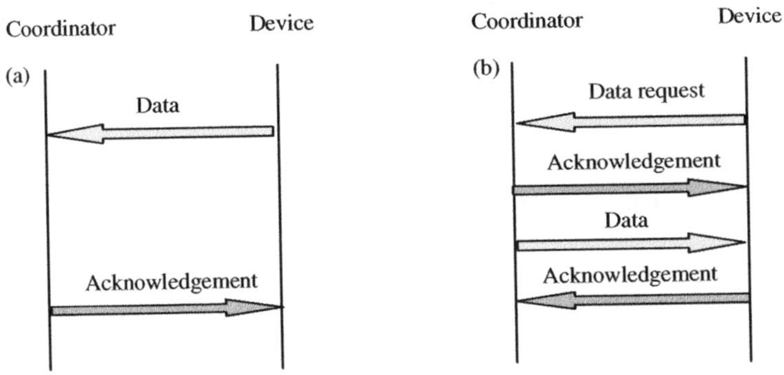

Figure 11-22. (a). Sending data from the coordinator without beacon. (b). Receiving data from the coordinator without beacon.

A superframe is optional, thus data transmission is possible for two cases, namely, when a superframe is defined by the coordinator and when a superframe is not defined. Random access in the former case is provided by a slotted CSMA/CA mechanism, in which all transmission attempts are made on slot boundaries. In the case of no superframe, CSMA/CA is employed without time slots (asynchronously). Figures 11-21 and 11-22 show the three data transfer modes for the two cases, Figure 11-21 for the case of a superframe defined and Figure 11-22 without a superframe.

11.6.5. MAC Frames

LR-PAN MAC defines four frame types.

1. *Beacon frame* is used in superframe communication. It carries network identification and synchronization information. If a frame contains guaranteed time slots, it carries the information about their allocation.

2. *Data frame* is used for data transfer of all kinds.
3. *Acknowledgement frame* is a short frame that carries a sequence number and FCS fields to ACK data frames. The acknowledgement frames bypass the CSMA/CA mechanism and are sent with priority.
4. *Command frames* are used for various MAC commands. Its payload contains command type and data.

11.6.6. MAC Security

The MAC sublayer also provides a secure mode of transmission in which four functions provide secure exchange of frames between peer MAC entities. These functions are: (1) access control, (2) data encryption, (3) frame integrity and (4) sequential freshness.

11.7. SUMMARY

Even though Bluetooth has an identifiable set of applications not clearly handled by WLANs, it has two problems, both relating to its scalability. First, it is too slow for many short distance (personal operating space—POS) appliances used in every home, such as connections between video and television. Second, it is too complex for simple toys and sensors. The first problem is fixed by IEEE 802.15.3, high data rate WPAN by defining PHY and MAC in tens of Mbps. The UWB option to be specified in 802.15.3a is going to raise the rate further up to hundreds of Mbps in addition to the capability of passing through walls from one room to another. The latter, that is, scaling down capability, is to be provided by IEEE 802.15.4 low-rate PAN, which are to provide from below 25 kbps to above 200 kbps. These rates are provided by a low-complexity, ultra-low power standard suitable for toys and sensors.

Work on UWB will continue well beyond IEEE 802.15.3a because of the variety of ways it can be applied in its own scope and as an additive to other existing wireless technologies. In reference to 802.15.3a, the latest on the DS-UWB proposal has come in the form of a compromise to let DS-UWB and multiband-OFDM PHY co-exist with a common signaling mode (CSM). The IEEE 802.11 has tried a similar concept, by allowing three PHYs in the original standards, and has resulted in having only one dominant candidate, the DS-SS. However, in IEEE 802.11 DIFR and FH-SS don't co-exist with DS-SS. So the exact balance of DS-UWB/MB-OFDM is unknown at this time.

REFERENCES

[1] Shan Tsung Wu, 'Multicarrier CDMA and its applications to high-rate UWB communications', available from http://my.nthu.edu.tw/~cuicp/file02.ppt

[2] G. Racherla, Ellis, J.L., Furuno, D.S., and Lin, S.C. 'Ultra-wideband systems for data communications', *IEEE International Conference in Personal and Wireless Communications*, December 2002.

[3] Jeffrey R. Foerster, 'Ultra-wideband technology enabling low-power, high-rate connectivity (invited paper)', available from http://dsp.jpl.nasa.gov/cas/full/forester.pdf

[4] Joy Kelly (*Presenter*), 'Multi-band OFDM Physical Layer Proposal Update', *IEEE 802.15-04/0122r4*, March 2004.

[5] Reed Fisher, Kohno, Ryuji, Ogawa, Hiroyo, Zhang, Honggang, and Takizawa, Kenichi, 'DS-UWB Physical Layer Submission to 802.15 Task Group 3a', *IEEE P802.15-04/0137r00137r00137r0*, March 2004.

[6] 'foreign', 'Ultra Wide Band (UWB)', home.ee.ntu.edu.tw/~lab554/chapter/slide/Ultra Wide Band.pdf

[7] Safecom, 'Emerging Wireless Technologies: Ultra Wide band Communications', www.safecomprogram.gov/admin/librarydocs8/Emerging_Wireless_Technologies-Part4.pdf

[8] IEEE, 'Medium access control (MAC) and physical layer (PHY) specifications for wireless personal area networks (WPANs)', *IEEE Std 802.15.1 2002.*

[9] Brent A. Miller and Bisdikian, Chatschik, *Bluetooth Revealed: The insider's guide to an Open Specification for Global Wireless Communications*, Prentice-Hall PTR, Upper Saddle River, NJ 2001.

[10] Jennifer Bray and Charles F. Sturman, *Bluetooth: Connect without cables*, Prentice-Hall PTR, Upper Saddle River, NJ 2001.

[11] IEEE, 'Wireless medium access control (MAC) and physical layer (PHY) specifications for high rate wireless personal area networks (WPANs)', *IEEE Std 802.15.3 2003.*

[12] IEEE, 'Co-existence of wireless personal area networks (WPANs) with other wireless devices operating in unlicensed frequency bands', *IEEE Std 802.15.2 2003.*

[13] IEEE, *IEEE 802.15 WPAN High rate alternative PHY Task Group 3a (TG3a)*, www.ieee802.org/15/pub/TG3a.html

[14] IEEE, 'Medium access control (MAC) and physical layer (PHY) specifications for low-rate wireless personal area networks (WPANs)', *IEEE Std 802.15.1 2002.*

CHAPTER 12

BROADBAND WIRELESS ACCESS (BWA)

For an all-IP wireless network, covering an area as large as a country or a continent, the fixed wireline networks can provide an excellent backhaul network. As long as there is wireless access mechanism available to a wireless or to mobile wireless terminals, the rest of the detail is not important. The access network can be provided by WPANs within a very short area, WLANs within a much larger area than the WPANs, and by cellular access architecture for the wide area case. The wireless access, however, is not required for mobile user only. Due to the fact that wireless access networks can be deployed with much less infrastructure than the wired counterparts, they are a prudent choice even for fixed user locations. Much of the network cost in all telecommunications networks is due to the subscriber access network. The cost is enhanced due to the requirement of reaching each subscriber individually. Areas with sporadic population density are especially hard to provide access services through wire because of the distances among various subscribers and the consequent reduced return on the infrastructure cost.

Wireless local loops (WLLs) have been used for PSTN to substitute for wired subscriber's loop in this situation. These loops have also been termed as last-mile technology[1]. These loops have been designed by using all kinds of infrastructure technologies, ranging from new terrestrial technologies, through the cellular networks to satellite loops. For broadband Internet access, a tech-

[1] The fact of the matter is that this is not just last mile, but also the first mile.

Wireless and Mobile Data Networks, by Aftab Ahmad

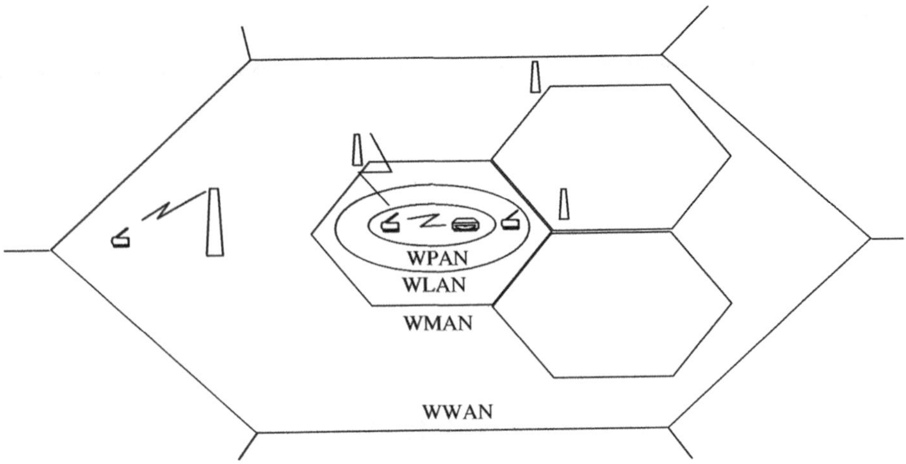

Figure 12-1. Wireless network hierarchy in terms of coverage.

nology like the WLL is imperative because the only other wireless alternative, that is high-speed cellular networks, have been designed for more generalized applications, with stringent requirements for mobility management. Also, due to the fact that the access networks for cellular systems have evolved from GPRS and other 2.5 G systems, the call control mechanisms are too complex for a simple point-to-point (uplink) and point-to-multipoint (downlink) connections. The wireless metropolitan area networks[2] (WirelessMAN™s) are projected to fill this spot. The IEEE Working Group (WG) 802.16 was set up to recommend specifications for WirelessMAN in a rather wide rage of microwave spectrum of 10–66 GHz, covering many existing standards, for example, LMDS (local multipoint distribution system). The group came up with the MAC and PHY specs for this microwave spectrum as IEEE 802.16. Later on, an extension group IEEE 802.16a added specifications for the WMAN at 2–11 GHz by appropriately amending the MAC and PHY specifications of the IEEE 802.16. This covered some other existing standards, or standards in the waiting, such as MMDS (multichannel multipoint distribution system). Figure 12-1 shows the hierarchy of various wireless networks in terms of coverage.

Earlier, the European DAVIC (digital audio visual council) had published specifications for some broadband digital loops [1] including LMDS at 28 GHz for line-of-site applications. Also, MMDS at frequencies below 3 GHz has been projected to provide digital access streams to broader distances. Besides the EU, other regulating agencies in countries, including the United States, South Korea, and Canada also allocated spectrum for LMDS and MMDS. Additionally, ETSI's broadband access infrastructure has HIPERACCESS speci-

[2] WirelessMAN™ is a trademark of the IEEE.

fied for the broadband access family of standards called BRAN (broadband radio access network).

In this chapter, we will have a look at the IEEE 802.16 and 802.16a specifications for BWA. A Working Group, IEEE 802.20, is also looking up at the mobile WBA (MWBA) at 3.5 GHz for speeds exceeding the vehicular range. We will have a few words about that, too.

12.1. LINE-OF-SITE (LOS) AND NON-LINE-OF-SITE (NLOS) SYSTEMS

The division of spectrum from 2–66 GHz between two parts (2–11 and 10–66 GHz) is mainly referred to as requiring LoS (latter) or not requiring LoS (former) communications. This is more of a consequence of having short wavelength in the case of frequencies above 10 GHz. Due to the short wavelengths, signals at these frequencies don't get reflected from buildings and foliage. Therefore, if there is no direct LoS communications, the signal is unlikely to reach the receiver. At these frequencies, the signal is also absorbed relatively more by humidity. In fact, the path loss is directly proportional to the α power of the frequency, where α is called path loss exponent. Thus, if the loss is proportional to $10\alpha \log(f)$, the signal coverage is severally limited at higher frequencies. As shown in Figure 12-2, for a range of frequencies between 2 GHz to 64 GHz, the term $20\log(f)$, with $\alpha = 2$, has a range of 30 dB. This translates directly into the coverage area. A second implication, due to lack of reflections is shown in Figure 12-3. In this figure, a signal reaches the receiver through three ways, direct (LoS), reflection from a wall and reflection from a tree.

This allows for a rake receiver design that will collect a given number of multipath signals and combine them into one (*multipath diversity*). At lower part of the spectrum (2–11 GHz), multipath diversity is possible to exploit due to reflections, while at higher parts, it is not. Therefore, designing systems and network protocols for wide ranges of frequencies is challenging and one should expect more than one solution for such a wide range of spectrum.

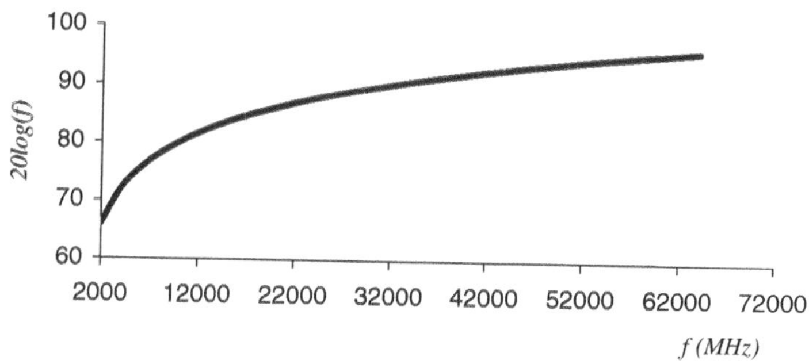

Figure 12-2. Increasing frequency can incur significant loss.

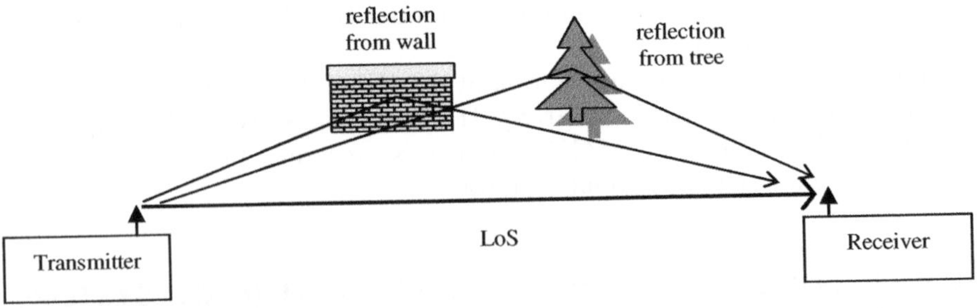

Figure 12-3. Multipath results in more signal energy arriving at the receiver.

TABLE 12.1. Channel Models for Various Antenna Heights and Directivity

Antenna Directivity	Antenna Height	Channel Model
Omni-directional at the receiver (at least)	Below obstacles.	Rayleigh
Omni-directional at the receiver (at least)	At least one above tobstacles, LoS exis	Ricean
Directional high-gain	LoS	Lognormal
Either	LoS and multipath	Nakagami-*m*

For higher spectrum values, absence of multipath means that the received signal will be weaker in general and the coverage area shorter. However, there are benefits for not having multipath diversity and the incapability of providing mobility related service. These benefits are seen in reduced receiver complexity at the subscriber side as well as base station. With no mobility-related infrastructure, a base station can easily manage more than one cell, resulting in savings as compared to using one base station per cell.

12.2. EFFECT OF ANTENNA TYPE

The received signal statistics can also be a strong function of antenna type and positioning. As reported in [2], by varying the directivity and height of antenna for LMDS system, one can get a variety of channel statistics. Table 12-1 lists various channel models resulting from changes in antennas.

Each of these channel models requires a specific type of receiver design for a given performance level.

Among other things, the above discussion shows that the higher level of spectrum (above 10 GHz) is not suitable for wireless systems that require mobility of the user. Therefore, this spectrum has been allocated for fixed wireless access where very high data rates are required, such as digital video broadcast systems.

TABLE 12.2. MMDS Allocation

Country/Region	MMDS Spectrum (GHz)
USA	2.5–2.686
Europe	3.3–3.6
Japan	3.4–3.6

TABLE 12.3. Worldwide Allocations for BWA[4] [4]

Country/Region	MMDS spectrum (GHz)	Comment
USA	24 (DEMS), 28, 31, 38	Canada too except 25–27 for 31.
Europe (Germany, Netherlands Norway, Spain)	26 (ETSI)	France: 27.5–29.5, UK: 10 GHz (ETSI)
Japan and South Korea	25–27	Japan also 38 and 18–24.

12.3. BWA SPECTRUM

Bandwidth for MMDS has been allocated at the lower part of the 2–66 GHz 'slab'. Table 12-2 shows the allocations as per Ref. [3]. The coverage area for this spectrum ranges anywhere up to 50 km[3].

The allocations for LMDS in the United States are primarily in 28 GHz and 31 GHz locations of the RF spectrum. Table 12-3 lists worldwide allocations for LMDS. HIPERACCESS, which is one of the four BRAN standards, also covers a broad spectrum. As mentioned in a letter of Liaison [5] by the BRAN Chair to the Chair of ETSI-ERM, the major spectrum requirement for HIPERACCESS (now this part is referred to as HIPERMAN) is around 3 GHz, but small pieces are required in a much broader range. The various categories of HIPERACCES spectrum as laid out in [7] are shown in Table 12-4.

12.4. BRAN VERSUS WIRELESSMAN™

The BRAN spectrum of 2–66 GHz, too, is viewed as 2 parts, one for the LoS access (HIPERACCESS) and the other, lower band, for NLoS access (HIPERMAN). The IEEE and the ETSI are working on harmonizing the two suites of standards. Figure 12-4 shows a comparison as per [8]. Figure 12-5 shows their scope in the ETSI BRAN family.

[3] The actual coverage is highly subject to the propagation environment and condition. It may go well above 50 km [6] claims a range of 100 km for MMDS.

[4] This table should be checked for any changes.

TABLE 12.4. BRAN Broadband Access Spectrum

Spectrum (GHz)	Comment
3.41–3.6	The big chunk. Draft FWA recommendation.
10.15–10.3/10.5–10.65	(Two way) Draft FWA recommendation.
24.5–26.5	Draft FWA recommendation.
27.5–29.5	Draft FWA recommendation. LMDS band
31.8–33.4	Draft FWA recommendation. (Possible candidate)
40.5–42.5	Draft revision of ERC/DEC/(96)05.
42.5–43.5	Draft revision of ERC/DEC/(96)05. sharing with radio astronomers.

HIPERMAN	802.16.1a		802.16.1	HIPERACCESS	
Variable packet size	Variable packet size		Variable packet size	Fixed packet size	MAC
OFDM, OFDMA	OFDM (TDMA), OFDMA	SC2	Single carrier (SC) modulation	Single carrier (SC) modulation	PHY

Figure 12-4. Feature comparison of IEEE and ETSI BWA standards.

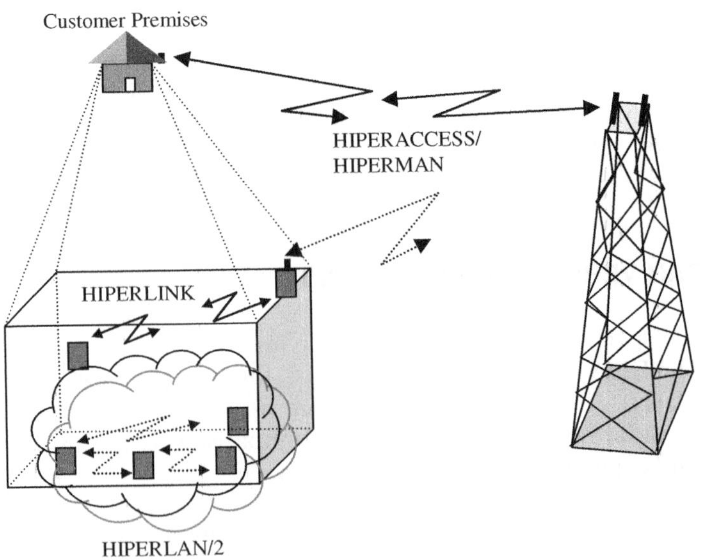

Figure 12-5. HIPERACCESS and HIPERMAN form part of ETSI BRAN family.

TABLE 12.5. IEEE 802.16.1 versus ETSI HIPERACCESS [9]

Standard Attribute	HIPERACCESS	802.16.1
Access	FDD (primary), TDD (for unpaired bands) TDMA/TDM (downlink)	Mode A: FDD Mode B: FDD/TDD TDMA + DAMA (uplink) TDM/TDMA (downlink)
Canalization schemes	28 MHz 14 MHz possible in uplink	20–40 MHz or 25–50 MHz
Modulation schemes	4/16-QAM + 64-QAM as option (downlink) 4-QAM + 16-QAM as option (uplink)	Mode A: QPSK + 16-QAM as option Mode B: QPSK/16-QAM + 64-QAM as option
Capacity (Mbps unless otherwise specified)	60 (downlink), 30 (uplink) Symmetrical capability although instantaneous traffic need not be symmetrical.	16, 20, 32, 40 Mbauds for 20, 25, 40 and 50 MHz channels, respectively.

The HIPERMAN and IEEE 802.16a were developed with reasonable coordination. However, HIPERACCESS and IEEE 802.16 standards do not share enough to be easily interoperable.

Since the interoperability is not at issue due to the disjoint nature of the access networks, the standardization and liaison efforts are centered around international trade and manufacturing realties rather than the technical issues. There is more coordination at the 2–11 GHz level (802.16.1a versus HIPERMAN). Table 12-5 [9] highlights some more differences between 802.16 and HIPERACCESS.

Harmonization of IEEE 802.16 efforts have resulted in representations from many countries in meetings of the IEEE 802.16 WG and its task groups. An interoperability forum for IEEE 802.16 (WiMAX) is set to make interoperability profiles.

12.5. IEEE WIRELESSMAN™

The IEEE WirelessMAN consist of two parts, a base standard [10] for the 10–66 GHz range and an enhancement [11] for 2–11 GHz. In this chapter, we may also use the notation WMAN for IEEE 802.16 WirelessMAN and eWMAN for its enhancement (2–11 GHz) specified as IEEE 802.16a. We will discuss the network components, layout, architecture, the medium access control (MAC) and associated sublayers, the physical layer (PHY). For details on hardware level, reader is referred to Ref. [12], available on the web.

12.5.1. WirelessMAN Station Types

The WMAN defines two types of stations, a base station (BS) and a subscriber's station (SS).

12.5.1.1. Base Station (BS). The base station controls and manages the connection. It sends data on the downlink in channels allocated to various subscribers. A base station can cover multiple cells (sectors) with the help of sectored antennas. In a point-to-multipoint (PMP) configuration, the downlink is multipoint. Each base station is configured with a 48-bit MAC address. The first 24 bits of this address identify the operator.

12.5.1.2. Subscriber's Station (SS). A subscriber's station is a terminal that communicates with the base station (BS). It sends data on the uplink, which is point-to-point in a PMP network configuration and either point-to-point or point-to-multipoint in a mesh topology. All SSs within the same sector and frequency channel receive the same downlink information. The 48-bit IEEE 802 MAC address uniquely identifies a SS. An SS could be a packet data or multimedia terminal with a range of transmission rate capabilities.

12.5.2. Network Topologies

The WMAN has a point-to-multipoint (PMP) configurations, in which a BS can send data and control information to many SSs while an SS can communicate to BS through the allocated or contention based resources. In the optional mesh topology of eWMAN, SSs can communicate directly, thus paving the way for multihop communications. Figures 12-6 and 12-7 show the two topologies.

In addition to the simplified Figures 12-6 and 12-7 layouts, the 802.16 network has many other potential applications. As mentioned in [12], these applications range from fractional T1 carrier for small businesses, DSL to residential and home offices, full T1 for businesses, backhaul network for cellular systems and wireless backhaul for a constellation of Wi-Fi hotspots. The cell radius could vary anywhere from one km to more than 10 km, depending on the carrier frequency.

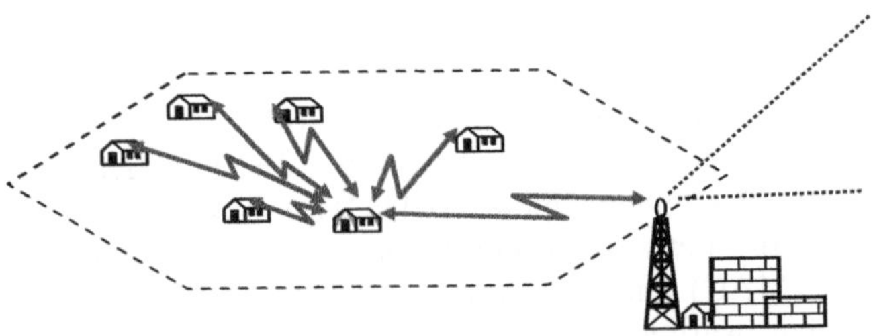

Figure 12-6. Mesh topology showing connections of just one user premises.

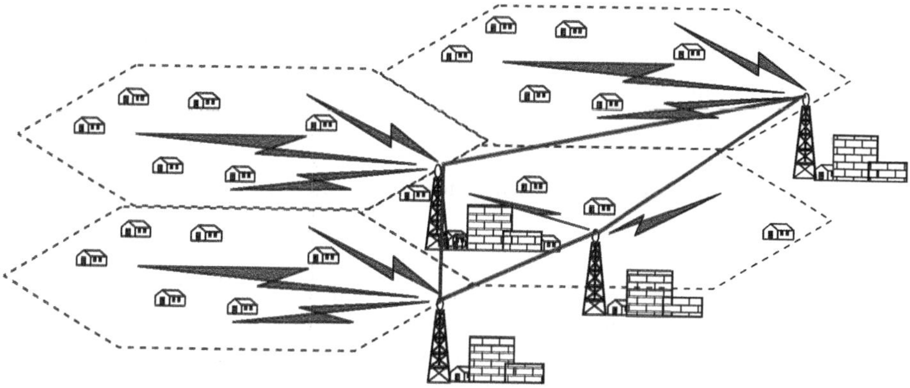

Figure 12-7. Point-to-multipoint network. Solid lines are wired connections between the BS towers.

The SS, when initialized goes through the following steps in a PMP topology:

1. Scan for downlink channels.
2. Establish synchronization with the BS.
3. Obtain transmit parameters from a downlink map.
4. Perform ranging.
5. Negotiate basic capabilities.
6. Performance authorization and key exchange for encrypion.
7. Perform registration.
8. Establish IP connectivity.
9. Establish time of day.
10. Transfer operational parameters.
11. Setup connections.

In the mesh topology, the concept of neighborhood (one-hop distance) and extended neighborhood (neighborhood of neighbors) and the need for coordination (even for BS) makes communication infrastructure both, more powerful and more complex. The SS, when initialized in a mesh topology goes through the following initialization steps:

1. Establish coarse synchronization by scanning the network.
2. Obtain network parameters.
3. Open a channel called *sponsor channel.* The sponsor channel is a temporary schedule setup by a sponsoring neighborhood node for the initializing (candidate) node.
4. Go through node authorization.

5. Perform registration.
6. Establish IP connectivity.
7. Establish time of the day.
8. Transfer operational parameters.

The standard employs the following novel concepts for flexibility and high resource utilization.

12.5.2.1. Bandwidth Stealing. Bandwidth stealing is done by a subscriber station (SS) when it is allocated a bandwidth for a connection (to exchange data), but uses this bandwidth in whole or a fraction for making a signaling request (to get more bandwidth).

12.5.2.2. Adaptive Modulation. Adaptive modulation is the ability of a station to send and receive signals by selecting a suitable modulation scheme from different modulation schemes. Using adaptive modulation, a receiving station can receive from a station using multiple burst profiles. Also, as a transmitter a station using adaptive modulation can send data to stations using different burst profiles.

12.5.2.3. Adaptive Antenna System (AAS). A receiving station is said to have AAS if it can receive signal from multiple antennas adaptively in order to increase the capacity and improve coverage.

12.5.3. WirelessMAN Protocol Architecture

The WMAN and eWMAN standards have been designed to meet several explicit and implicit objectives. Independence from the upper layer protocols in one such objective, security is another. In fact, the medium access control (MAC) sublayer has been designed to coexist with multiple PHYs. Figure 12-8 shows a protocol architecture reference model.

In the following, we will discuss briefly the main functions of the PHY and MAC sublayers.

12.5.4. MAC Sublayer

The core MAC functions are provided by the MAC common part sublayer. Broadly speaking, these functions include channel access and multiple access in the uplink and point-to-multipoint operation for WMAN in the downlink. For eWMAN, the option for mesh topology, if implemented, requires additional procedures for channel resource management. Five different scheduling mechanisms are available for SSs in a WMAN to use the uplink bandwidth. The actual combination of these mechanisms is left to the operator and should

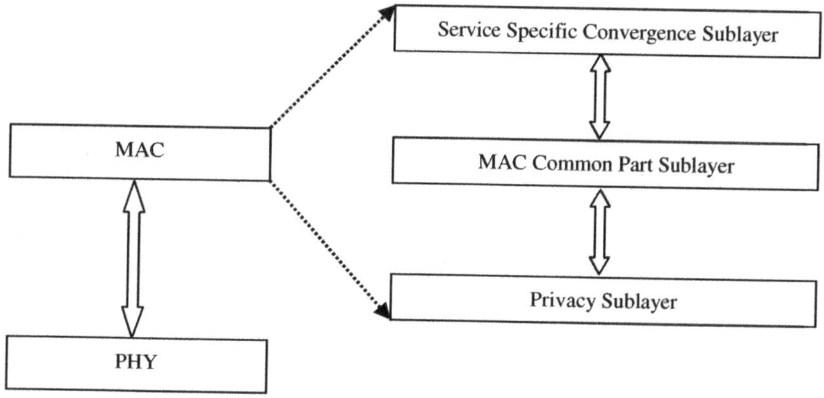

Figure 12-8. Reference model for IEEE 802.16*a*.

ideally be a function of traffic distribution as a function of bandwidth allocation mechanisms; including unsolicited allocations, polling and random access. Central to the bandwidth allocation is the concept of *service flow.*

12.5.4.1. Service Flow. The IEEE 802.16 MAC is connection oriented. A 16-bit connection ID (CID) is used to identify all connection, limiting the total number of connections to 2^{16}. Higher layers data can be multiplexed on a single connection, using the same CID. The quality of service (QoS) attributes of the data exchanged over a connection define a service flow. Alternatively, a service flow has QoS parameters associated with it over a connection. Service flows are set up at the time of subscription. As soon as a SS registers after installation, connections are associated with the service flows depending on subscription profile of the subscriber. A service can be one of the four types: unsolicited grant service (UGS), real-time polling service (rtPS), non-real-time polling service (nrtPS) and best effort (BE). All of these services, including the BE uses the CID of the MAC connection, even if the convergence sublayer is providing service to a connectionless protocol, such as IP. A MAC SDU constitutes a service flow. A MAC SDU could consist of one or more MAC PDUs (using fragmentation) and vice versa (using packing). A service flow is identified by a 32-bit service flow ID (SFID) and an ASCII name descriptive of the QoS represented by the flow.

12.5.4.2. MAC PDU. The MAC PDU carries the CID and the service flow is implicit from the PDU. Figure 12-9 shows the general format of the MAC PDU.

12.5.4.2.1. MAC Header. The 48-bit MAC header is of two types; a bandwidth request (BR) header, to request bandwidth and a generic MAC header, carrying from less than one to more than one MSDU. The generic MAC

Generic MAC Header (6 bytes)	Payload (0–2042 bytes)	CRC (0 or 4 bytes)

Figure 12-9. MAC PDU for IEEE 802.16.

HT (1)	EC (1)	Type (6)	RSV (1)	CI (1)	EKS (2)	RSV (1)	LEN msb(3)
LEN lsb(8)			CID msb(8)				
CID lsb(8)			HCS (8)				

Figure 12-10. Generic MAC header fields along with number of bits.

HT (1)	EC (1)	Type (6)	BR msb(8)
BR lsb(8)			CID msb(8)
CID lsb(8)			HCS (8)

Figure 12-11. Bandwidth Request header fields along with number of bits.

header carries management information of convergence sublayer data. Figures 12-10 and 12-11 show the two types of the generic MAC header.

Table 12-6 lists the various fields of the generic MAC header fields and their functions.

MAC Subheaders

The 'Type' field in the MAC header may indicate the presence of MAC subheaders. Table 12-7 lists the three types of supported subheader.

12.5.4.2.2. CRC. If a flow service requests for a CRC, an IEEE 802 CRC (32-bit) is generated over the encrypted MAC header and payload and is included in every PDU.

TABLE 12.6. Generic Header Fields and Their Functions

Field	Function
HT (Header type)	Define 2 header types.
EC (Encryption control)	Whether encryption used or not.
Type	Payload type, including subheaders.
RSV (Reserve)	Reserved.
CI (CRC indicator)	Whether CRC appended.
EKS (encryption key sequence)	Index of the key.
LEN (Length)	Length of MAC PDU in bytes including header.
CID (Connection identifier)	MAC connection identifier.
HCS (Header check sequence)	Header error check.
BR (bandwidth request)	Indicate the requested bandwidth (number of bytes).

TABLE 12.7. MAC Sub-headers

Sub-header Type	Scope	Function and Scope
Fragmentation sub-header	MAC PDU	Indicates the state of fragmentation and the sequence number of the fragment.
Grant management sub-header	MAC PDU	Used by SS to covey bandwidth needs to the BS. Has about 30 message types.
Packing sub-header	MAC SDU	Preceded every MAC SDU if multiple SDUs are packed into a single PDU. This is the inverse of fragmentation.

TABLE 12.8. Generic Header Fields and Their Functions

Function	Description
Concatenation	Multiple PDUs in a single transmission. Works for uplink an downlink.
Fragmentation	Fragment a large SDU into smaller PDUs. Could b done by SS or BS.
Packing	Multiple SDUs per PDU. Packing of fixed length as well as variable length SDUs possible. Can be combined with fragmentation.
Encryption	Use exchange of key sequence for data cipher. Authentication mechanism not specified, but authentication itself specified.
CRC	If requested. No retransmissions for WMAN (10–66 GHz).

12.5.4.3. Transmission of MAC PDU. The MAC service data units are transmitted in the payload of the MAC PDUs. The standard provides a number of functions relating to the transmission of PDUs. Table 12-8 lists these functions along with a brief account of each.

12.5.4.4. QoS Provisioning. The 802.16 MAC provides four types of scheduling services with respect of QoS differentiation of service flows. Table 12-9 lists these service types and their characteristics. Each of the four types correspond to a different scheduling mechanism.

Additionally, the WMAN includes several QoS-related concepts, such as dynamic service establishment and two-phase activation model.

12.5.4.5. Distributed and Centralized Scheduling in eWMAN. In mesh topology, the concept of distributed scheduling allows the SSs to make sure that transmissions do not depend on BSs. In the distributed scheduling, the SSs have a three-way handshake with the neighbors to reserve channel resources. The three-way handshake consists of request, grant, and confirm grant paradigm, in which the first and last messages are sent by the requesting SS.

This distributed scheduling could be either in a coordinated fashion or uncoordinated fashion. The uncoordinated distributed scheduling is performed using contention based mechanism for sending requests and responses to requests. A backoff mechanism is used in the uncoordinated distributed scheduling to avoid collisions. In the coordinated distributed scheduling, all requests and responses use allocated resource, thus avoiding collisions. Figure 12-12 shows a handshake using mesh(MESH)-distributed(D)-scheduling (SCH)-related messages MESH-DSCH:request, MESH-DSCH; Grant and MESH-DSCH:Grant. The figure shows two neighbors of the requesting SS.

In the distributed scheduling, the BS exerts no control over the scheduling. However, in a centralized scheduling, the BS determines the resources assigned from the requests made by SSs. The centralized scheduling is similar to the point-to-multipoint scheduling except that in the case of latter all SSs are directly in communication with the BS. In mesh topology (centralized scheduling), not all SSs are directly connected to the BS.

12.5.4.6. Duplexing Techniques. The MAC supports several duplexing techniques for framed and unframed PHYs. For the current framed PHYs, the MAC CPS (common part sublayer) provides FDD and TDD support. The FDD frame (burst) could be used to provide full-duplex and half-duplex modes. In full-duplex mode, a SS can receive signals on the downlink burst and, at the same time, transmit at the uplink burst. The two modes (FDD and TDD) are shown in Figures 12-13 and 12-14.

In TDD mode, transmission and reception occur at different times and at the same carrier frequency. The TDD frame is of fixed duration just like the FDD frame. It is divided in two ways; first it is divided into *physical slot* (PS), and then into two subframes A physical slot is the smallest unit of bandwidth that can be managed (not shown in the Figure 12-14, but one can imagine each small rectangle of the TDD frame as a single PS). The two subframes are used

TABLE 12.9. QoS Related MAC Services

QoS Service Type	Purpose	Service Parameters	How
Unsolicited grant service (UGS)	For real-time, fixed size regularly transmitted packets, e.g., voice codec.	Unsolicited grant size, the grant interval, grant jitter, request/transmission policy.	The BS provides fix size Data Grant Burst periodically.
Real-time polling service (rtPS)	For real-time variable size regularly transmitted packets, e.g., MPEG video.	Polling interval, polling jitter, request/ transmission policy.	The BS provides SS the opportunity to request bandwidth on a regular basis.
Non-real-time polling service (nrtPS)	For non-real-time service flows, requiring variable size, regular Data Grant Burst.	Polling interval, minimum reserved traffic rate, maximum sustained traffic rate, request/ transmission policy, priority.	The BS provides SS opportunity to request bandwidth using unicast and contention methods.
Best effort service flow	E.g., FTP.	Minimum reserved traffic rate, maximum sustained traffic rate and priority.	The BS allows SS to use all available mechanisms for transmission requests.

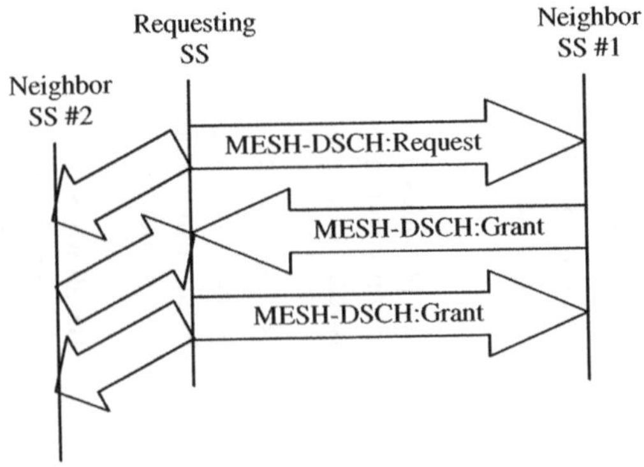

Figure 12-12. Handshake for distributed scheduling. Possibility of collisions exists in uncoordinated handshake.

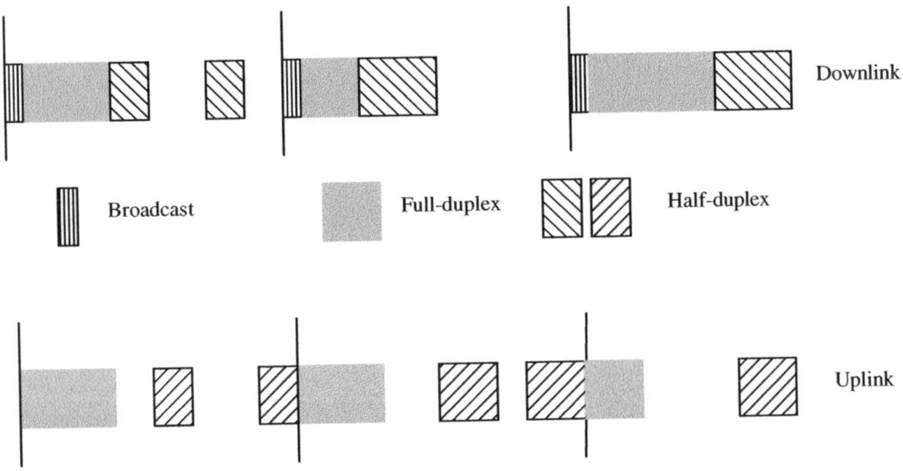

Figure 12-13. Example of full-duplex and half-duplex transmission with FDD.

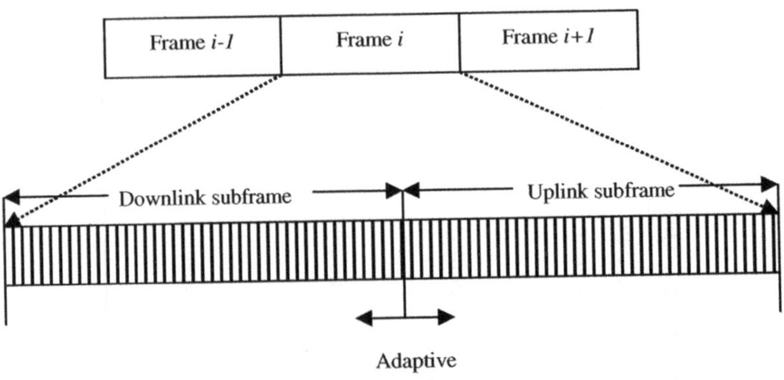

Figure 12-14. Example two-way transmission with TDD mode.

for uplink and downlink transmissions. The boundary between the uplink and downlink subframes can be changed, thus allocating different bandwidths in the two directions.

12.5.4.7. Bandwidth Management. The access in the uplink and downlink is granted through the uplink map (UL-MAP) and downlink map (DL-MAP). The resolution of the bandwidth is in terms of physical slots, but the allocation unit in the uplink is in *minislots*. A minislot consists of n PSs, where n is an 8-bit number (0–255). The uplink map (UL-MAP) is used to allocate the minislots to SSs. Similarly, the downlink map (DL-MAP) is used to allocate bandwidth to the SSs in the downlink for burst mode PHYs. Minislots are organized into *transmission opportunities* (TO). A TO is a set of minislots in which a SS can transmit. Even though the MAC is connection-oriented, collisions still may occur during the Initial Maintenance and Request intervals. The collision resolution is done in two steps. These steps involve a binary exponential backoff mechanism. As step number one, a SS that has data to send will set a backoff timer at a random value selected with the help of DL-MAP. It transmits after the backoff timer expires and waits for the BS response. If it does not get a response in a specified time, the SS, as a second step, doubles the maximum limit of the random backoff and repeats the process. If it gets a response from the BS, the contention has been resolved and uplink and downlink resources may have been allocated.

12.5.4.8. Adaptive Antenna Systems (AAS). The 2–11 GHz eWMAN specifications provide a way to employ the adaptive array antenna technology by AAS. There are many benefits of using AAS including:

- Improve range and capacity through increased spectral efficiency.
- Strong reception at target users by coherently combining signals from multiple antennas.
- Interference reduction by steering the beam nulls in the direction of interferers.

In case an AAS is used, an SS will not have BS antennas beam pointing at it all the time. This results in the possibility of the SS making a bandwidth request and having the request lost due to the BS not pointing toward it. To avoid this situation, the BS in an AAS system instructs the SSs whether to make bandwidth requests or not.

12.5.4.9. Dynamic Frequency Selection (DFS). Since eWMAN contains license-exempt bands, some channels may be used by other 802.16~ and non-802.16~compliant networks and devices. If the interfering networks or devices are the designated primary users for the frequency band in question, then an SS is required to change the channel. DFS function helps the SS get a new channel with less interference. The DFS results in a uniform load distribution over the available channels in addition to avoiding unnecessary interference with the primary users. The DFS involves steps shown in Table 12-10.

12.5.4.10. Other MAC Sublayers. The service specific convergence sublayer (SS-CS) and the privacy sublayer are part of the WMAN MAC. CS provides the opportunity for MAC to be independent of the upper layer protocols. For example, as shown in Figure 12-15, a MAC can interact transparently with ATM and IP because the CS hides the details of the ATM and IP networks from MAC and prepares MAC SDUs that are acceptable to the MAC sublayer. Similarly, when CPS delivers a data unit to the CS, it converts it back to the upper network format (ATM cell or IP datagram).

TABLE 12.10. Steps for Using Dynamic Frequency Selection (DFS)

Step	Detail
Testing channels for primary users	During start up testing period and operating test period.
Discontinue operations after detecting primary users	Stop transmission of MAC PDUs containing management message within 'management operations period', and MAC PDUs carrying data within 'max data operations period'.
Detecting primary users	Method not specified.
Scheduling for channel testing	A BS may ask an SS to measure a channel interference.
Requesting and reporting of measurements	SS prepares a detailed report of its findings on the channel requested for measurement by the BS.
Selecting and advertising a new channel	The BS, based on the measurements, may decide so. Algorithm for this not specified.

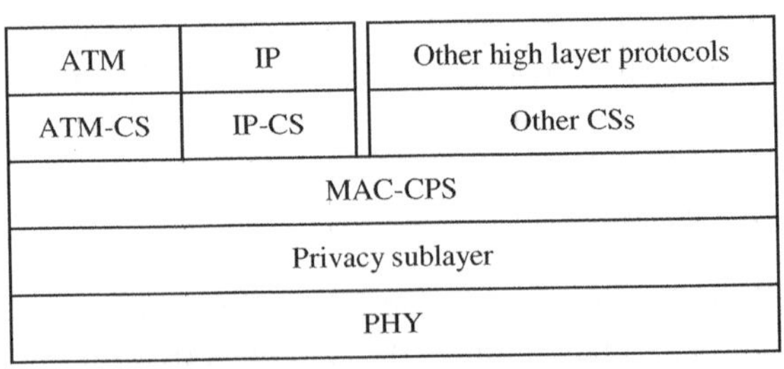

Figure 12-15. Each upper layer protocol requires a convergence sublayer to conform back and forth with MAC sublayer.

The privacy sublayer protects the 802.16 network against unauthorized SSs through authentication and protects data through encryption. The privacy sublayer has two major components.

1. A data encapsulation protocol that includes a set of cryptographic suites and rules to apply them. A security association (SA) between the BS and SS identifies these suites.
2. A secure key distribution protocol that allows a BS control the distribution of key to SSs in a secure way. X.509 v 3 certificates are used by SSs for digital signature.

12.5.5. WirelessMAN PHYs

The medium access control sublayer of the IEEE 802.16 has been designed to work over multiple PHYs. A transmission convergence part (TC) of the PHY hides the underlying detail of the PHY from MAC. A physical medium dependent (PMD) part of the PHY arranges the signals from the PHY into logical groups for the TC for transparent handling by the TC. Just like a unique CS is needed above MAC CPS for each upper layer protocol, every PHY requires a unique TC. A general PHY uses three sets of primitives to provide various services to MAC. These primitives relate to data transmission, management functions and sublayer-to-sublayer interaction related to layer control. We will look at the characteristics of PHY for WMAN (10–66 GHz) and eWMAN (2–11 GHz) in the following.

12.5.6. WMAN PHY (10–66 GHz)

In order to accommodate a range of spectra allocated for BWA (e.g. for various bands of LMDS) the PHY design has been kept flexible. This layer provides modulation, error control and burst-oriented transmission in the downlink and uplink. The PHY supports two types of duplexing, FDD and TDD. For FDD, different carrier frequencies are used for uplink and downlink connections. The SS and BS can use the uplink and downlink carriers simultaneously for a full-duplex connection or in alternate fashion for a half-duplex connection mode. The downlink transmissions are organized in a single TDM stream with TDMA to support half-duplex FDD. On the uplink, SSs use TDMA and DAMA (demand assignment multiple access) mechanisms to access channel resources. Both uplink and downlink operate in burst mode. One of the important feature regarding the flexibility of PHY is adaptive burst profiling. In adaptive burst profiling, the burst parameters, including modulation and error control coding, can be changed from burst-to-burst.

12.5.6.1. PHY Frame. The PHY supports three frame sizes, 0.5, 1 and 2 ms. Both TDD and FDD modes operate in frame (burst) form. Figures 12-13 and

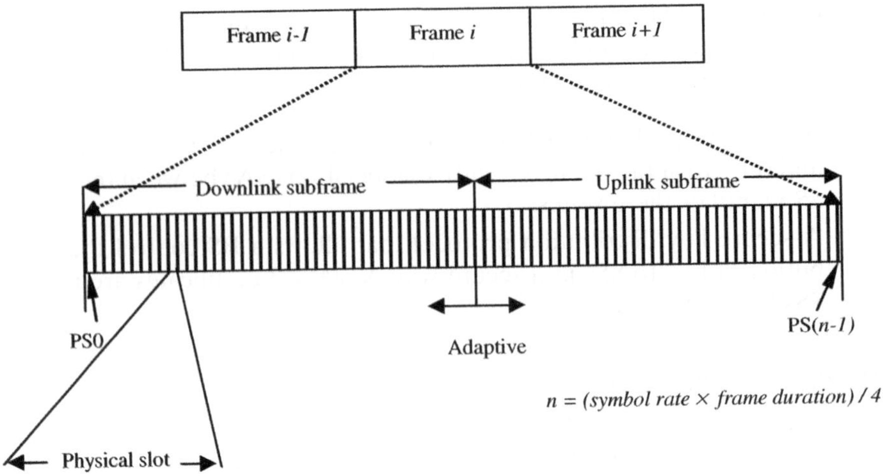

Figure 12-16. Example transmission with TDD mode.

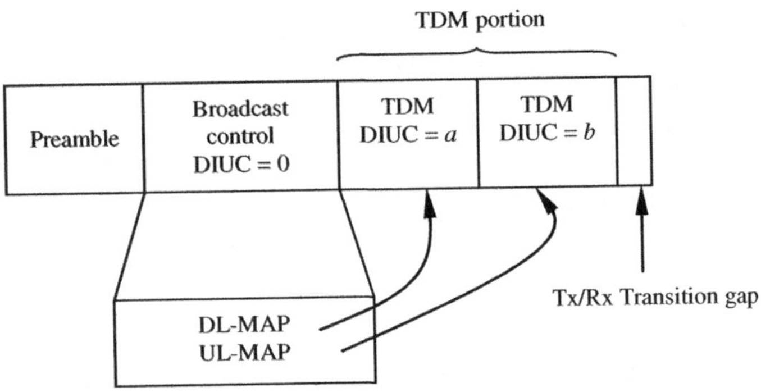

Figure 12-17. Downlink TDM frame format.

12-14 above show the burst format for both. We have discussed them under the MAC 'Bandwidth management', because the bandwidth, though defined at PHY, is managed by MAC. Figure 12-16 shows the physical slots (PSs) for the TDD frame. The downlink interval usage codes (DIUC)s dictate the burst profile that can be used in this interval. for example, a DIUC = 0 in the broadcast control field of a downlink frame implies the use of QPSK.

12.5.6.2. Downlink Frames. Figures 12-17 and 12-18 show the downlink frames for TDD and FDD. The TDD downlink frame consists of four parts, the function of each is explained in Table 12-11.

The FDD downlink frame has a TDM portion for full-duplex operation and a TDMA portion to support half-duplex portion. It has a preamble for both frame parts. Just like the TDM part is divided into usage intervals, the TDMA,

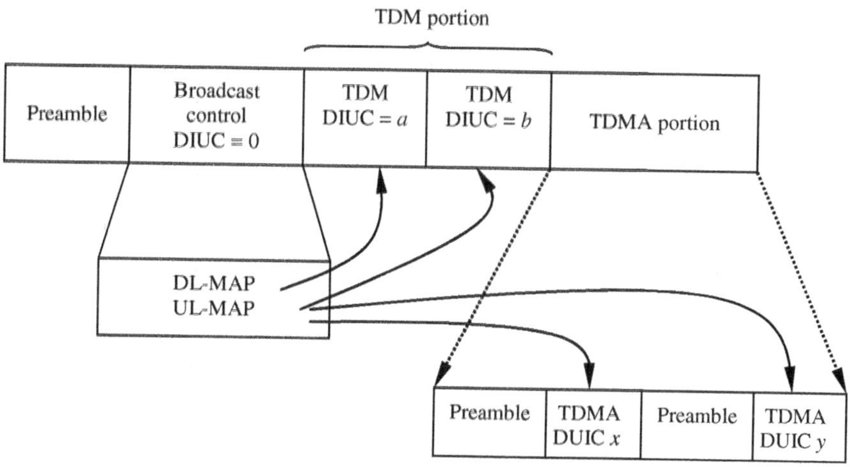

Figure 12-18. Downlink FDD frame format.

TABLE 12.11. Various Fields of the Downlink Frame in 802.16 10–66 GHz PHY

Field	Contents	Function
Preamble	45° rotated constant-amplitude, zero-correlation (CAZAC) sequence.	Start of frame detection.
Broadcast control	Information for SSs.	DL/UL-MAP and other control information.
TDM slots	User information	For data and control
Tx/Rx transition gap	No information.	To allow Receiver sufficient time to become transmitter for full-duplex operation (opposite of Tx/Rx Transition gap).

too, is divided into usage intervals, with the difference that each TDMA interval is preceded by a TDMA preamble.

The transmission convergence (TC) sublayer composes the TC packets to encapsulate the MAC PDUs in the TDD and FDD bursts. In TC packets, MAC PDUs that are not aligned to the boundaries of the TC packet. Instead of aligning MAC PDUs with TC packets, the TC sublayer inserts a pointer at the beginning of the packet with the location of the first PDU in this TC packet. This is shown in Figure 12-19 and it allows to break a PDU between two or more TC packets, thus increasing the link utilization.

12.5.6.2. Uplink PHY Frame. The uplink provides contention-based and grant modes to the subscriber stations (SS)s. There are two types of contention

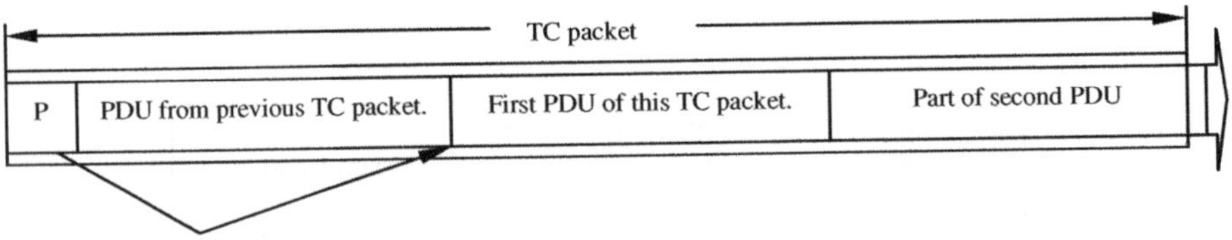

Figure 12-19. The TC sublayer uses a pointer (P) to identify the start of first PDU in the TC packet.

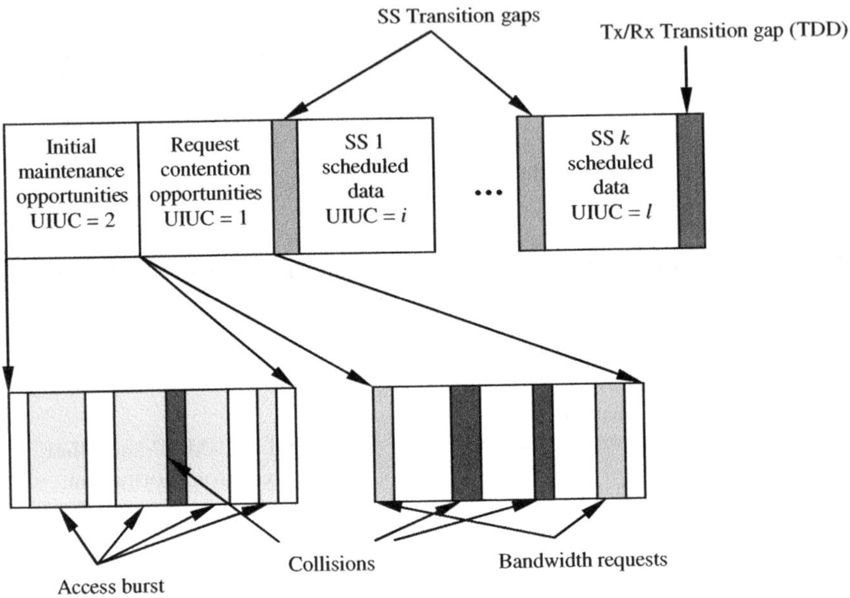

Figure 12-20. The uplink subframe structure.

based transmissions, one for initial maintenance and the other for response to multicast and broadcast. These three (two contentions and a grant-based) transmission mechanisms are defined as three burst classes within the uplink frame. The actual organization of the three within an uplink burst depends on traffic conditions and resource allocation strategy of the operators, and can be done to maximize utilization. The actual burst profile is specified by the BS in every burst and a SS acts accordingly to use the uplink. Since the connection on the uplink is a multipoint-to-point type, a ramping down of burst from one SS and synchronization with another is required for the BS for correct reception. For this purpose, the BS inserts SS Transition Gaps between uplink subframes. Figure 12-20 shows the format of the uplink frame. The purpose of the uplink interval usage code (UIUC) is the same in the uplink as that of the DIUC in the downlink.

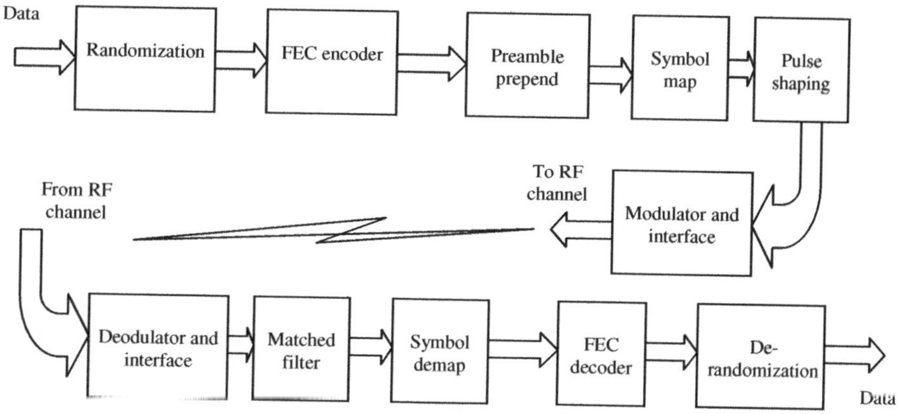

Figure 12-21. Transmitter and receiver blocks for 802.16 10–66 GHz PMD sublayer.

TABLE 12.12. PMD Characteristics for the 802.16 PMD (10–66 GHz)

Function	Purpose		
Randomization/ Derandomization	Using a scramble seed with generator polynomial $x^{15} + x^{14} + 1$		
Pulse shaping	Raised cosine pulse $s(t) = I(t)\cos(2\pi f_c t) + Q(t)\sin(2\pi f_c t)$		
FEC codec	Reed-Solomon over GF9[256] without and with inner code (24,16) conv. code. (9,8) Parity/check optional. Block Turbo code optional.		
Modulations	Modulation symbols from QPSK, 16-QAM, 64-QAM		
Channel size (MHz)	20	25	28
Symbol rates (MBauds)	16	20	22.4
Number of PSs /frame	4000	5000	5600
Power control	Recommended not specified. (should support 10 dB/s fluctuation rate, and 40 dB depth).		
Channel models	3 (with 1, 2 and 3 Taps).		
Max transmitted power	14 dBw/MHz (BS) and 30 dBw/MHz (SS)		

12.5.6.3. Physical Medium Dependent (PMD) Sublayer. The PMD performs the core PHY functions, such as mapping data to physical layer symbols, transmission of symbols and an array of error control signal conditioning functions. Figure 12-21 shows block diagrams for the transmitter and receiver.

Table 12-12 describes the some of the PMD characteristics.

Table 12-13 describes the PHY characteristics for Wireless-MAN-SCa (single carrier) for the 2–10 GHz band. Figure 12-22 shows a transceiver block diagram.

TABLE 12.13. Summary of the Wireless-MAN-SCa PHY for 2–11 GHz

PHY Attribute	Value
Bandwidth	For licensed bands 1.25 MHz times the power of 2.
Duplexing	TDD/FDD
Access	TDM (DL), TDMA (UL)
Randomization/ Derandomization	Using generator polynomial $x^{15} + x^{14} + 1$
Error control	Concatenated FEC (Reed-Solomon outer and TCM inner) Mandatory for QPSK and 16-QAM, Optional for 256-QAM, Mandatory in BS for 64-QAM and in SS for UL for BPSK.
Modulations	QPSK, 16-QAY (Mandatory at SS and BS), BPSK (at SS), 64-QAM (at BS), 256-QAM (optional).
Pulse shape	Square-root raised cosine.
Power control	Algorithm not specified but uplink control for maximum fade depth of 10 dB and a max change rate of 30 dB/s specified. Step size of 1 dB, +0.5 dB, –0.5 dB.
Channel quality control	BS may ask SS to measure and report channel quality.

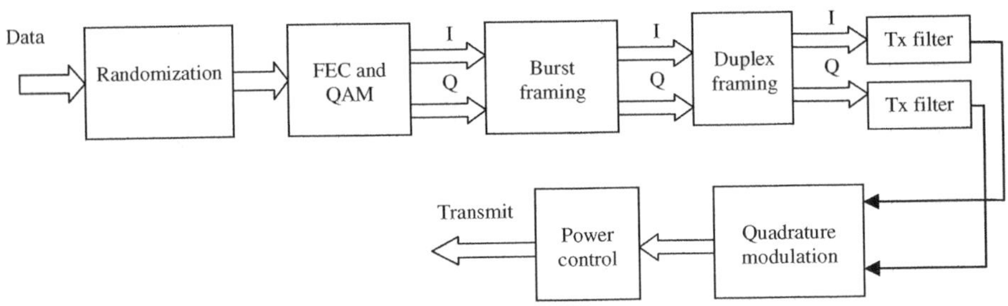

Figure 12-22. Transmitter and receiver blocks for 2–11 GHz WirelessMAN-SCa PHY.

Table 12-14 summarizes some of the characteristics of the WirelessMAN-OFDM PHY for 2–11 GHz.

Table 12-15 summarizes some of the characteristics of the WirelessMAN-OFDMA PHY for 2–11 GHz.

12.6. IEEE 802.20 MOBILE BROADBAND WIRELESS ACCESS (MBWA)

Traditionally, wireless broadband access has been associated with being fixed. However, two developments are destined to change this scenario. These are, the IEEE 802.16e, the task group of the same Work Group that specified

TABLE 12.14. Summary of the Wireless-MAN-OFDM PHY for 2–11 GHz

PHY Attribute	Value
OFDM number of carriers	256 (200 used)
Duplexing	TDD/FDD
Randomization/ Derandomization	Using generator polynomial $x^{15} + x^{14} + 1$
Error control	Concatenated FEC (Reed-Solomon outer and 1/2 rate conv. inner), Turbo optional.
Modulations	QPSK, 16-QAM (Mandatory), 64-QAM (Optional).
Power control	Algorithm not specified but uplink control for maximum fade depth of 10 dB and a max change rate of 30 dB/s specified. Step size of 1 dB, +0.5 dB, –0.5 dB.
Channel quality control	BS may ask SS to measure and report channel quality.

TABLE 12.15. Summary of the Wireless-MAN-OFDMA PHY for 2–11 GHz

PHY Attribute	Value
OFDMA number of carriers	2048
Duplexing	TDD/FDD
Randomization/ Derandomization	Using generator polynomial $x^{15} + x^{14} + 1$
Error control	Concatenated FEC (Reed-Solomon outer and 1/2 rate conv. inner), Turbo optional.
Modulations	QPSK, 16-QAM (Mandatory), 64-QAM (Optional).
Power control	Algorithm not specified but uplink control for maximum fade depth of 10 dB and a max change rate of 30 dB/s specified. Step size of 1 dB, +0.5 dB, –0.5 dB.
Channel quality control	BS may ask SS to measure and report channel quality.

TABLE 12.16. Comparison of the IEEE 802.20 and IEEE 802.16e

Item	802.20	802.16e
Bandwidth	3.5 GHz or lower	Same is WirelessMAN™
Target speeds	Ultra-vehicular (>200 kmph)	Lower than MBWA
Data rates	In excess of 1 Mbps	Much higher than MBWA

WirelessMAN, and another Work Group IEEE 802.20. Table 12-16 compares the objectives of the two groups.

IEEE 802.20 Work Group is scheduled to complete the draft standard by September 2005 and have the final approved document by December 2006

TABLE 12.17. IEEE 802.20 Targets as Proposed in [14]

Attribute	Value
Sustained spectral efficiency	1 b/Hz/s/cell
Peak aggregate data rate per cell (UL)	800 kbps
Peak aggregate data rate per cell (DL)	4 Mbps
Duplexing	FDD and TDD
Security	AES (advanced encryption algorithm)
Bandwidth	Rates for 1.25 MHz one way.
QoS	IPv4 and IPv6 architectures
Latency	10 ms–10 s
Packet error rate	10^{-8} to 10^{-1}
Handoff support	Link level and IP level

[13]. The following summary of the requirements for the IEEE 802.20 is based on [14].

12.6.1. Objectives

The main objective of the 802.20 standard is to have specifications for the PHY and MAC for mobile broadband wireless access systems for packet exchange between the 802.20-enabled mobile terminal and external networks (e.g., IP network) or another 802.20-enabled terminal.

- The target frequency spectrum is below 3.5 GHz.
- It should be optimized for IP data transport.
- Peak data rates per user in excess of 1 Mbps.
- Vehicular speeds up to 250 kmph supported.
- Spectral efficiency, sustained user data rate and number of active users to be higher than any current mobile system.

Other targets proposed in the document [14] are summarized in the Table 12-17.

12.7. CELLULAR AND SATELLITE NETWORKS AS WIRELESS LOCAL LOOPS (WLLs)

The WLL function generally implies telephone local loop. Much of the Internet access is provided to homes and small businesses over the telephone loops too. Therefore, when cellular networks are to be used for WLL service, the main challenge is to emulate the PSTN signaling that could be used by the handset [15]. However, a study on the application of cdma2000 (1xEV-DO) has shown [16] that the number of terminals can be doubled if the network is used for fixed wireless service at 9.6 kbps. This can be an incentive for the 'CellCo's (Cellular Companies) to deploy WLL with the mobile technology. Having said that, the cellular networks perhaps will not be able to compete with the BWA systems, such as WirelessMAN due to a variety of factors, data rates being one critical factor. Satellite systems have provided WLL capability and are candidate for the growing broadband access market [17]. The non-geostationary satellites could provide data rates as high as 64 Mbps, but perhaps not at a price of terrestrial fixed networks. All these technologies may find their niche due to one reason or the other and justify their co-existence.

REFERENCES

[1] George Hendry, 'DAVIC standards overview', *Stanford Wireless Broadband Inc.*

[2] Wei Zhang and Moayeri Nader, 'Classification of statistical channel models for local multipoint distribution service using antenna height and directivity', *IEEE 802.16.1pc-00/07*, January 2000.

[3] Pramote Srisuksant, 'The frequency reallocation for broadband wireless access', *APEC TEL Broadband Workshop #3*, March 2004.

[4] Roger B. Marks, 'Technical consensus in broadband wireless access technology', *IEEE 802.16 Working Group (Presentation).*

[5] EP-BRAN—HIPERACCESS standard area, grouper.ieee.org/groups/802/16/liaison/bran/80216lb-99-02.pdf

[6] Sanjay Moghe, 'Commonality between MMDS and LMDS standards', *Broadband Wireless Access, ADC Telecommunications,* November 1998.

[7] Aldo Bolle, 'HIPERACCESS status and plan', *BRAN 312, HA/N-WEST Joint Session*, January 1998.

[8] OFDM Forum WG-1, 'Progress in BWA standardization', www.ofdm-forum.com/presentations/WG-1.pdf

[9] Xilinx, 'IEEE 802.16 WirelessMAN solutions', www.xilinx.com/esp/networks_telecom/wireless_networks/collateral/wirelessMAN_esp.pdf

[10] IEEE 802.16, 'Part 16: Air interface for fixed broadband access systems', *IEEE Std 802.16,* December 2001.

[11] IEEE 802.16.1a, 'Part 16: Air interface for fixed broadband wireless access systems—Amendment 2: Medium access control modifications and additional physical layer specifications for 2–11 GHz', *IEEE Std 802.16a (amendment to IEEE Std 802.16.1)*, April 2003.

[12] Hassan Yaghoobi, '802.16 Broadband Wireless Access: the next big thing in wireless', *Intel Broadband Wireless Division, Wireless Networking Group*, available from www.intel.com/idf/us/fall2003/presentations/F03USWNTS111_OS.pdf

[13] IEEE, 'IEEE 80-2.20 Project development Plan', http://grouper.ieee.org/groups/802/20/P_Docs/IEEE 802.20 PD-07.ppt

[14] IEEE 802.20, 'System requirements for IEEE 802.20 mobile broadband wireless acces system—version 14', *IEEE 802.20-03/15*, July 2004.

[15] 3GPP2, 'Wireless Local Loop: Stage 1 Description', *3GPP2 S.R0024*, Version 1, September 2000.

[16] Eduardo Esteves, Gurelli, Mehmet I., and Fan, Mingxi, 'Performance of Fixed Wireless Access with cdma2000 (1xEV-DO', *IEEE Vehicular Technology Conference (VTC)*, October 2003.

[17] Leslie A. Taylor, 'Terrestrial and Satellite Wireless solutions', www.lta.com, May 2000.

APPENDIX

OVERVIEW AND GUIDE TO THE IEEE 802 LMSC

September 2004

DISCLAIMER

This guide assembles in one place some info to make life easier for first time attendees of the LAN/MAN Standards Committee (LMSC) meeting (and also for habitual attendees). It draws on the work of other folks in the committee and acknowledges their contribution to this guide.

INTRODUCTION

LMSC (or IEEE Project 802) develops LAN and MAN standards, mainly for the lowest 2 layers of the Reference Model for Open Systems Interconnection (OSI). LMSC coordinates with other national and international standards groups, with some standards now published by ISO as international standards. There is strong international participation, and some meetings are held outside the U.S.

The material in this appendix is presented with thanks to and permission from the IEEE 802.2 Committee. It is available at the URL http://grouper.ieee.org/groups/802/802%20overview.pdf.

Wireless and Mobile Data Networks, by Aftab Ahmad

With work ongoing on a number of standards, the overall picture can be confusing at first. The material handed out at registration, and the overview of meetings and organization in this guide, will help make some sense of the many parallel activities. At the plenary session week, be sure to attend the opening plenary meeting on Monday from 11 am to noon. The Working Group chairs and others give a status report and tell what will be happening during the rest of the session. Arrive early for the best seats—with 1500 or so attendees; the room can fill up fast!

Shortly after the opening plenary, Working Group meetings start and then continue each day through Friday, concluding with a meeting of the LMSC Executive Committee on Friday afternoon. You'll find a wide range of attendance and topics in the Working Group meetings—for instance, in the early stages of a standard there is more technical presentation and less editorial work. In between plenary sessions, work continues over the four months until the next plenary session: most Working Groups hold interim sessions and continue discussions and document editing by E-mail. The personal contacts you can make during the week will be very helpful, so take time to meet people as well as study the documents.

CONTENTS

- PLENARY SESSIONS (registration, schedule, how to find the right room)
- HISTORY (how did we get here?)
- ORGANIZATION (who does what, WGs / SGs / TAGs, IEEE Standards Office)
- STANDARDS PROCESS (inception, liaison, consensus, ballots, & publication)
- DOCUMENTS (drafts, mail, standards, and how to get them)

PLENARY SESSIONS

In March, July, and November of each year, all the subgroups of LMSC meet together at one location. These plenary sessions are scheduled 1–2 years ahead and are attended by about 1500 people. For information about upcoming sessions, visit the 802 web site at **www.ieee802.org**. You can contact the plenary session planners, via E-mail at **802info@ieee.org**.

Pre-registration by credit card is offered to help reduce the wait for registration at the session. A registration fee is collected for each person, and helps cover the cost of the session (document copies, wireless and wired LAN,

Internet connection, refreshments, and Wednesday social). Pre-registration information is available at www.ieee802.org. The registration office is open 5–8pm Sunday and from 8am to 5pm Monday for registrations and information. When you register you'll be given a packet (read it carefully!) including the room schedule for the week and a list of documents and standards that can be ordered.

The overall schedule for the plenary meeting week is as follows:

Sunday:	all day: some Working Group meetings* 5:00 pm–8:00 pm: registration
Monday:	8:00 am–10:30 am: Opening Executive Committee meeting 8:00 am–10:30 am: some Working Group meetings* 11:00 am–12:00 noon: Opening Plenary meeting 1:00 pm–6:00 pm Working Group meetings 6:30 pm–9:30 pm: tutorials 8:00 am–5:00 pm: registration
Tuesday:	8:00 am–6:00 pm Working Group meetings 6:30 pm–9:30 pm: tutorials 8:00 am–5:00 pm: registration
Wednesday:	8:00 am–12:00 noon Working Group meetings 6:30 pm–9:30 pm: Social Reception 8:00 am–5:00 pm: registration
Thursday:	8:00 am–6:00 pm: Working Group meetings
Friday	8:00 am–12:00 noon: Working Group meetings 1:00 pm–6:00 pm: Closing Executive Committee meeting

* these WG meetings are actually interim meetings since they're before the opening plenary

NOTE: Executive Committee meetings are open to any interested observers from 802

The evening tutorials on Monday and Tuesday are often used to publicize work on a potential new standard—the topics are listed in the material handed out at registration. The Wednesday evening social is an opportunity to talk with people in other groups, as well as being known for free hor d'ouvres! The real work of the plenary session happens in the individual Working Group meetings, with some Working Groups split into several subgroups during part of the week. It is not possible to see what's happening in all groups—best to concentrate on one or two. Since the detailed schedules for each group may change during the week, check the bulletin board at the registration office for the latest room assignments & meetings.

HISTORY

The first meeting of the IEEE Computer Society 'Local Network Standards Committee', Project 802, was held in February of 1980. (The project number, 802, was simply the next number in the sequence being issued by the IEEE for standards projects). There was going to be one LAN standard, with speeds from 1 to 20 MHz. It was divided into media or Physical layer (PHY), Media Access Control (MAC), and Higher Level Interface (HILI). The access method was similar to that for Ethernet, as well as the bus topology. By the end of 1980, a token access method was added, and a year later there were three MACs: CSMA/CD, Token Bus, and Token Ring.

In the years since, other MAC and PHY groups have been added, and one for LAN security as well. The unifying theme has been a common upper interface to the Logical Link Control (LLC) sublayer, common data framing elements, and some commonality in media interface. The scope of work has grown to include Metropolitan Area Networks (MANs) and higher data rates have been added. An organizational change gave us the 'LMSC' name and more involvement in the standards sponsorship and approval process.

ORGANIZATION

LMSC is organized in a number of Working Groups (WGs) and Technical Advisory Groups (TAGs) as well as a Sponsor Executive Committee (SEC).

802.0 **Sponsor Executive Committee**

Chairman—Paul Nikolich E-mail: **p.nikolich@ieee.org**

1st Vice Chairman—Mat Sherman E-mail: **matthew.sherman@baesystems.com**

2nd Vice Chairman—Howard Frazier E-mail: **mailto:hfrazier@sbcglobal.net**

Executive Secretary—Everett O. Rigsbee E-mail: **everett.o.rigsbee@boeing.com**

Recording Secretary—Bob O'Hara E-mail: **bob@airespace.com**

Treasurer—John Hawkins E-mail: **jhawkins@nortelnetworks.com**

ACTIVE WORKING & TECHNICAL ADVISORY GROUPS

802.1 High Level Interface (HILI) Working Group

Chairman—Tony Jeffree E-mail: **tony@jeffree.co.uk**

802.3 CSMA/CD Working Group

Chairman—Bob Grow E-mail: **bob.grow@intel.com**

802.11 Wireless LAN (WLAN) Working Group
Chairman—Stuart Kerry E-mail: **stuart.kerry@philips.com**

802.15 Wireless Personal Area Network (WPAN) Working Group
Chairman—Bob Heile E-mail: **bheile@ieee.org**

802.16 Broadband Wireless Access (BBWA) Working Group
Chairman—Roger Marks E-mail: **r.b.marks@ieee.org**

802.17 Resilient Packet Ring (RPR) Working Group
Chairman—Mike Takefman E-mail: **tak@cisco.com**

802.18 Radio Regulatory Technical Advisory Group
Chairman—Carl Stevenson E-mail: **carl.stevenson@ieee.org**

802.19 Coexistence Technical Advisory Group
Chairman—Steve Shellhammer E-mail: **stephen.j.shellhammer@intel.com**

802.20 Mobile Wireless Access Working Group
Chairman—Jerry Upton E-mail: **Jerry1upton@aol.com**

802.21 Media Independent Handover Working Group
Chairman—Ajay Rajkumar E-mail: **ajayrajkumar@lucent.com**

HIBERNATING WORKING GROUPS (standards published, but inactive)

802.2 Logical Link Control (LLC) Working Group
Chairman—David E. Carlson E-mail: **dcarlson@netlabs.net**

802.5 Token Ring Working Group
Chairman—Bob Love E-mail: **rdlove@ieee.org**

802.12 Demand Priority Working Group
Chairwoman—Pat Thaler E-mail: **pat_thaler@agilent.com**

DISBANDED WORKING GROUPS (all standards withdrawn or did not publish a standard)

802.4 Token Bus Working Group
Chairman—Paul Eastman E-mail: **paul@rfnetworks.com**

802.6 Metropolitan Area Network (MAN) Working Group
Chairman—James F. Mollenauer E-mail: **jmollenauer@technicalstrategy.com**

802.7 BroadBand Technical Adv. Group (BBTAG)

802.8 Fiber Optics Technical Adv. Group (FOTAG)
Chairman—J. Paul 'Chip' Benson, Jr. E-mail: **jpbenson@lucent.com**

802.9 Integrated Services LAN (ISLAN) Working Group
Chairman—Dhadesugoor R. Vaman E-mail: **dvaman@megaxess.com**

802.10 Standard for Interoperable LAN Security (SILS) Working Group
Chairman—Kenneth G. Alonge E-mail: **alonge_ken@geologics.com**

802.14 Cable-TV Based Broadband Communication Network Working Group
Chairman—Robert Russell E-mail: **rrussell@knology.com**

Each project approved within an existing group is assigned a letter, for example 802.1D for MAC Bridges in the High Level Interface WG. A Study Group (SG) is formed when a new area is first investigated for standardization. The SG can be within an existing WG or TAG, or it can be independent of the WGs. A new project in an existing group is developed by a Task Force, while a new independent project creates a new WG.

Membership in LMSC is by WG/TAG, with voting rights after attending two of the last four sessions. The interim sessions of a WG/TAG may be counted under some circumstances. Attendance means you must be present for at least 75% of a meeting and attend at least 75% of the meetings in a WG/TAG session. Attendanceis tracked by sign-up sheets. Credit is given for attendance at only one group per plenary meeting.

The working style of each WG/TAG depends on the number of members and the subject at hand, with some topics decided informally while others are subject to letter ballots. The Chair of each group is given latitude to set the procedure for the group. In all cases, approval of a draft standard by the WG/TAG requires a letter ballot and an effort to resolve any 'No' vote.

In addition to the volunteer members of LMSC, there is a professional staff at the IEEE Standards office that supports our work and gets standards published. The IEEE Standards Board and Standards Staff are responsible for a wide range of standards activities beyond the LAN/MAN standards in LMSC. The IEEE Standards office can be contacted at (732)562-3800 or **http://stdsbbs.ieee.org/**.

LMSC also relies on a meeting management firm (802info@ieee.org) to administer the arrangements for each meeting, including registration. They can be contacted for hotel and transportation information and meeting pre-registration.

STANDARDS PROCESS

Each standard (or recommended practice, or guide) starts as a group of people with an interest in developing the standard. A Project Authorization Request (PAR) is normally submitted for approval within 6 months of the start of work. In LMSC, new projects require supporting material in the form of '5 criteria' to show that they meet the charter of LMSC. The draft PAR is voted on by the SEC, and then goes to the IEEE Standards Board New Standards Committee (NesCom) which recommends it for approval as an official IEEE Standards project. Part of the PAR identifies which outside standards groups there will be liaisons with, for instance ITU for some international standards. The liaisons help avoid conflicts or duplication of effort in one area.

Proposals are evaluated by the WG, and a draft standard is written and voted on the by the WG. The work progresses from technical to editorial / procedural as the draft matures. When the WG reaches enough consensus on the draft standard, a WG Letter Ballot is done to release it from the WG. It is next approved by the SEC and then goes for Sponsor Letter Ballot. In the past, the sponsor group was the Technical Committee on Computer Communication, so the sponsor ballot is still referred to sometimes as a TCCC ballot, even though LMSC now is a sponsor and conducts its own Sponsor Letter Ballots.

After the Sponsor Letter Ballot has passed and 'No' votes are answered, the draft Standard is sent to the IEEE Standards Board Standards Review Committee (RevCom). Once recommended by RevCom and approved by the Standards Board, it can be published as an IEEE standard. Most draft standards in LMSC are also sent to ISO at or before the time they go to Sponsor Letter Ballot. A parallel approval path is followed in ISO JTC1/SC6 (Joint Technical Committee 1, Subcommittee 6—responsible for LANs) that leads to publication as an ISO standard. The process from start to finish can take several years for new standards, and less for revisions or addenda.

DOCUMENTS

The main work in LMSC sometimes appears to be generating documents, and it's certainly true that we generate a lot of text to develop a standard. In the early stages, a WG will have a number of proposals and drafts, and these will be copied to the WG for their work. You can get copies of other WG's documents via the web at www.ieee802.org.

To reduce paper use and speed the process, the WGs use E-mail and the Internet and LAN distribution for proposals, discussions, minutes, and drafts to varying extents. Check with the WG chair for whether this is available and how to use it. Also, a central LMSC web site is maintained at www.ieee802.org. This web site includes meeting information and other general info, as well as some WG areas.

Once a standard has been published, you may order it from IEEE Document Sales, (800)678-4333 (USA only) or (732)981-0060 (voice), (732)981-9667 (fax) for the first six months after publication. After the document has been published for six months, it is available at no charge through the Get IEEE 802 program http://standards.ieee.org/getieee802/.

The following is being included in this overview on a trial basis, to 'fill in the blanks' beyond what's covered in the rules of IEEE 802 or the Working Group. Because the 'culture' of each WG is distinct, there are exceptions and differences from the following list in each WG.

Please note that some WGs are developing their own rules, and also be aware that IEEE and the Computer Society have rules to help reach consensus on each standard (including appeals process and patent policies).

These guidelines are meant to explain what might otherwise take some trial and error to learn—please help improve them by getting your feedback to one of the WG Chairs or other SEC members.

GUIDELINES FOR IEEE 802 PARTICIPANTS

Based on the LMSC operating rules, the function of IEEE 802 Working Groups is as follows:

'The function of the Working Group is to produce a draft standard, recommended practice or guideline. These must be within the scope of the LMSC, the charter of the Working Group and an approved PAR, or a PAR under consideration by the IEEE Standards Board, as established by the Executive Committee. After the approval of the Working Group's standard, recommended practice or guideline, the function of the Working Group is to review, revise, and affirm its documents.'

Within this framework, the following guidelines serve to outline the responsibilities of Working Group participants.

Observer

- Pay applicable session fees: Pay IEEE 802 Session Fee for each plenary session you attend; pay the Session Fee for each Interim session you attend.
- Sign the attendance book when and only when you will be participating in that Working Group for the majority of that session. For each session, you should sign the attendance book of no more than one WG.
- If you are aware of any patents that pertain to the work of the subcommittee you should so advise the chair.
- Initially, mostly listen. Learn about the issues and procedures of the Working Group to effectively contribute to the Working Group's progress and to gain the Working Group's respect and attention when you participate in its discussions.

- Contribute to the progress of the Working Group. Since standards are based on consensus, focus on win-win solutions when you are seeking to establish or change a Working Group position. Vote on Straw Polls during the plenary and interim meetings.

When you are on a WG E-Mail reflector

- Read the E-Mail and keep abreast of the issues.
- Respond to the E-Mail when you have appropriate input.
- Get documents from the WG FTP site to review as appropriate.
- Review Working Group drafts out for ballot, and submit your comments on or before ballot closing date.
- Familiarize yourself with Robert's Rules of Order.
- Review the meeting minutes.
- Comment on required changes as appropriate if you have attended the last meeting

Working group member

- Same as Observer with the following additional responsibilities:
- The Working Group members decide technical issues by vote
- Vote on Official Working group Motions
- Vote on Working group Drafts circulated for Ballot.
- Vote for Working Group officers as appropriate.
- Familiarize yourself with the 'Policy and Procedure of IEEE Project 802, LAN MAN Standards Committee (LMSC)'. We must follow those rules.
- Consider volunteering to host interim sessions (See the chair for details)
- Consider volunteering to serve the Working Group in additional capacities

INDEX

Wireless and Mobile Data Networks, by Aftab Ahmad

Printed and bound by CPI Group (UK) Ltd, Croydon, CR0 4YY

16/09/2026

14963442-0004